DESIGN AND DECISION SUPPORT SYSTEMS IN ARCHITECTURE

Design and Decision Support Systems in Architecture

edited by

HARRY TIMMERMANS
Urban Planning Group,
Faculty of Architecture, Building and Planning,
Eindhoven University of Technology,
Eindhoven, The Netherlands

KLUWER ACADEMIC PUBLISHERS
DORDRECHT / BOSTON / LONDON

Library of Congress Cataloging-in-Publication Data

Design and decision support systems in architecture / edited by Harry Timmermans.
p. cm.
Proceedings of a conference held in Mierlo, the Netherlands in July 1992 and organized by the Faculty of Architecture, Building, and Planning, Eindhoven University of Technology, Eindhoven, the Netherlands.

1. Architectural design--Data processing--Congresses. 2. Computer-aided design--Congresses. I. Timmermans, H. J. P. II. Technische Universiteit Eindhoven. Afdeling der Bouwkunde.
NA2728.D48 1993
720'.285'53--dc20 93-24413

ISBN 978-90-481-4313-9

Published by Kluwer Academic Publishers,
P.O. Box 17, 3300 AA Dordrecht, The Netherlands.

Kluwer Academic Publishers incorporates
the publishing programmes of
D. Reidel, Martinus Nijhoff, Dr W. Junk and MTP Press.

Sold and distributed in the U.S.A. and Canada
by Kluwer Academic Publishers,
101 Philip Drive, Norwell, MA 02061, U.S.A.

In all other countries, sold and distributed
by Kluwer Academic Publishers Group,
P.O. Box 322, 3300 AH Dordrecht, The Netherlands.

Printed on acid-free paper

TABLE OF CONTENTS

PREFACE

The papers collected in this volume were originally presented at the conference on Design and Decision Support Systems in Architecture and Urban Planning that was held in Mierlo, the Netherlands in July 1992. This conference was organized as one of the events celebrating the 25th anniversary of the founding of the Faculty of Architecture, Building and Planning at Eindhoven University of Technology.

The organizing committee had a strong feeling that many interesting developments in this area were emerging within different institutional frameworks and informal networks that do not interact frequently. For example, scholars working on architectural problems are not particularly familiar with computer applications in urban planning. Likewise, although many computer-aided design systems claim to be based on principles of design methodology, serious discussions on the methodological underpinnings of such systems are relatively scarce. Consequently, we may have little opportunity to learn how scholars in closely related disciplines approach specific design or planning problems.

The DDSS conference was organized in an attempt to build bridges between these divergent research communities, and it succeeded in attaining this goal: approximately 100 scholars from many different disciplinary backgrounds and countries attended the meetings. Participants came from far and near: from Japan and New Zealand, from Argentina and Canada, and even from some African countries; almost all European countries were represented. Many topics were discussed extensively. As this volume contains a selection of papers across topics and countries, it should provide an interesting overview of current research directions in this area. Other papers presented at the conference will be published in special issues of various international journals. It is unfortunate that, due to space constraints, I had to leave out many interesting papers. While this volume is intended to be a major reference text for students, researchers and practitioners, I hope it also reflects some of the excitement of the conference.

Of course, the success of the conference was largely due to the efforts of the organizing committee. I am grateful to Marlyn Aretz, Thijs Bax, Roel Daru, Geert Smeltzer, Leo van Veghel, and Andreas van Wagenberg for their help in shaping the conference in this potentially risky format and for chairing many of the sessions. I am also very grateful to Mandy van Kasteren, who saved me from missing important deadlines by helping out in the preparation of the programme and during the conference itself.

Most publications are the result of teamwork, and this book is no exception. Many people helped, directly or indirectly, to bring it to completion. I would particularly like to express my thanks to Nancy Smyth van Weesep, who acted as copy editor, and Toos van Raaij, who typed most of the manuscripts. I feel very fortunate to have been able to call upon the drafting skills of Ton Davits, Ton van Gennip, Jac de Kok, and Bert Lammers, all of Eindhoven University of Technology, who prepared the final drawings for the book. Most of these activities took place during the time of year that I work in Edmonton, Canada. Therefore, I feel deeply obligated to Ilse Vermeulen, who coordinated the project during my absence and, as always, did a marvellous job.

Harry Timmermans
January 1993

INTRODUCTION

It is generally recognized that the degree of sophistication in architectural and urban design has, at least in principle, increased considerably over the last decades. The discipline has seen the emergence and growth of various specialisms. Each of these has accumulated knowledge and built models about particular domains. As the designs become more complicated, they require the expertise of many consultant professionals. Each one makes a more or less independent contribution to the design in a particular domain. Consequently, the potential for disagreement is greatly enhanced. This has led to increased efforts to build project teams with common goals and information bases. To some extent, this common ground can be established by setting particular standards.

Part of the discipline has also attempted to apply scientific principles to the design process. Design theory has sought to develop frameworks, classifications and models to describe or prescribe the design process and provide a rational basis for assessing design quality and performance. According to some schools of thought, architects are expected to justify their designs in terms of implied goals and objectives. These may pertain to various dimensions, such as cost, user evaluation, physical aspects, construction, flexibility, etc. Such explication could enable the user or contractor to assess whether the ultimate design meets a set of a priori standards or criteria. Alternatively, it could provide an argument to involve members of a design team in a design process whereby the options and variants are successively evaluated until a final design decision is reached. Partly due to this increased level of complexity and sophistication, but also because the architectural profession is not shielded from harsh reality - productivity, efficiency and market orientation -, the last decades have also witnessed a search for better, more efficient working methods and means of communication.

Each of these developments has potentially strong ramifications for the role and importance of design systems. In fact, many believed that the new computer technology, coupled with a strong design theory, would be the panacea for most, if not all, of the problems mentioned above. It is easy to show that computer technology has indeed provided professionals with a wide variety of tools and systems which support their design or decision-making process. Obviously, these are very effective tools for communicating ideas and concepts to other members of the design team or to the ultimate users. The field of computer graphics is rapidly approaching a turning point. The user will be able to create such realistic images that they will be able to evaluate the aesthetic quality of designed spaces. Virtual reality systems will even allow one to walk through non-existent environments. Multi-media systems are very powerful in linking quantitative, verbal and image information with sound. This too opens entirely new avenues for communication as well as information storage and retrieval. Traditional drawing and simple computer-aided systems are increasingly augmented with specialized knowledge. Sometimes, this is implemented in terms of simple rules. However, knowledge engineering and artificial intelligence are gaining ground, and the first expert systems are now available. Models for generating an optimal or quasi-optimal building layout, subject to architectural constraints, are currently in use as well. Models for calculating building costs, visual impact assessment, thermal analysis, multi-criteria evaluation, and so on and so forth are now widely available; often the user has many to choose from. These are only a few examples of the strides made during the last decade.

Notwithstanding these achievements, it is abundantly clear that design and decision support systems have their limits. It is not my intention to go into too much detail on this issue. Such systems are not as widely employed in the building industry as in many other industries. This is partly the result of its idiosyncrasies. But their sparse implementation also stems from the fact that their proponents merely seek to replace well understood and perfectly acceptable practices or media with new computer technologies which, for the time being, are not necessarily providing better quality or greater flexibility. This is not to say that no progress can be made in the future, or that an emphasis on computer technology should be avoided. On the contrary, I feel it is imperative to develop and test design and decision support systems. However, I believe that, after the rapid initial progress, new developments will be less easy. They should be based on a thorough understanding of current and future design practices, advanced models of the phenomena under investigation, superior design theory and realistic expectations.

This volume contains a set of papers which make a potential contribution in this respect. Many address more than one topic, which makes their sequence somewhat arbitrary. We have chosen to start with contributions to design theory, because this constitutes the premise of all computer systems. These are followed by chapters which describe improvements or extensions of more traditional computer aided design systems and chapters which concentrate on integrated decision support systems. Next, a few contributions are included on product modelling and related developments. Finally, there is a series of chapters which focus on improvements in or applications of evaluation methodology. In some cases, the authors concentrate on methods; in other cases their prime concern is empirical analysis of building performance. The examples have been chosen to highlight the many different aspects of building: cost, energy consumption, construction, layout, user behaviour, etc.

A frequently heard criticism of many currently available computer-aided design systems is that they force the architect to develop different working styles and practices. Computer-aided design systems often lack flexibility and thus cannot accommodate different styles. Consequently, an important item on the research agenda concerns the development of computer-aided design and decision support systems which allow the user to stick to his or her accustomed style. Such research presumes some basic understanding of the variety of styles in designing. *Van Bakel*'s chapter offers some interesting ideas on this topic. He uses an interactive personality model to explain and interpret strategic design behaviour. He proposes a distinction between site-oriented, brief-oriented and concept-oriented behaviour which, when combined with a priority hierarchy, leads to a classification of six different working styles. This could be an important building block for design systems, yet to be developed.

Computer-aided design has traditionally focused on computer graphics for the creation and production of two- or three-dimensional visual representations. *Koutamanis* and *Mitossi* point out the importance of research on making historically significant documents or archives accessible to CAD systems. They suggest using automated recognition of digitized architectural drawings to reduce storage size, improve recall and manipulation time, and enhance compatibility with CAD systems. Such recognition systems should identify geometric primitives, recognize building elements, and evaluate designs with respect to a set of requirements.

Smeets sketches a functional specification of a decision support system developed to assess the rentability of housing estates. His system draws heavily on concepts of strategic management. He identifies the information that is required to perform quality control and advocates using this system for early warning, evaluation, planning and dialogue. An interesting aspect of the proposed system concerns the incorporation of both quantitative information, focusing on a rentability index, and qualitative information. The idea to link the system to a SWOP analysis will probably also improve its usefulness for managers.

The next chapter, written by *Pullen, Wassenaar, van Hettema, Toonen Dekkers, Janszen, Boender, Tas* and *Stegeman* elucidates the BOAST system. It is used to match available real estate with housing need, in this case that of government organizations. Their effort to match supply and demand is thus an aim they have in common with Smeets' study. The difference between the two papers is that Pullen and his colleagues developed the decision support for the public sector, whereas Smeets' system pertains to the private sector, although it does not necessarily exclude government organizations. Moreover, the two approaches differ considerably. Smeets basically uses a qualitative approach, together with some simple indices. In contrast, Pullen et al. apply multi-criteria evaluation techniques and optimization algorithms for matching supply and demand of building accommodation.

Like Smeets' study, the integrated facility management system described by *De Scheemaker* has a strong management focus. She outlines a system to be developed for the Dutch government. In addition to traditional information storage and retrieval functions, that system contains a monitoring component and a sensitivity analysis module. An interesting feature of her chapter is the way it links management control and style to the functional specification of the decision support system.

Communication between members of a design team tends to be a major problem in architectural design. This is largely because members typically develop partial designs or solutions specific to their own particular domain. Within this general context, product modelling becomes increasingly important. *Drach, Langenegger,* and *Heitz* take the product modelling approach one step further. They describe an attempt to combine fundamental structures for product modelling with design concepts which are dependent on a designer's knowledge and interpretation. Their system, called "a+", differs from many related ones in that user control is a leading principle. The user is not only in control at the level of product modelling, that is in finding appropriate design solutions, but also at the knowledge level, through interactive definitions of design consistency and support. It seems that such an approach may be especially valuable for highly complicated, ill-structured design problems. It is there that traditional expert systems tend to fail because the rule structures are very difficult to maintain.

Gauchel, Hovestadt, van Wyck and *Bhat*, who are working in the same tradition, also point out that the implementation of a unified building model with traditional database management systems has proven to be difficult. They therefore propose an alternative. It is based on a modular concept of building models that can be shared, but also customized, by members of the design team, according to their specific needs.

An interesting approach related to this problem is presented in *Lucardie*'s paper. Specifically, he poses that the theory of functional classifications is a powerful means to integrate knowledge technology and database technology. As we can see in some of the other papers, the integration of knowledge and data is often established by transferring data structures and processing techniques. Lucardie focuses, however, at the semantic level. He argues that such an emphasis provides the flexibility we are looking for. The need for flexibility is a result of rapidly changing products, major difficulties in defining and maintaining logical consistency, and the variety of user demands. The quintessence of functional classification is the goal-oriented reconstruction process: objects are classified into functionally equivalent classes. The potential of this approach is made clear by *De Gelder* and *Lucardie* in the next chapter. There, they describe a knowledge-level analysis using the theory of functional classifications. This analysis concerns the requirements for a product model of a portal structure in steel.

Perhaps one of the most important aspects of performance assessment in the design process is building costs. This is one of those areas where considerable progress has been made lately. These developments typically lend themselves to inclusion in decision support systems. *Poortman* illustrates how translation and calculating ratios, using characteristic quantity figures, may improve the reliability of one's building cost estimates and assist in cost control evaluations. One of the major obstacles in this type of evaluation is collecting reliable information. In contrast, the mechanics of the evaluation and the automation aspects are typically straightforward. Often, they can be performed with cheap, readily available computer software. This is nicely illustrated by *Gilleard* and *Olatidoye*, who demonstrate that Lotus 1-2-3 can be used together with AutoCAD and AutoLisp to determine elemental square foot costs and cost significance for residential construction. This procedure allows non-expert CAD users to work with a relatively complicated cost model.

Although this is not explicitly mentioned, the model developed by *Prins, Bax, Carp* and *Tempelmans Plat* is also available, albeit as a stand-alone spreadsheet program. Their approach differs from the previous two chapters in that they incorporate building change and flexibility into their model. Their model is not really capable of simulating future costs, and it is debatable whether it is possible to obtain reliable forecasts. However, their attempts push this area one step forward. This approach should be drawing more attention in industrialized countries, where an increasing proportion of building is adapting to changing user needs.

Shabha's paper also examines the issue of measuring flexibility, in this case of school buildings. He concentrates on operational measures of the flexibility of school buildings in use. The proposed scheme specifies indicators of organizational change, indicators of change in activity characteristics, indicators of adaptation and the cost thereof, and indicators for measuring flexibility. This framework could be used in a decision support system.

Designs are complex in nature of designs and are subject to many potentially conflicting interests. Therefore, the formal evaluation methods may play an increasing role in design evaluation processes. Many different methods have been developed over the years, although often in areas other than architecture, to assist the designer in this process. These methods include applications of utility theory, regime analysis, conjoint measurement, multi-criteria analysis to name just a few. *Friedman* recommends the use of decision theory techniques. To illustrate how expected monetary value method and weighted objective decision analysis can be used for this purpose, he describes their use to the problem of flexible internal partition options in multi-unit housing in Montreal, Canada.

One alternative evaluation method already has a good track record in urban planning and environmental impact assessment: multi-criteria analysis. *Mortola* and *Giangrande* apply concordance analysis to a design entered in competition sponsored by Strathclyde University for European students in architecture. This evaluation model was a module of an 'interface for designing' computer system.

Quantitative evaluation such as decision theory has the potential advantage of scientific rigour. It should be noted that this is often only a superficial advantage: it can be very difficult to quantify evaluation criteria and weights reliably. One way of dealing with this drawback is to use qualitative multi-criteria analysis, or mixed data analysis. Alternatively, one might use graphic evaluation methods. These may be more appealing to architects, who are often graphically inclined. An example is presented in the paper by *Winteraecken-Bruls*, who describes an interactive graphic heuristic procedure, based on Bertin's work in cartography, as applied to a space-planning problem.

Kim and *Brawne* take a different approach to evaluation. They concentrate on the design of exhibition spaces in art galleries. They suppose that the quality of visitor experiences is influenced by the route followed through the exhibit. This in turn, is assumed to be a function of the spatial configuration. They investigate the relationship between spatial organization and visitors' behaviour. Information on behaviour was collected by the trailing technique. These observations were correlated with a set of physical variables, including geometric proportion, perimeter, length and area. Some moderately strong positive relationships were found. The authors did not explicitly position their study as a decision support system. Yet, it is evident that models like this one could be incorporated in the model base of such systems. Alternatively, they could be translated into decision rules and be used to complement traditional computer-aided design systems.

Their evaluation method is thus explicitly based on empirical analysis. This holds true for *Hensen*'s paper as well. He describes recent developments in simulating building and plant energy in the context of decision support and design or building performance evaluation. Generally speaking, the developments in this area are very similar to those in other fields. First, there is a tendency to build increasing sophisticated models. These range from simple deterministic steady state models to dynamic models. Second, there is a tendency to incorporate expert systems in these systems in order to assist non-expert users.

The final chapter in this volume is a paper by *Bourdakis* and *Fellows*, who are concerned with building performance evaluation. They assess the effects of structural systems on the environment, the activities, the resources and culture in sports hall and swimming pool buildings.

Overall, this collection of papers seems to reflect the diversity of leading-edge research in design and decision support systems in the field of architecture. Evidently, some of these contributors take advantage of progress in particular subareas. Equally innovative are those who sketch the contours of new systems yet to be developed. Technological progress is fast; future achievements and developments will probably be restricted only by the limits of our imagination. But regardless of the technology, we should not forget that the implementation of any design and decision support system ultimately depends on human factors. Scholars working in this area have an obligation to demonstrate that their "toys" are indeed capable of supporting design activities and can help create better or more intelligent designs. The time will tell whether the new generation of systems can actually move the field beyond the simple standardized design activities typical of most currently used systems.

Harry Timmermans,
January 1993

PERSONALITY ASSESSMENT IN REGARD TO DESIGN STRATEGIES

A.P.M. VAN BAKEL
Department of Architecture, Urban Planning and Management
Eindhoven University of Technology
P.O. Box 513, 5600 MB Eindhoven
The Netherlands

ABSTRACT. This paper discusses some preliminary results of several knowledge-acquisition and documentation-structuring techniques that were used to assess the working styles of architects. The focus of this assessment was on their strategic design behaviour. Hettema's Interactive Personality Model (Hettema 1979, 1989) was used to explain and interpret these results. The methods used to acquire the necessary data are protocol analysis, card sorting and interviews. The results suggest that at least three parameters can be used to explain and differentiate the strategic design behaviour of architects. These parameters are S (site-oriented), B (brief-oriented) and C (concept-oriented). A priority hierarchy of these parameters reveals six major distinguishable working styles. These results are captured in a new design model that can be used in data bank implementations.

1. Introduction

This introduction reviews recent developments in research on architectural design as well as the reasons why this research was conducted. Then the personality model is presented as a framework for understanding the implications of the data derived from the experiment. After a description of the experiment and its results, their implications for future research and software development are discussed.

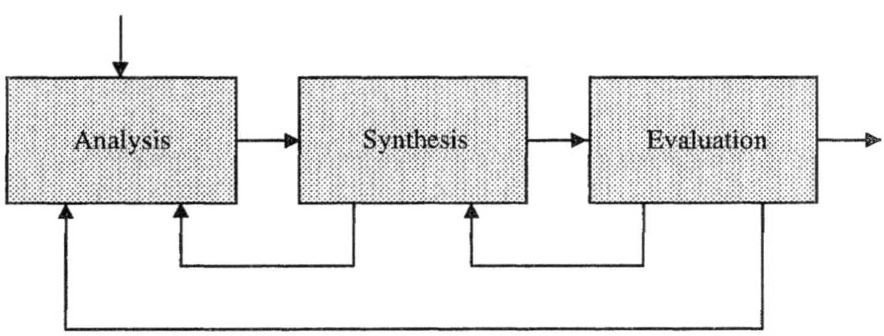

Figure 1. A classic model of the general design process

1.1 CURRENT DESIGN MODELS

Many theories about design, architectural design in particular, describe the design process as a sequence of activities like analysis, synthesis and evaluation. Figure 1 shows a classic model of these three activities. There is a problem, however. Models of this kind are partly tautological. They do not lead to better insight in the complex design process, nor to fruitful development of a design-support system, unless they are extended and expanded with special features. On the basis of research, these models were augmented to generate more elaborate models like, for instance, the one designed by Archer (1969) shown in Figure 2.

H. Timmermans (ed.), Design and Decision Support Systems in Architecture, 1–16.

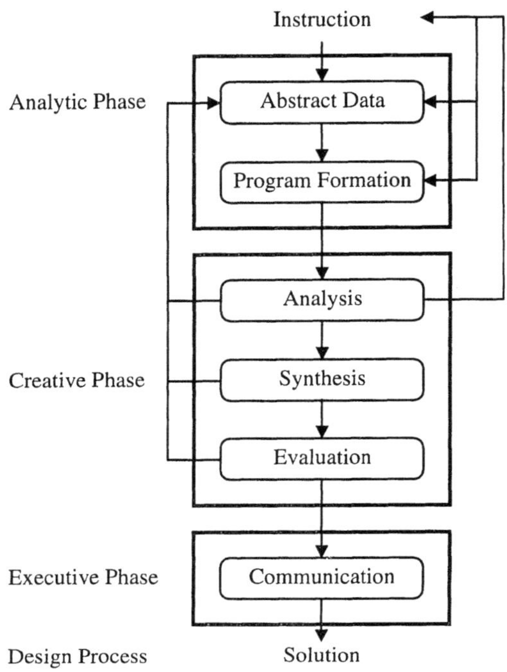

Figure 2. A more elaborate design model by Archer

The problem with these more elaborate models is that they often describe the ideal design situation. That is, they show how the design process should take place (prescriptive models) instead of describing how it actually does take place (Van Bakel 1991).

1.2. CURRENT DESIGN TOOLS

Unfortunately, computer scientists use the models described above as a basis to set up an artificial process dispatch to support designers in their work. The problem is that when we talk about CAAD or Design Decision Support Systems (DDSS), the emphasis seems to lie on the role of the computer instead of the user (Daru 1984). Design programs used to be (and sometimes still are) developed without using the explicit and implicit knowledge of the designer. Fortunately, program developers have recognized these shortcomings; we now have some very useful design tools at our disposal. These are not merely number-oriented drawing tools. Using interview and observation techniques, the investigation revealed that these programs are very promising. Instead of just being used as a drawing tool, they can be applied as a design tool in the preliminary phase of the design process. A few examples of forms synthesized during the initial stage of a design process are shown in Figures 3 and 4.

These shapes were created in the sketch phase of the design process by running one of these programs on an Apple Macintosh computer. It took just a few minutes to create this conceptual image on the computer. Without a computer, it would take several hours to reproduce different representations of this visual image that can be rotated in space. Unfortunately, most modelling and rendering programs are developed with an emphasis on the description or representation of a design instead of the generation of a design. Also, to a lesser extent they are suitable to support sub-processes like calculating the costs, the requirements of energy, air-conditioning, etcetera.

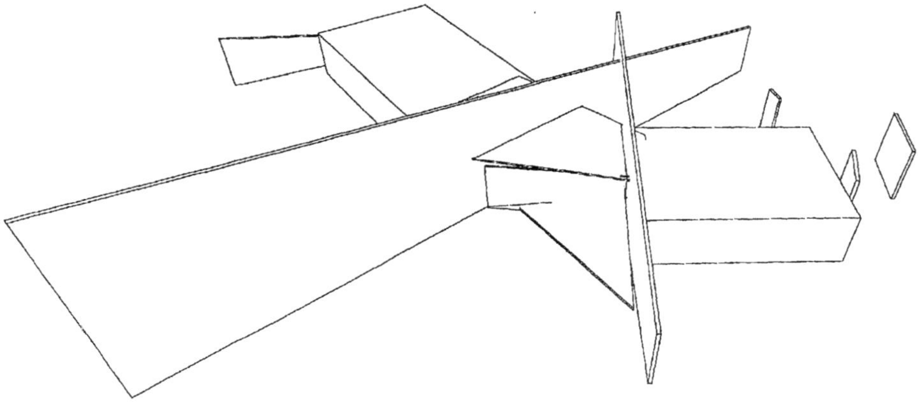

Figure 3. An early phase design concept by one of the subjects

But it is difficult or sometimes even impossible to use these programs in the design stage instead of the decision stage.

From the perspective of artificial intelligence, however, these developments are not surprising (Simon 1970a). At present, we are able to write complex algorithms to execute some specific problem-solving task. These algorithms usually pertain to generic tasks. A generic task is one that cannot be divided in sub-tasks (Chandrasekaran 1988). But we can ask ourselves whether architectural design is a problem-solving activity (Simon 1970a, 1970b; Newel and Simon 1972). We can also ask whether we may conceive of design and DDSS as a fixed number of sequential generic tasks to be executed. In the development of expert systems, it became common practice to involve a knowledge technologist with a background in the cognitive sciences. His task is to use several Knowledge Acquisition and Documentation Structuring techniques to map the expertise without being restricted to implementation formalisms. KADS is such a method. It enables us to map domain-specific knowledge into a conceptual model without being restricted to implementation formalisms (Breuker et al. 1987). This conceptual model can, in turn, be used for the implementation of a design-supporting system (DSS). This is the main reason why we suggest shifting the focus of the attention. Instead of talking about a computer-aided design approach, we would like to emphasize a different approach, namely, a designer-aided computer point of view (Daru 1984).

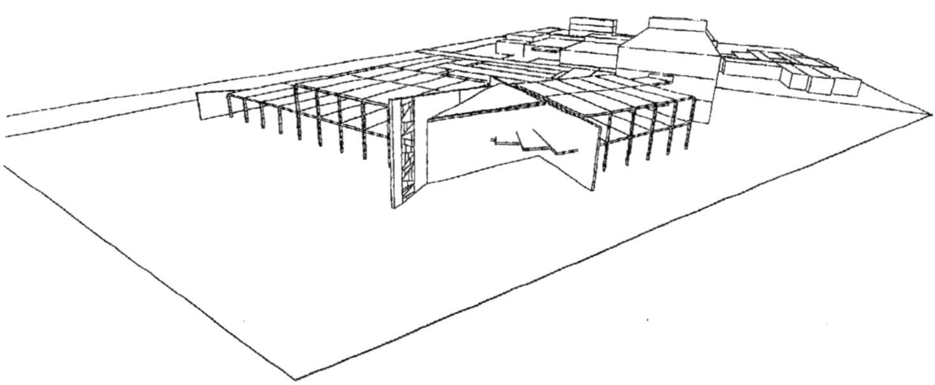

Figure 4. A more elaborate design concept in the sketch phase by one of the subjects

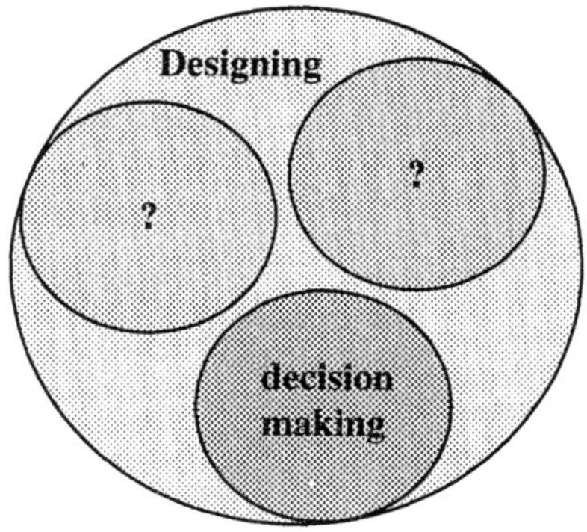

Figure 5. Decision-making as a sub-process of design

Notice that we refer to the approach as design support and not decision support. We believe that decisions follow design and are a sub-domain of design, not vice versa. This relationship is schematically represented in Figure 5. Making a decision means that you have to choose an action or solution out of a set of options based on one or more criteria like cost, relation between rooms and buildings, amount of available space, etcetera. Some preliminary design activities will already have taken place before a designer has to make a choice between alternatives. Most existing tools are suitable for decision support but not for design support. This is probably because such tools are not attuned to how the designer works in general, or at a specific moment in the design process in particular. The most obvious way, then, to support design processes is to study the design process at a higher level of abstraction.

1.3. DESIGN CHARACTERISTICS

Before we continue, we should clarify a basic question: what are the characteristics of design processes? Many theorists describe design processes in terms of flexible, creative, undetermined, multidirectional, rough activities. Designers often use paper and pencil to create an outline using global and sometimes specific information about the type of building, the site, the brief, and so on. Therefore I think flexibility, multidirectionality, roughness, etcetera should also be features of a good design support tool for these preliminary design activities.

This may be an important reason why decision-making tools are not often used in the first phases of design. There is simply not enough specific information available at the time to decide upon. Even more plausibly, the designer may want to be free of these restrictions for a while. Before we can implement a useful design support tool, we need to know what is to be analyzed, synthesized and evaluated. We have to know how this is done and in what order. In other words, we need to know more about the working style of the architect.

Style, in the context of this paper, has two aspects: a strategic and a tactical aspect. Strategy pertains to the general approach or plan a designer uses to come to a design. Tactics pertain to activities a designer executes in case his design strategy fails to be successful. We are not alone in assuming that the design processes of different designers can be distinguished from one another. Several other theoretical publications make the same assumption. Archea makes a distinction between designers who perform problem-solving activities in contrast with puzzle-making activities (Archea 1987). Perhaps a combination is possible, as suggested by Van Bakel (1991a, 1991b). Broadbent (1973) makes another interesting distinction between pragmatic, analogic, canonic, and iconic design activities. According to the definitions of these working strategies, it is very likely that the designer is going to use the available tools and information-processing systems in a different way for each strategy. As mentioned earlier, we are not alone in raising the possibility of distinguishing design processes. We are unique, however, in our approach. We use the Interactive Personality Model of Hettema, which in fact is a general model. It is used to predict and interpret human behaviour based on knowledge about specific situations (Hettema 1979, 1989). The relation of this model with respect to design is represented in Figure 6. To understand how the model can be used in this study, it will be introduced briefly in the next section.

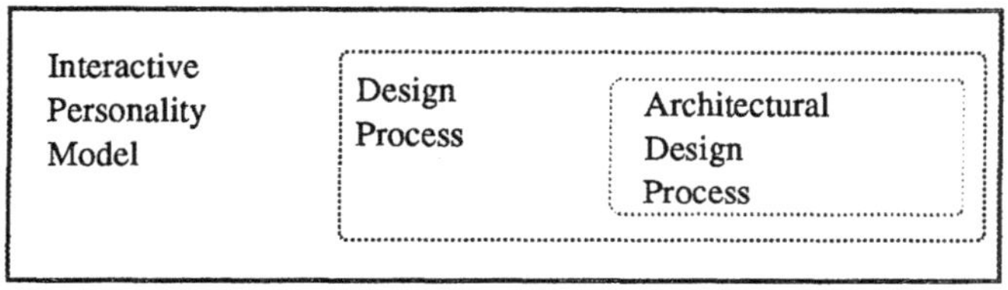

Figure 6. Relation between different levels of behaviour

2. Interactive Personality Theory

The interactive personality theory of Hettema (1979, 1989) describes personality as "an active, intentional entity that has considerable impact on situations." Furthermore, it states that "persons and situations reciprocally affect each other, so that persons obtain feedback on their actions from the situations in which they act" (Hettema 1989, pp. 18-19). The theory uses a three-layer model to describe the personality system. This model consists of a cognitive-symbolic level, a control level and a sensorimotor-operational level. A schematic representation of this model is shown in Figure 7, where it is tailored to design processes in general and architectural design in particular.

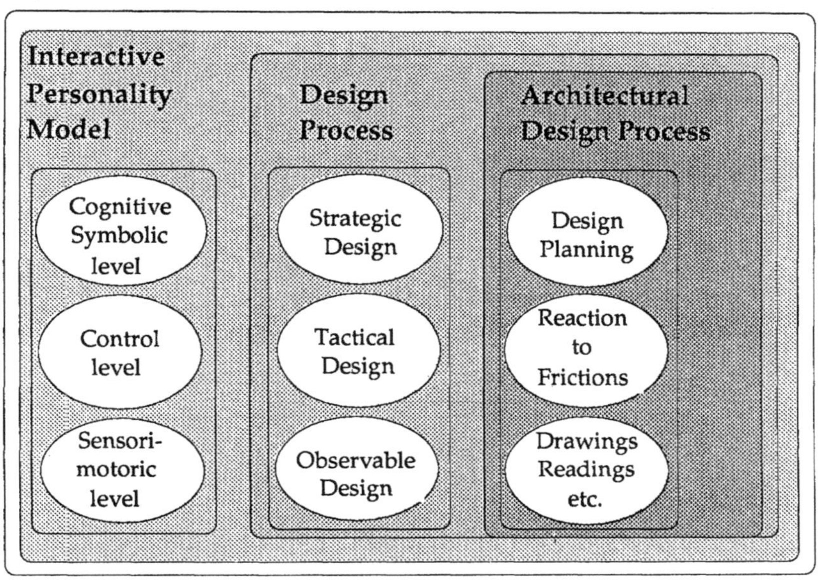

Figure 7. A three-layer hierarchy of design actions

2.1. COGNITIVE-SYMBOLIC LEVEL

At the cognitive-symbolic level, a person can make use of situation concepts and transformation rules. People have a conception of the present situation (Sy1), a conception of the desired situation (Sy2), and, according to the theory, preferences for particular transformation rules in specific situations (Ry). These preferences are based on social learning and private experiences and are summarized in the strategic subsystem of personality. There is an interesting parallel with the personality model of Hettema (1989) and the ideas presented by Simon (1970a). The latter says that there are only a few 'intrinsic' characteristics of the inner environment of thinking man that limit the adaptation of his thought to the shape of the problem environment. All else in his thinking and problem-solving behaviour is artificial; that is, learned and thus subject to improvement.

Translated to the domain of architectural design, this means that an experienced architectural designer uses a more elaborated practical experience to enter the design situation. Accordingly, an experienced designer might have a much riper strategy to execute a design task than a novice designer would have.

2.2. SENSORIMOTOR-OPERATIONAL LEVEL

The effect of the strategic system becomes manifest at the sensorimotor-operational level. That is, people observe the actual situation (Sx1), they execute specific actions (Rx), and they can observe the final situation (Sx2) when Sx1 is transformed. Behaviour specified by the strategic system may or may not be effective in transforming the environment as intended by the individual. Translated to the domain of architectural design, a plan of activities may not be practicable because of a shortage of time or because the techniques are not suitable for a solution.

2.3. CONTROL LEVEL

At the control level, processes have the capacity to alter the state of elements at the other two levels. At the sensorimotor-operational level, they can mould the environment to conform to cognitions. They do so by way of three possible state-transition mechanisms: exploration, substitution and persistence. An architect could look around to find support for his solution of the design problem (explore). He could execute a few other activities (substitute). Or he could try to convince his employer that his solution is the only possible in this case (persist).

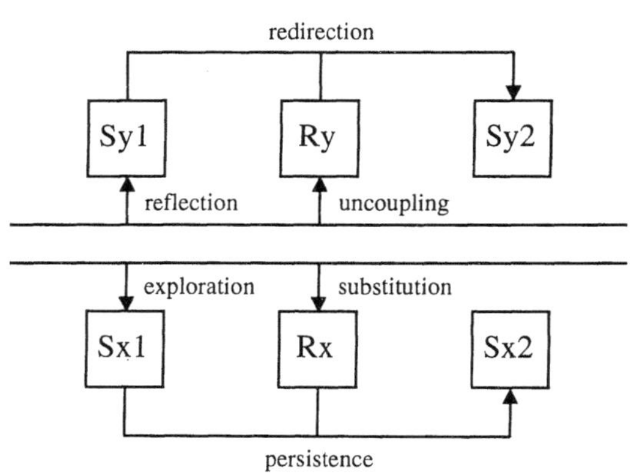

Figure 8. Relation between six state-transition mechanisms

At the cognitive-symbolic level, however, processes can replace cognitions to conform to the environment. They could do so by the use of three other state-transition mechanisms: reflection, uncoupling and redirection. An architect could alter his conception of the design problem (reflect). He could replace a few transformation rules and generate a modified plan (uncouple). Or he could decide to aim for a complete new solution (redirect).

The six state-transition mechanisms, represented schematically in Figure 8, belong to the tactical subsystem of personality and are used to maximize control. Tactics are also assumed to be largely innate mechanisms, acquired on the basis of heredity (Hettema 1989, pp. 170-171). However, in case the state-transition mechanisms fail to restore control, the disruption mechanisms - avoidance, extinction and distraction - are started to minimize non-control.

3. Experiment

3.1. BASIC MODEL

As far as the conceptual design model is concerned, we are looking for a general model that, despite its generality, can be elaborated to include more detail without losing its readability. This model should enable us to distinguish variations in the design process due to individual preferences based on personality characteristics and experience. In other words, the initiation phase should not be predetermined by theoretical assumptions of how design should take place, in order to make it possible to distinguish design strategies. Furthermore the number of recursions and switches between phases must also be flexible to make it possible to distinguish design tactics. Finally, the model must be flexible, allowing the addition of new information (and control) to the design process and permitting recourse to automated processes. If we can find a model that meets these conditions, we can use the model to distinguish design styles. Thus, we would be able to support a designer on the basis of his personal preferences. The assumption was that a design style as a function of human behaviour is determined by personality characteristics. It is also determined by foreknowledge based on culture and social learning (Bandura 1977; Mischel 1977; Pervin 1984), as well as by the environmental influences under which activities take place. This should preferably lead to an integration of parameters in the model that correlate with these aspects. The model should thus consist of three processes: (i) an integration of a constantly changing environment; (ii) an interaction possibility between the environment and the individual; and (iii) the integration of variable as well as consistent parameters. IDEF0 diagrams enable us to meet these conditions (IDEF0 1981). If we describe the components discussed above in the top level of an IDEF0 scheme, the model looks like the one represented in Figure 9. The figure represents the top layer of the model.

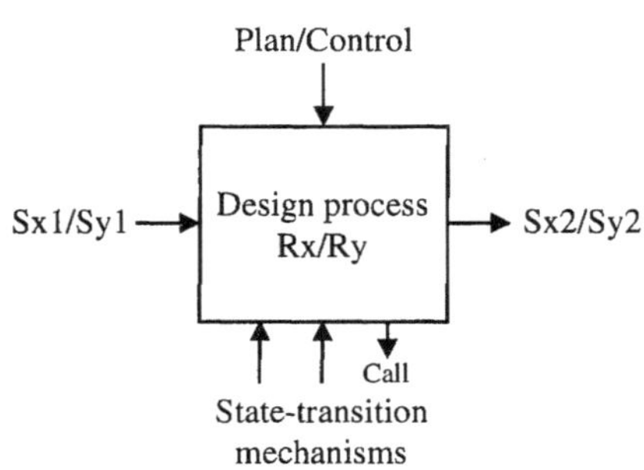

Figure 9. Top level of the design support model

We must realize that the number of design solutions is not finite. This is partly because architectural design concerns an ill-structured problem (Simon 1970b). We believe that the difficulty goes even beyond this statement. We agree with Archea that architectural design is not only about problem-solving but also about puzzle-making (Archea 1987; Van Bakel 1991a, 1991b). The consequence is that we do not have a finite number of solutions, not even when the designer structures the design task and creates sub-tasks to be solved as suggested by Simon (1970a). This is only true for the design constraints that a designer imposes to himself. An important fact is that these constraints are very personal. The restructuring of the design task takes place by calling up certain constraints (Ry), sub-goals, and generators (Rx) from his long-term memory. These are factors not specified in the problem statement. The formulation of the task is therefore dynamic (Coyne et al. 1990). Ry and Rx are thus very important parameters in assessing the design styles of architects.

3.2. WHAT DATA?

Our main concern in this investigation is to identify the characteristics of the strategic design styles of architects. In other words, we concentrate on the strategic level of the personality model. We want to know which Rys are preferred by individual designers. What kind of research do we need to conduct in order to elaborate the model? It might be helpful to consider developments in the cognitive sciences. These sciences are in a pre-paradigmatic stage. Therefore, the best way to study design processes from that perspective is to observe them, for design consists of cognitive as well as sensory processes (Derksen 1985). These observations can, in turn, lead to a useful implementation-independent model to interpret design activities at a high level of abstraction. Hettema's Mode * Media Matrix offers an opportunity in this direction.

*TABLE 1. Mode *media matrix*

		mode		
		Self report	Psycho-physiological measurement	Observation
medium	Verbal	I personality questionnaire	II lie detection	III verbal intelligence
	Pictorial	IV projective measures	V film stress studies	VI perform. intelligence
	Natural	VII interview	VIII telemetry	IX natural observation

3.3. SUBJECTS

The subjects were 11 male architects in the age range of 45-55 who responded to an announcement in a monthly journal for architects (B.N.A. 1990). The journal is distributed by the B.N.A., which is an abbreviation for Bond van Nederlandse Architecten (Union of Dutch Architects). The announcement asked for experienced architects who would be willing to take part in an investigation of design style.

3.4. EXPERIMENTAL DESIGN

The focus of this paper lies on the strategic aspect of design styles. These are represented at the cognitive-symbolic level. Design strategy pertains to the plan a designer has when he is asked by a client to design a building. At this point, the only necessary information concerns the type of building to be designed. We can ask the architect what he would like to do to arrive at a solution. In making this plan of activities, there is no cause to change a preferred approach to finding a design solution. Therefore, cells I and VII of the Mode * Media Matrix will be used. Four techniques were used to acquire the necessary data. The first phase of knowledge acquisition took place at the private design studio of the architect. The other three took place in a laboratory situation.

3.4.1. *Knowledge Acquisition 1.* In the first phase, all subjects were visited at their own design studio. Notes were taken listing the furnishings of the room, such as books, drawing Tables, drawing tools, computers, etcetera. During this visit a structured interview was also conducted, using a questionnaire as a guideline. The questionnaire included items like: "What do you do if you have difficulty finding the solution you are looking for?", "How much communication is there between you and your client?" and "What architectural style do you like very much and what style do you dislike very much?" These interviews were recorded on audiotape.

3.4.2. *Knowledge Acquisition 2.* In this session, the architect was asked to describe the activities he would preferably carry out if he were asked to design a day-care center. In other words, what is his plan to reach a solution? That was the only information provided about the hypothetical task at the beginning of the session. If the architect wanted additional information about a topic, this was given to him. So it was not really necessary for the subject to design the building if he did not like to design. The session took place in a laboratory situation specially designed for this research. Most of the necessary equipment that was available at a subject's own studio was also available here. Also materials concerning the design task were available, such as pictures of the scenery, a map, a brief, information about the client, municipal destination of the area, etcetera. When the subject asked for additional information, it was given to him. These protocols were recorded on audiotape.

3.4.3. *Knowledge Acquisition 3.* In the third session, a card-sorting technique was employed, using a Macintosh II computer with a 19-inch monochrome screen, Design/2.0 software to execute the task, and Screen-Recorder 2.0 to tape all the screen activities. The concepts displayed on the screen were derived from the structured interview of the first session, augmented with concepts from the NNI, which is the Dutch Standardization Institute. The subjects were asked to talk aloud and the protocols were recorded on audio tape. The task to be executed was formulated as follows:

> In this task several concepts are displayed randomly at the bottom of the figure. There is also a matrix in which the concepts can be ordered. The matrix has two columns and several rows. The left column should contain concepts regarding all design tasks as they are executed in practice. The right column can contain concepts that are sometimes related to a design task. The top row contains concepts that are relevant at the beginning of the design task as executed in practice. The bottom row contains the concept pertaining to the end of a design task. It is your task to order these concepts in the matrix. You are free to remove, rename or add concepts at your discretion.

3.4.4. *Knowledge Acquisition 4.* In the last session, subjects were again asked to employ a card-sorting technique, using the same equipment as described for Session 3. The task, however, was different. The task to be executed was formulated as follows:

> In this task you again see a number of concepts randomly displayed at the bottom of the screen. You are asked to split these concepts into two categories. Each of the categories has to be labelled. You can do this by using a concept included in the group or one that does not appear on the screen. Repeat this activity with each group until you have a maximum of two or three concepts in the resulting sub-group. As soon as you have only groups of two or three elements, you'll have to regroup the members of these small groups using labels displaying why they belong together. You can use an existing concept as a label or you can add a new one. Repeat this activity until you are satisfied with the result.

3.5. RESULTS

There are two types of results. The first type consists of the protocols recorded during each session. The second consists of the material produced by the subjects, either on paper or on the computer screen. The results will be discussed in the order of the sessions.

3.5.1. *Results Session 1.* Some architects described the architectural design process as a spiral like the one discussed by Zeisel (1981). According to the description they gave, the design process is like a spiral starting very wide and gradually narrowing down as it reaches the design solution. Ten out of the 11 subjects mentioned that communication is a very important activity that takes place during the general design process (Column 2 of Table 2). And four of them even mentioned the importance of communicating with the client while creating the concept or image of the design (Column 3 of Table 2). Communication with the proper authorities and other people involved influences the design process. This involves gathering extra information, discussing propositions, but also negotiating about elements described in the brief. Many subjects also gain some extra information by reading magazines or books about the topic, and some even plan a vacation to get new impressions for a particular design task.

3.5.2. *Results Session 2.* In the analysis of the protocols recorded during this session, it appeared that subjects had different ideas about which elements play an important role in the design p rocess. The number of elements and activities differed. One interesting fact was that all subjects mentioned three particular elements they considered important in every design task. These are the brief, the situation, and the imaginary solution or concept (Column 4 of Table 2).

The situation concept refers to the location, urban plan, site, etcetera. The distinction between the brief and the situation regards the flexibility of their elements. The brief can be changed more easily than the situation. Whereas you can easily change the number of rooms in the brief, it is more difficult to change the existing surrounding buildings on the site. Of course, according to the subjects, there are exceptions.

The brief is described as containing the conditions that the solution should address, such as the volume, the height, the number of rooms, the organization that is going to use the building, etcetera. Some of the subjects even make a distinction between qualitative and quantitative aspects of the brief. The quantitative aspects are the ones mentioned above. Qualitative aspects concern how the building should look or what it should represent. Typically, some of the architects who make this distinction also mention that this aspect is explored by communicating with the client. The other subjects who do not make this explicit distinction claim to find these qualitative aspects in the situation and the brief not.

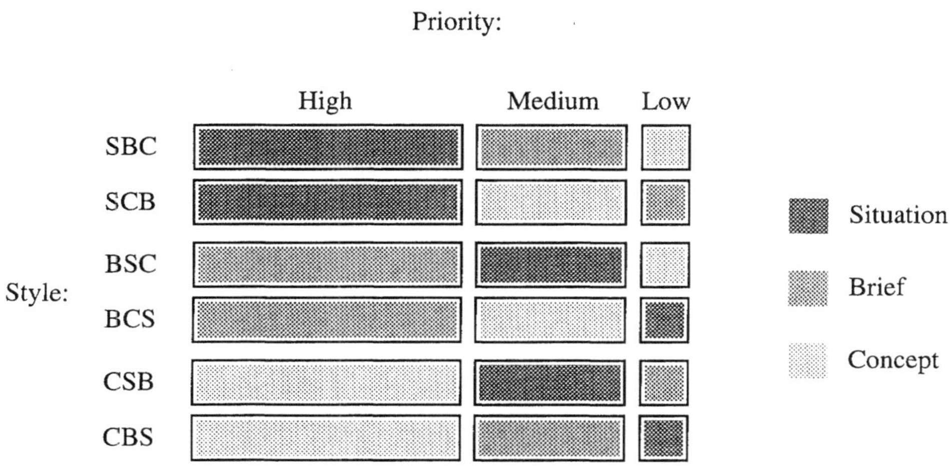

Figure 10. Six style types using three parameters

The imaginary solution or concept is in fact a visual image that the subject has upon receiving the task or after gaining some extra information. It is not necessarily a complete solution but includes what the building should look like and/or self-imposed constraints. In this session, subjects had to describe a plan of activities after receiving the globally formulated design task. An interesting fact was that the order in which these elements were discussed differed among the subjects. Also the importance of the concepts as described by the subjects differed. Another even more interesting fact is that all possible 3*2*1=6 sequences appeared. This is schematically represented in Figure 10.

Each bar shows the priority of the concepts according to the subjects. The widest part in the bar corresponds to the aspect that was given the highest priority. The smallest part of the bar corresponds to the aspect that was given the lowest priority. This priority was estimated by the sequence in which the aspect was discussed, the amount of time spent on it, and by direct questions. Some of the architects explicitly mentioned that this preference of activities cannot always be carried out in practice, for instance because of time pressure. The distribution of these 11 architects across the six styles is 2,2,3,1,2,1, as shown in Figure 11.

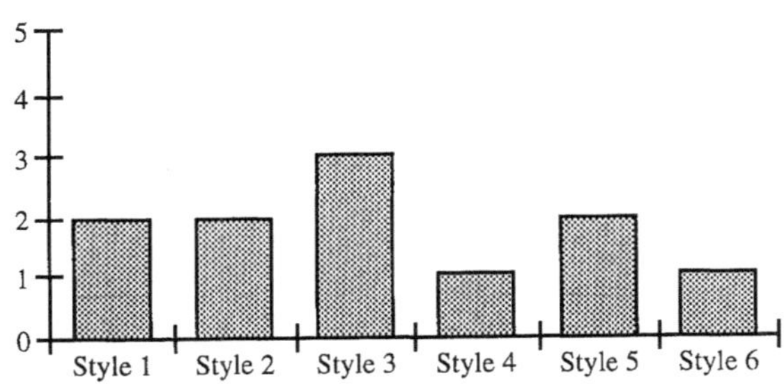

Figure 11. Style distribution of knowledge acquisition 2

3.5.3. *Results Session 3.* Some remarkable differences between architects are displayed in Table 2. As shown, the three aspects of design discussed under the results of the previous session appear again in the Table (Column 5 of Table 2). However, the order in which they appear differs for three subjects with respect to the results of the previous session. This new frequency, as obtained by this technique, is displayed in Figure 12. According to Figure 12, style 2 (situation, imaginary solution, brief) and 4 (imaginary solutions, situation, brief) don occur in practice for this population. Some architects emphasize finances as an important determinant of design, whereas others don (Column 6 of Table 2). The use of alternative designs, variants and models is also variable (Columns 7, 8, 9 of Table 2).

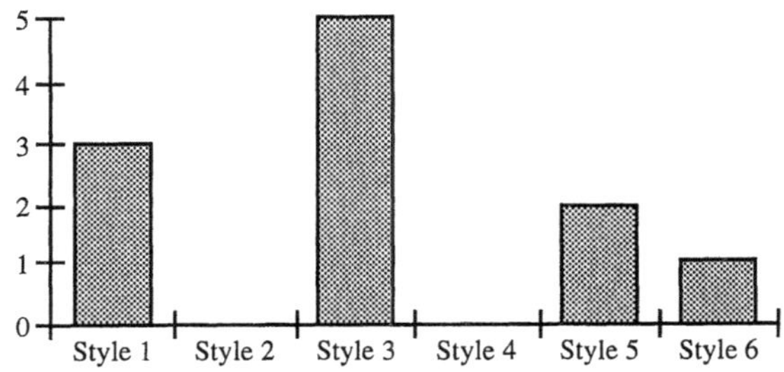

Figure 12. Style distribution of knowledge acquisition 3

3.5.4. *Results Session 4.* Every structure obtained by the subjects differed from those of the other subjects. Summarizing these results, we can say that architects have different ideas about the relations between actions and concepts.

TABLE 2. Quantitative results of the knowledge acquisition sessions

	Session 1			Session 2		Session 3			
Sub ject	Comm	Comm conc	SBC	SBC	Strategy shift	Finan- ces	Alterna- tives	Variants	Model
1	•		123	123			sometimes	sometimes	sometimes
2	•		123	123		•	sometimes	always	sometimes
3	•		312	312			sometimes	sometimes	always
4	•		132	123	•		always	always	sometimes
5	•		213	213			sometimes	always	sometimes
6			312	312			never	never	always
7	•		132	213	•		always	always	sometimes
8	•	•	213	213		•	always	always	sometimes
9	•	•	231	213	•	•	always	always	sometimes
10	•	•	321	321			sometimes	always	sometimes
11	•	•	213	213			sometimes	always	sometimes

4. Discussion

Looking at the results of the different techniques, we can say that interaction and communication are two fundamental aspects of the architectural design process. All the architects mention the importance of communication with the proper authorities and the client to gain the necessary and relevant knowledge to reach a design solution. To some architects this interaction with the client is even desirable to get more insight in the qualitative aspects of the brief. This means that some of the architects hold discussions with their clients about what the design should look like. The implications for the design-supporting system to be developed are that it should allow communication at different levels with the client and the proper authorities.

Architects mention that in practice design is exposed to changes at different levels. Changes can take place at the qualitative level by changing the imaginary solution, using another detailing because of financial problems, etcetera. But changes can also take place at the quantitative level, for instance by fundamental changes in the brief concerning the size and height of the building, the number of rooms, etcetera. The design-support system should enable us to implement these changes easily and as soon as possible at the different levels involved with these changes.

If we compare the results of Figure 11 and Figure 12, there seems to be a contradiction in the data obtained (see Figure 13). They are, in fact, not contradictory because of the formulation of the tasks being executed. In the second session, subjects were asked to describe a plan of activities they would prefer to execute during a design task, while in the third session they were asked to describe the activities as they take place in design practice. This was confirmed by asking these three subjects to explain the differences between the results. This was done by first explaining the assumption that all design tasks consist of at least three elements: namely, situation, brief, and imaginary solution. They were then asked to rank these with respect to preferable importance. The results were in accordance with those of Session 2.

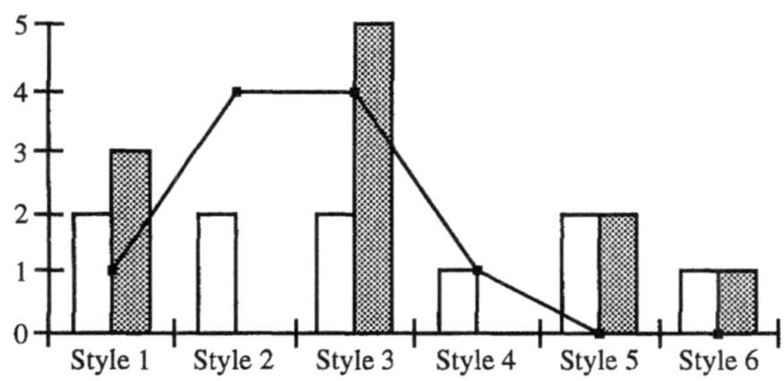

Figure 13. Non-linear correlation between strategy and tactics

Then they were asked whether this could always be achieved in practice, and to rank these elements with respect to design practice. This time the results were in agreement with those of Session 3. Interpreting these results with respect to the Interactive Personality Model of Hettema, this means that indeed there is a distinction between design strategies and design tactics in practice (Rx, Ry). The line in Figure 13, which displays the variance between design plan and design practice, is fortunately not straight. This means that architects can indeed have a design plan that in practice needs to be adjusted because of particular frictions that occur.

Also on the basis of the elements distinguished by the IPM and the elements retrieved by these experiments we can say that the model might be useful in distinguishing strategic design styles. If we refine the basic IDEF0 scheme of Figure 9 by using these results and the IPM, the next layer looks like the one shown in Figure 14. The model enables the user to follow his design strategy by providing flexibility in the initiation phase and by allowing different recursions. The only restriction is that each phase should be encountered at least once in order to come to a design solution. Depending on the tactical mechanisms (that need to be assessed in future research) preferred by the user, a strategy shift can take place. Furthermore a critical analysis of design tools (Daru 1992) can help us in suggesting a specific tool for a specific situation. For instance, if a designer prefers to begin with the imaginary solution, the system might suggest the use of a tool such as Architrion. If this does not lead to a satisfying conceptualization, the system might suggest the use of a different tool, depending on the design strategy and design tactic of the user.

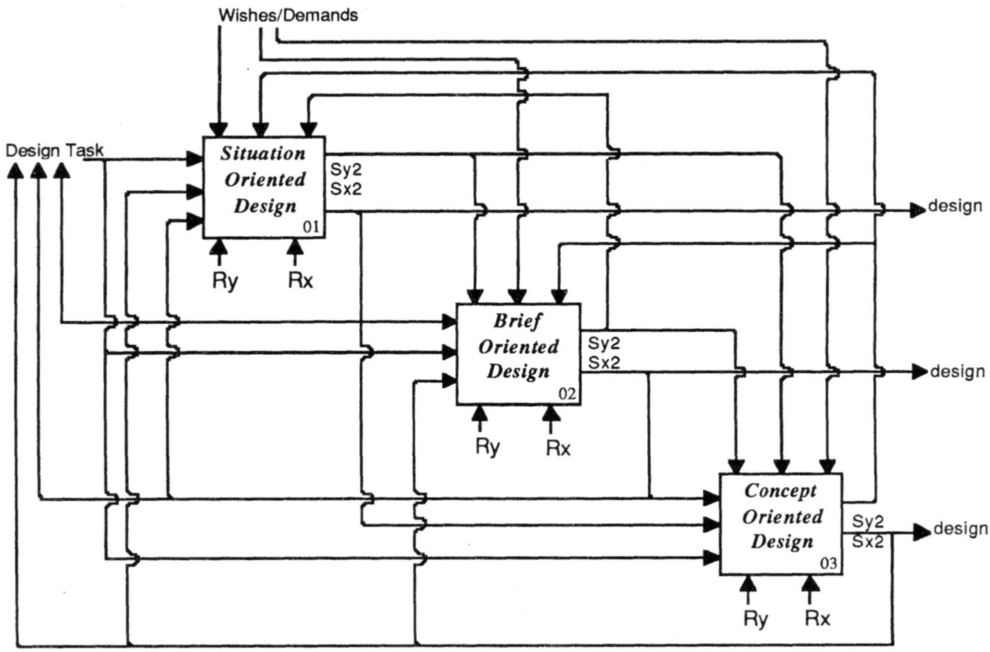

Figure 14. Architectural design style model

To test the usefulness of this model, future research is aimed at the development of an SRS personality questionnaire and the assessment of psycho-physiological measurements. The SRS questionnaire will be used to assess a design strategy in a guide and easy way. SRS is an abbreviation for Situation-presented, Response, and Situation-transformed. This questionnaire should be specifically developed for the domain of architectural design. A description of the construction of such a questionnaire can be found in Hettema (1989). At present, we are working on a prototype. The use of this questionnaire, if it is balanced concerning the representation of domain-specific situations, can probably lead to a first fruitful attempt to create a useful design-supporting system. In case the design-support model enables us to distinguish design strategies and design tactics, it is very likely that the system based on this model can make a suggestion about which tool to use in a specific problem situation based on personal preferences of the designer, which have been assessed before implementation. The advantage of using SRS questionnaires is that we are not restricted to the use of fictitious design tasks, executed in a laboratory (Teyken 1988; Hamel 1990). The psycho-physiological measurements will be used to assess the design tactics. This knowledge, and the use made of the model in previous design tasks, can be assembled in a hypercard stack. This stack should keep track of tactical manipulations and strategy shifts due to experience. As can be seen, a lot of work still needs to be done. But we can conclude that the approach discussed here can lead to a practical and useful design-supporting system based on personal preferences correlated with design styles.

Acknowledgements:
First of all I would like to thank Tom van der Zanden for letting me use his complete documentation of a child day-care center. I would like to thank Helmi Gerrits and Roel and Myriam Daru for their critical reviews of the text. I would like to thank Hugo de Haan for letting

me use his fantastic drawings in this paper and Phil Winteraeken-Bruls for helping me create IDEF schemes. And last but not least, I am grateful to all the architects and students who took part in this study. You know who you are!

References

Archea, J. (1987) "Puzzle Making: What Architects do when no one is looking", in E. Y. Kalay (ed.), *Computability of Design,* John Wiley and Sons, London, pp. 37-52.

Archer, B. (1969) *The Structure of the Design Process*, Clearinghouse, London.

Bandura, A. (1977) *Social Learning Theory*, Prentice-Hall, Englewood Cliffs, New Jersey.

B.N.A. (1990) *Mededelingen* **8**, 4.

Breuker, J.A., Wielinga, B., Van Someren, M., De Hoog, R., Schreiber, G., De Greef, P., Bredeweg, B., Wielemaker, J., Billault, J-P., Davood, M., and Hayward, S. (1987) Model-Driven Knowledge Acquisition: Interpretation Models, ESPRIT Project P1089, Deliverable D1 (Task A1), University of Amsterdam and STL Ltd.

Broadbent, G. (1973) *Design in Architecture*, John Wiley and Sons, London.

Chandrasekaran, B. (1988) "Toward a Functional Architecture for Intelligence Based on Generic Information Processing Tasks", in J. Boose and B. Gaines (eds.), *Knowledge Acquisition Tools for Expert Systems*, Academic Press, London, pp. 1183-1192.

Coyne, Richard D., Radford, A.D., Balachandran, M., and Gero, John S. (1990) *Knowledge-Based Design Systems*, Addison-Wesley, Sydney.

Daru, R. (1984) "Designer Aided Computing in Tegenstelling tot Computer Aided Designing", *Bouwadviseur* **26**, 11-20.

Daru, R. (1992) "The Performance of Design Support Tools", paper presented at the DDSS Conference, Mierlo.

Derksen, A.A. (1985) *Wetenschap of Willekeur*, Coutinho, Muiderberg.

Hamel, R. (1990) *Over het Denken van de Architect*, AHABooks Art History, Amsterdam.

Hettema, P.J. (1979) *Personality and Adaptation*, North-Holland, Amsterdam.

Hettema, P.J. (1989) *Personality and Environment: Assessment of Human Adaptation*, John Wiley and Sons, Chichester, England.

IDEF0 US Airforce/Soft Tech Inc. (1981) IDEF: Integrated Computer-Aided Manufacturing (ICAM). Architecture; Part II, Vol. IV, V, VI, Function Modelling Manual USAF, Report Number AFWAL TR-81-4023, Wright Patterson AFB, Ohio.

Mischel, W. (1977) "Self Control and the Self", in T. Mischel (ed.), *The Self: Psychological and Philosophical Issues*, Rowman & Littlefield, Totowa, New Jersey, pp. 31-64.

Newel, A. and Simon, H.A. (1972) *Human Problem Solving*, Prentice-Hall, Englewood Cliffs, New Jersey.

Pervin, L.A. (1984) *Personality: Theory and Research*, John Wiley and Sons, London.

Simon, H.A. (1970a) *The Sciences of the Artificial*, MIT Press, Cambridge, Massachusetts.

Simon, H.A. (1970b) "Style in Design", in J. Archea and C. Eastman (eds.), *Proceedings of the 2nd Annual Environmental Design Research Association Conference*, Dowden, Hutchinson and Ross, Stroudsbury, pp. 1-10.

Teyken, C. (1988) *Ontwerpen als Proces*, Zwets & Zeitlinger, Amsterdam.

Van Bakel, A.P.M. (1991a) "Design Styles of Architects", paper presented at the First International Conference on Style in Philosophy and the Arts. 3-5 April, Amsterdam.

Van Bakel, A.P.M. (1991b) "Stylistic Description instead of Methodological Prescription of Architectural Design", in L. Bay Tseng (ed.), *Proceedings of the Second IES Conference*, Singapore, pp. 108-116.

Zeisel, J. (1981) *Inquiry by Design: Tools for Environment-Behavior Research*, Brooks/Cole Publishing Company, Monterey, California.

ARCHITECTURAL COMPUTER VISION: AUTOMATED RECOGNITION OF ARCHITECTURAL DRAWINGS

A. KOUTAMANIS and V. MITOSSI
Faculty of Architecture
Delft University of Technology
Berlageweg 1, 2628 CR Delft
The Netherlands

ABSTRACT. Computer vision offers the ability to transform digitized drawings into documents that can be used with computer systems. Recognition of digitized drawings can occur at the levels of (a) geometric elements, (b) building elements, and (c) spatial articulation. The last two levels apply not only to digitized images but also to computer-produced ones. The enormous burden placed on the user for inputting and manipulating CAD drawings suggests that automated recognition can add to the capabilities of CAD by making the computer more flexible with respect to inputting design information and more responsive to the actual concerns of the designer.

1. Digitized Drawings

One of the pressing problems in the computerization of architecture is the transfer of existing architectural drawings to the computer. This is done either to preserve and disseminate historically significant documents or to make the (usually massive) archives of a professional design firm accessible to computer-aided design (CAD) systems. The traditional manual digitization of drawings using tracing devices like graphic tablets to input existing drawings to CAD systems is too labour-intensive and unreliable to offer a viable large-scale solution. Optical digitization of architectural drawings resolves the problem to a degree only. Digitized drawings are reasonably accurate and detailed copies of the original. Their resolution and portability are adequate for any practical purpose such as reproduction with laser printers and plotters. However, digitized images may also demand enormous storage capacity and are often slow to recall and manipulate. In particular, digitized images are more difficult to modify than CAD documents, while they are incompatible with design analysis and evaluation programs.

One way to reduce storage sizes, improve recall/manipulation times, and enhance compatibility/utilization with CAD and other computer programs is the automated recognition of digitized architectural drawings. Recognition of digitized drawings can occur at three levels. The first is recognition of geometric primitives such as lines and shapes in a drawing, which amounts to a direct transformation of a digitized image into a CAD document. The second level is recognition of building elements, such as doors, walls and columns. This represents a transformation of the digitized drawing into a more abstract, comprehensive and knowledgeable description that remains compatible with CAD and other computer programs. The third level is recognition of the more implicit spatial articulation of a design. This level is the most relevant to architectural design thinking, as it facilitates analysis and evaluation from several points of view, as well as the execution of corresponding design manipulations.

2. CAD Drawings

Another problem that architects will sooner or later have to face concerns the arduous nature of

H. Timmermans (ed.), Design and Decision Support Systems in Architecture, 17–25.

current computerized drafting, which offers little more than a digital equivalent of the drafting table. In computerized drafting, the user has to input the description of a design in great detail and with high precision. The computer facilitates certain drafting operations such as copying and pasting or repetition of a standard element far better than the adhesive sheets of symbols and drafting templates used in hand drawing. Moreover, it allows what appears to be extremely effortless and versatile manipulation of drawings. Any individual line and shape can be modified in position or dimensions with great ease and accuracy, while grouping capabilities simplify such operations on more than one graphic element.

All these, together with the unprecedented compactness of storage, paint a very positive and optimistic picture of computerized drafting. This picture certainly conveys all the advantages but fails to acknowledge some of the problems that arise in practice. One such problem is currently obscured by the employment of specialized staff to run the computer facilities instead of the architects who design a project. This problem is that effectiveness, efficiency, reliability, and flexibility in the modification or other manipulation of a computer-produced drawing is primarily dependent upon the structure of the graphic elements which comprise the drawing. Placing a single element on the wrong layer, in the wrong group, or connecting by mistake the dimensions of unrelated building elements does not simply reduce the speed of manipulation. It may also render useless the documents which were automatically produced on the basis of the drawing, such as bills of materials and cost estimates. In order to avoid the frustration and financial loss caused by such events, it is imperative that the structure of a CAD document is always impeccable.

The enormous intellectual burden placed on the user for inputting and manipulating computer-produced drawings can be reduced through the automated recognition of the building elements and spatial characteristics in the configuration of lines and shapes input by the user. This effectively means that the computer interprets the user input and automatically elaborates/structures the representation of a design with respect to both general graphic/geometric aspects (Pavlidis 1988) and the specific domain-dependent issues discussed in the present paper. The computer can recognize the building and spatial elements in the configuration of lines input by the user and allocate each identified element to the correct layer and group, as well as establish the appropriate connections between identified elements. The user can therefore input the drawing more freely and flexibly, more or less as he would present the design to a skilled draftsman. This type of computer support to user interaction represents a logical and fruitful evolution of drawing software toward the established goals of design automation. In fact, recognition of building elements and spatial arrangement in a CAD document is far easier than in a digitized image because of the former's reliance on vector graphics as opposed to the latter's pixel basis.

3. Geometric Primitives

Recognition of geometric primitives such as lines and shapes is essentially similar to what has already been attempted for the automated conversion of engineering drawings to CAD documents (Hofer-Alfeis 1987; Joseph 1989). Chain coding and edge following are employed in order to identify the straight and curved line segments in a drawing and subsequently also the two-dimensional shapes (mostly closed loops) these line segments form. Figure 1 is an example of the application of chain coding and edge following techniques to the recognition of geometric primitives in a floor plan. First, the appropriate chain code is allocated to each black pixel of the image and then uninterrupted sequences with the same (or, for curves, related) coding are grouped together. Further grouping is possible on the basis of rule systems that examine the relationships between connected or adjacent groups of pixels (i.e., line segments) against prototypical or typical line configurations in standard symbolizations of doors and windows, for instance. In architectural drawings, the width of lines varies more than in schematic or mechanical engineering drawings. This variety poses more problems for the already tenacious line following (problems which can be remedied by thinning the image prior to line following), but it also allows for greater precision in the identification of shapes composed by the recognized line segments, because in many cases only intersections of segments with the same width are considered.

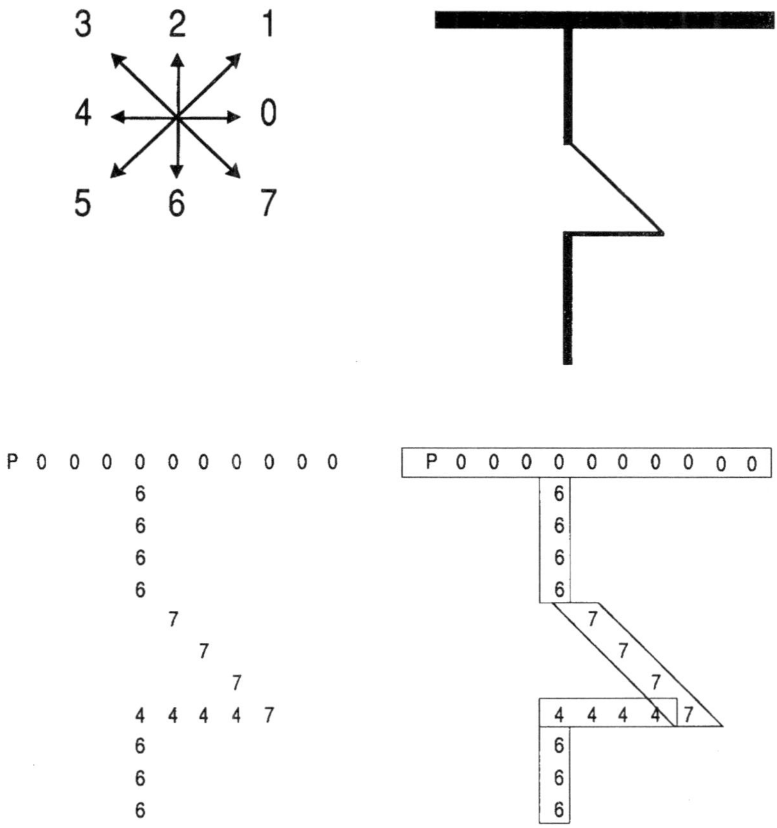

Figure 1. Recognition of geometric primitives using line following: an example
top left: chain coding scheme; top right: floor plan part; bottom left: coding of pixels (starting from an arbitrary point P); bottom right: grouping of pixels

In orthographic projections (floor plans, sections, elevations) recognition of geometric primitives transforms the digitized image into a collection of two-dimensional shapes that is fully compatible with two-dimensional drafting software. In perspective projections (which may even include shadows and shading), three-dimensional shapes formed by the line segments can be recognized by adaptations of the technique used for the recognition of line drawings of scenes with trihedral objects (Waltz 1975). These recover the volumetric composition of the scene and thus render the digitized image compatible to three-dimensional CAD software. Recognized axonometrics and isometrics normally require few modifications and corrections, while other perspective projections may require extensive dimensional corrections.

Recognition of lines and shapes effectively makes the digitized document compatible with

CAD programs, as these also employ geometric primitives to describe a design. Recognition at this level is therefore extremely useful for the computerization of existing drawings and their subsequent manipulation with computer programs. However, as related research into the transfer of other types of engineering drawings has shown, it is also computationally tenacious and expensive (Hofer-Alfeis 1987; Joseph 1989). Recognition techniques and algorithms are in several cases only poorly served by the widely available computer technology, and the current patterns of change do not suggest great optimism for the immediate future. Moreover, the utilization of a CAD document is always dependent upon its structure (subdivision of graphic elements into layers and groups, annotations). This means that in order to use the recognized drawing for anything beyond simple graphic operations, it is necessary to develop its structure interactively. This is a time-consuming and labour-intensive process. In conclusion, it appears that the main benefit from this level of recognition is compacting the size of a digitized image while at the same time improving the usability of the image with the types of software most often used by architects and designers.

4. Building Elements

The second level at which recognition can be applied to architectural drawings is that of building elements such as walls, columns, doors or windows. Most of these elements have a more or less standardized appearance in each type of architectural drawing, as manifested by the intensive use of predefined symbols in architectural drafting (on adhesive sheets or drafting templates in hand drawing and in the form of menus and 'CAD libraries' in CAD). Such standardization invites the use of *recognition* (as opposed to drafting) *templates* (Figure 2). Template-based recognition of building elements in architectural drawings is essentially similar to template-based optical character recognition (OCR). Recognition is achieved by matching each segment of the image to a number of known examples or prototypes which represent the types of symbols possible in an image of a particular type (i.e., a floor plan or a section). As in OCR, a certain degree of training and adaptation is required, and recognition can be efficient only if the number of possible patterns is relatively low.

The number of basic types of building elements in each kind (mostly floor plans and sections) and scale (typically 1:200 to 1:50) of architectural drawings is indeed limited. But considerable variations in the symbolization of building elements can be caused by differences in the materials and form of the real elements, as well as by differences in drawing conventions. The situation is analogous to OCR of multi-font texts where similar variability is confronted by appropriate structuring of the template database and/or abstraction of the patterns (e.g. skeletonization of characters so as to eliminate font attributes). More effective in the case of architectural drawings and especially for floor plans, the most essential drawings for conveying information on building elements, is the use of elastic templates and templates with 'don't care' pixels (Persoon 1988). Such templates can easily account for local perturbations, small variations in orientation, and other imperfections caused either by real differences in the original image or by the process of digitization.

One of the key problems in the use of templates for the recognition of building elements in architectural drawings is that some elements do not occur in standard sizes. In a floor plan at the scale of 1:100, for example, one can expect that practically all door symbols in a residential building will be approximately 1 cm wide. The use of elastic templates or templates with 'don't care' pixels can account for small variations in size or type. The length of wall symbols, on the other hand, may vary from a few millimetres to several centimetres. The use of adaptive hierarchical partitioning techniques like quadtrees can resolve such problems through the subdivision of the image into parts containing a single building element symbol. Partitioning on the basis of a uniform grid determined by the size of key elements, such as doors, is also possible. In some cases (e.g. in drawings of buildings which make use of standardized elements or modular coordination), partitioning is more relevant to the structure of the particular design.

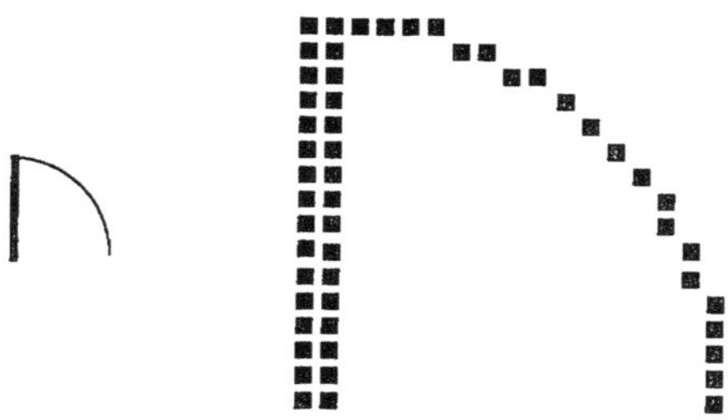

```
00000000000000000000
01111110000000000000
01100001100000000000
01100000011000000000
01100000000100000000
01100000000010000000
01100000000001000000
01100000000000100000
01100000000000010000
01100000000000010000
01100000000000001000
01100000000000001000
01100000000000000100
01100000000000000100
01100000000000000100
01100000000000000100
01100000000000000100
```

Figure 2. A door template
left: a common representation of a door in floor plans; middle: pixel array corresponding to the representation; right: a template for the recognition of the representation: ones match to black and zeros to white pixels

Recognition of building elements, being a knowledge-based process, is more demanding than that of geometric primitives. Still, recognition at this level can be more efficient and reliable, as well as more relevant to the designer user. The resulting description of the image as an array of building element symbols can always be transformed into a metrically precise and detailed drawing by substituting each instance of a symbol with an accurate drawing of the denoted building element. This technique is in fact more effective and efficient than the structuring of the elements produced by recognition at the level of geometric primitives, especially since it can accommodate different conventions or habits in computerized drawing (i.e., different approaches to the use of layers and grouping relations) and hence also ensures compatibility with a greater number of drafting software.

Another advantage of describing a drawing in terms of its building elements, even if the description is rather abstract, is that certain design analyses and evaluations (e.g. structural and thermal analyses) apply directly to the configuration of building elements. For these evaluations, the type of building elements and their physical/spatial relationships are generally more important than their precise dimensions and detailed appearance. In the framework of the above transformation of the abstract symbol array into precise and detailed drawings, these evaluations can even play a role in determining the precise geometric attributes of the building elements.

5. Spatial Articulation

The third and more advanced level at which automated recognition can be applied to visual architectural representations concerns the spatial articulation of a building and, through that, the most essential aspects of architectural design. On the basis of the spaces in a floor plan and the arrangement of these spaces with respect to geometry, adjacency and physical access, for instance, it is possible to evaluate the design with respect to design requirements concerning circulation (Koutamanis and Mitossi 1991) or even formal goodness (Koutamanis 1990).

Recognition of spatial articulation is best performed on the basis of *skeletonized* versions of the digitized image. The purpose of skeletonization is to reduce information existing in the image to the absolutely essential for the recognition of spatial characteristics. This is achieved by abstracting or even removing whole building elements or properties of elements that are only secondary for the overall spatial articulation of a building. These may be the number of stairs and the exact form of a staircase, the thickness of walls, the type and precise position of doors and windows, as well as the furnishing of a space. A skeletonized floor plan, for example, includes only the boundaries of each space in the building, reduced to unity thickness (Figure 3).

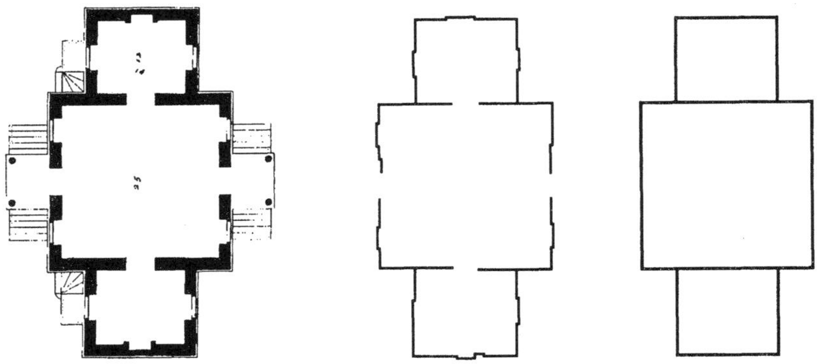

Figure 3. Skeletonization of a floor plan
left: plan; centre: the axes of its components; right: normalized skeleton

Skeletonization is achieved by first thinning the digitized image, whereby lines of unity thickness in the original digitized image (normally symbolizations of stairs, furniture, the swing of a door and similar elements) are eliminated, then filling small gaps that have emerged in the skeleton (normally openings in the boundaries of spaces or intercolumniation spaces), and finally normalizing the skeleton (on the basis of a small set of chain coding transformation rules, as in Bowie and Young (1977) so as to eliminate accidental or insignificant perturbations.

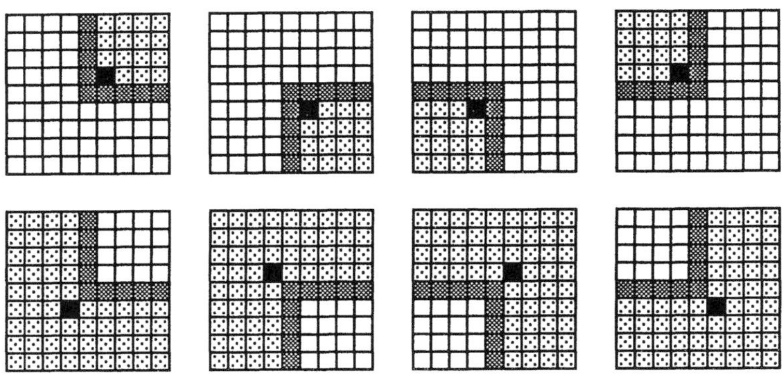

Figure 4. The typology of space corners in rectangular floor plans
top row (from left to right): types I, II, III, IV; bottom row: ¬I, ¬II, ¬III, ¬IV; dark cells represent black pixels (skeleton lines); light cells represent white pixels which belong to the space under investigation; black cells represent the corner under investigation

Line following can be used to identify boundaries of spaces in the skeletonized floor plan in the form of closed loops of line segments (Lawson and Riley 1982; Joseph 1989). An alternative, less tenacious, and conceptually more consistent alternative is to employ a variation of the technique used for the recognition of line drawings of three-dimensional scenes with trihedral objects (Waltz 1975), which also relates to the perception of illusory (subjective) contours and figures. According to this technique, recognition of spaces in a floor plan is based on (i) a typology of edge junctions (i.e., interior space corners; Figure 4), in the form of templates that are matched to the appropriate parts of the image, and (ii) the connectivity of different junction types, which allows the development of expectations concerning the type and position of the neighbours of each junction (i.e., other corners of the same space directly linked to it with edges; Figure 5) (Koutamanis 1990).

vertex	up	right	down	left
I	II,¬III	IV,¬III	*	*
II	*	III,¬IV	I,¬IV	*
III	*	*	IV,¬I	II,¬I
IV	III,¬II	*	*	I,¬II
¬I	III,¬II	III,¬IV	*	*
¬II	*	IV,¬III	IV,¬I	*
¬III	*	*	I,¬IV	I,¬II
¬IV	II,¬III	*	*	II,¬I

Figure 5. Connectivity of space corner types for rectangular floor plans

The conceptual advantage of the above technique is that, while edge following deals with the elements that *bound* a space, this technique concentrates on the space itself, in very much the same way that illusory shape perception operates. The resulting practical advantages include direct recognition of the shape of a space and relative freedom from local perturbations and missing parts in the boundary of a space. The latter can be of particular significance for the recognition of drawings of buildings with semi-unified or flow spaces.

The resulting description of a building in terms of its interior spaces forms the basis of most aspects of architectural design thinking and paves the way for the integration of the corresponding analyses and evaluations on the basis of standard design representations. This integration is traditionally one of the toughest problems for the development of coherent and comprehensive computer-aided design systems (Schmitt 1988). An example of the degree of integration allowed by automated recognition is that the description of a building's spatial articulation can be used to automatically create geometric and topological representations for the analysis and evaluation of dynamic design aspects such as the circulation of people, goods, or even information in a building (Koutamanis and Mitossi 1991). The topological representation is automatically derived from the (geometric) floor plan and correlated with programmatic requirements so as to determine the various circulation patterns (route and flow) at an abstract topological level. These patterns are then transferred back to the floor plan in order to obtain a more accurate representation of the route and its possible conflicts with the form of the building (Figure 6).

A more advanced case of analysis and evaluation is that of the formal goodness of a building, which includes automated recognition of grouping relationships between interior spaces (such as alignment, repetition and adjacency) and direct correlation of the resulting spatial configuration with architectural formal constraints. These make explicit in the drawing which elements and relationships pertain to a particular architectural formal system ('style'), such as classicism, and hence allow for an accurate and precise consideration of the design with respect to this formal system (Koutamanis 1990).

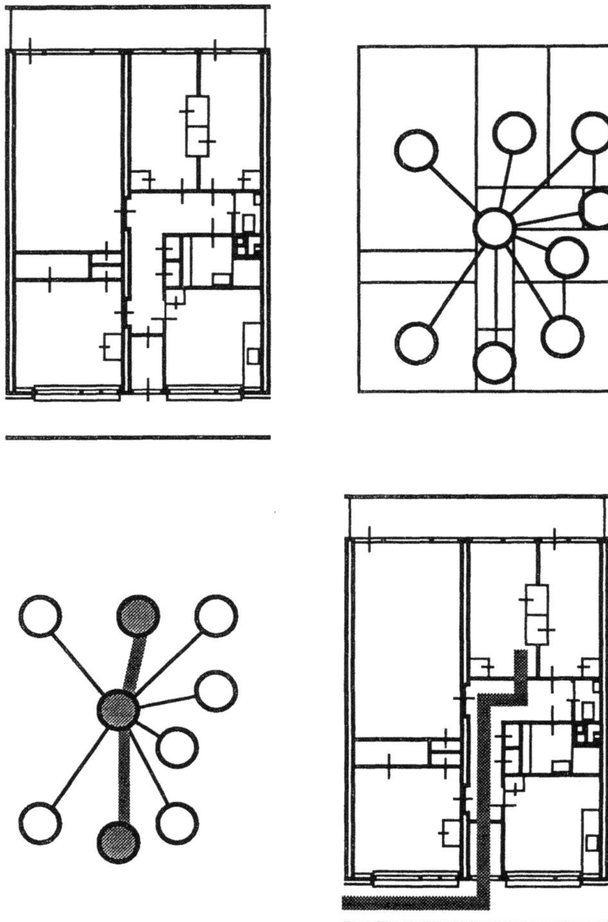

Figure 6. Transformation of a geometric design representation into a topologic one, for the analysis and evaluation of pedestrian circulation in dwellings

6. Extensions

Automated recognition of architectural drawings has obvious significance for the two problem areas we have already outlined: the transfer of existing architectural drawings to the computer, and the automated interpretation of user input to CAD systems. The transformation of the graphic elements comprising a drawing into a dual correlated network of building and spatial elements presents further possibilities which can have far-reaching extensions. One such extension concerns the use of autonomous robots in building construction.

The complexity of building tasks and the inability to use standard spatial set-ups like an assembly line on all sites suggest the potential of autonomous robots in building. Such autonomous robots can operate only if they have clear answers to what should be done and where. Architectural drawings can serve as the basis for the answer to both questions. The configuration of building elements clearly represents the final state of the building process. Moving a building component which arrives on site to its proper position involves the ability to recognize this component as part of a particular structure which has been specified somewhere in the drawings. Recognition of the building elements in a drawing can therefore provide answers concerning where something should be placed and in what relation to other objects.

The movement of a robot on a site, as well as in a finished building, represents another class of problems whose solution is at least facilitated by the recognition of the spatial elements represented in a drawing. The configuration of spaces in a floor plan obviously offers a measure by which the robot can judge the final product of its endeavours. At the same time, it offers a comprehensive spatial framework which facilitates understanding of not only how the robot can move from one space to another but also where exactly a building component should be positioned even if no related components are already in place. This spatial framework and its relationships with the building structure also facilitate the use of autonomous robots after construction, for instance in the diagnosis of structural and other faults, maintenance work and security supervision.

References

Bowie, J.E. and Young, I.T. (1977) "An Analysis Technique for Biological Shape —II", *Acta Cytologica* **21**, 455–464.

Hofer-Alfeis, J. (1987) "Automated Conversion of Existing Mechanical Engineering Drawings to CAD Data Structures: State of the Art", in K. Bø, L. Estensen, P. Falster, and E.A. Warman (eds.), *Computer Applications in Production and Engineering, CAPE'86* , North-Holland, Amsterdam, pp. 259–267.

Joseph, S.H. (1989) "Processing of Engineering Drawings for Automatic Input to CAD", *Pattern Recognition* **22**, 1–11.

Koutamanis, A. (1990) Development of a Computerized Handbook of Architectural Plans, Ph.D. Dissertation, Delft University of Technology.

Koutamanis, A. and Mitossi, V. (1991) "On the Representation and Analysis of Dynamic Aspects of Architectural Design", in F.H. Post and A.J.S. Hin (eds.), *Proceedings Second Eurographics Workshop on Visualization in Scientific Computing*, Delft.

Lawson, B.R. and Riley, J.P. (1982) "ISAAC: A Technique for the Automatic Interpretation of Spaces from Drawn Building Floor Plans", in A. Pipes (ed.), *CAD 82: 5th International Conference and Exhibition on Computers in Design Engineering*, Butterworths, Guildford, pp. 663–666.

Pavlidis, T. (1988) "Image Analysis and Graphics", in R.A. Earnshaw (ed.), *Theoretical Foundations of Computer Graphics and CAD*, Springer Verlag, Berlin, pp. 1177–1195.

Persoon, E. (1988) "A Pipelined Image Analysis System using Custom Integrated Circuits", *IEEE Transactions on PAMI* **10**, 110–116.

Schmitt, G. (1988) *Microcomputer Aided Design for Architects and Designers,* John Wiley and Sons, New York.

Waltz, D. (1975) "Understanding Line Drawings of Scenes with Shadows", in P.H. Winston (ed.), *The Psychology of Computer Vision*, McGraw-Hill, New York, pp. 19–91.

HOUSING TENANCY, DATA MANAGEMENT AND QUALITY CONTROL

J. SMEETS
Eindhoven University of Technology, Faculty of Building and Architecture
Department of Architecture, Urban Planning and Management
P.O. Box 513 Postvak 4, 5600 MB Eindhoven
The Netherlands

ABSTRACT. This paper deals with housing tenancy, data management and quality control. The proposed method is focused on quality characteristics of housing estates in view of rentability risks. It entails a cycle of registration, analysis and implementation of measures. The starting point is the behaviour of the housing consumer in a market-oriented context. The model is framed within theories of strategic management and marketing. Systematic registration and evaluation of consumer behaviour, by means of a set of relevant process and product indicators, can yield relevant information in the four phases of the rental process: orientation, intake, dwelling and exit. This information concerns the way in which the dwelling (characterized by product indicators) fits the needs of the consumer. The systematic analysis of the process and product indicators during the phases of the rental process makes a 'strength-weakness analysis' of housing estates possible. The indicators can be presented in aggregated form by way of a 'rentability index'. The 'strength-weakness analysis' steers the intervention in the quality characteristics of housing estates. The possibilities for readjustment, however, are different. The quality control system is not only an early warning system, but also has several other functions: evaluation, planning and communication. The method described here lays a solid foundation for a decision-support system in the area of housing tenancy.

1. Introduction

At present, few issues attract as much attention among housing managers as the quality of the housing stock. In the background, a number of developments on the housing market have given quality a high priority. The shift from a supply to a demand market in various segments and regions is especially pertinent. Due to these signs of saturation, more attention is being paid to stock management. When intervention in the housing stock takes place, the role of the market becomes more important. It involves the competitive position of dwellings and the extent to which they contribute to the preferred differentiation in type of housing and in price. The conditions in which management is taking place on the local level have been altered sharply as a result of deregulation and decentralization.

For a long time, the discussion about housing stock quality was concerned with the decline of housing estates as a consequence of problems in the post-war housing stock operated by non-profit landlords. The Prak and Priemus (1986) model of decline describes a relation between social, physical and financial decline. The interdependence between these clusters of factors leads to a spiral of further decay. Methods are also being developed within this model's framework to signalize decay. Spiral movements in the model have been recorded by means of indicators (Spaans and Hoenderdos 1989). This approach has some limitations, however. It only registers the weak points in housing estates, while strong points remain underexposed.

Our approach is not concerned with decline but with quality, specifically the appreciation of housing estates, in a negative as well as in a positive sense. Quality characteristics of housing estates and quality preferences of the dwelling client are central. Moreover, this client-oriented

H. Timmermans (ed.), Design and Decision Support Systems in Architecture, 27–37.

approach is embedded in a discussion of market-directed housing management (Smeets et al. 1989; Boekhorst and Smeets 1990).

2. From Product to Client

Under the influence of changing relations on the housing market and the government's policy of deregulation and deconcentration the environment in which housing managers operate is subject to radical change. This has far-reaching consequences for the organizations that are engaged in the management of housing. Housing managers are becoming increasingly aware that their product-oriented approach is no longer sufficient (Smeets et al. 1989).

With this product-oriented way of thinking and acting, the manager sets his goals and indicates appropriate working processes. The organization is structured along those lines (see Elffers and Vulperhorst 1991). To a large extent the product approach seeks to achieve internal goals and is thus an inside-out approach. The reason why managers insufficiently anticipate unexpected and/or undesirable developments can largely be attributed to this approach. In order to reduce environmental dependency (for instance by utilizing emerging opportunities and/or avoiding risk), a flexible future-directed (external) and innovative way of thinking and acting is required. The inside-out approach ought to be replaced by a more externally oriented, outside-in approach.

In his book Images of Organisation, Morgan (1986) makes a distinction between 'egocentrism' and 'the self-reflective evolution'. An 'egocentric organization' tries to preserve its own identity in a threatening outer world. It becomes preoccupied with its own importance and underestimates the meaning of the wider context. The expression 'self-reflective evolution' suggests that changes in an organization reflect an evolution of the own identity in relation with the environment. Morgan argues for more self-reflection and less self-centredness. The main characteristic of an externally oriented organization is that it puts the consumer (the market) at the centre. In the marketing literature, this interpretation is called the 'consumer-directed' mentality or marketing concept. With this interpretation, the manager will be able to enlarge his capacity to offer (prospective) tenants his services effectively and efficiently. Looking for a client-oriented form of organization is a fundamental and self-reflecting process: what can I do for the client?

A client-oriented strategy needs an organizational structure that clearly anticipates market instabilities. A solution might be a division into market segments, where client groups are identified for whom differentiated products and services are being developed, with freedom of choice for the client, and where each division is responsible for the satisfaction of the housing needs of a specific client group. Implementation of the marketing concept is vital to the survival of an enterprise. The marketing concept should be regarded as an instrument rather than a goal in itself. At the same time, it is a way of thinking that needs to be part of the corporate culture, which should be open to the marketing concept. When this is the case, the organization can be structured in such a way that its various departments are able to act in a market-oriented fashion.

3. Strategic Management and Quality Control

In the future, housing managers should be able to familiarize themselves with a more market-oriented way of thinking and acting. An aid to this end is strategic planning. Strategic planning means a systematic preparation of an answer to the question whether the organization will offer certain products to a specified public in a given market area and by which means and organizational services it will carry out all these tasks. The most important elements of this method are the formulation of objectives, the identification of target groups, the identification of product/market combinations, internal strength/weakness analysis, inventory of opportunities and threats and strategic choices (Sweringa 1989).

Strategic management includes the activities the manager expects to undertake in the long term in response to changes in his surroundings. On the basis of an inventory of the present situation, a vision of the future of the various sectors of the housing stock is necessary. The changing role of information supply should be pointed out. In view of the speed with which developments appear, a need for a systematic procedure arises whereby these trends and incidents can be quickly identified. Only then can an organization react in time. Information will become an increasingly important production factor, alongside factors such as finance, personnel and housing (Nauta and Smeets 1989). One of the activities concerned is the measurement of the quality of the housing services, i.e. the degree to which the housing services fulfil the client's needs (Suurmond and Fokkema 1989). After all, dwelling characteristics do not only concern the product, that is, a bundle of housing services. They also imply a coherent package of management services rendered during the tenancy process. In regard to quality control, several aspects can be distinguished (Figure 1; see also Born et al. 1991).

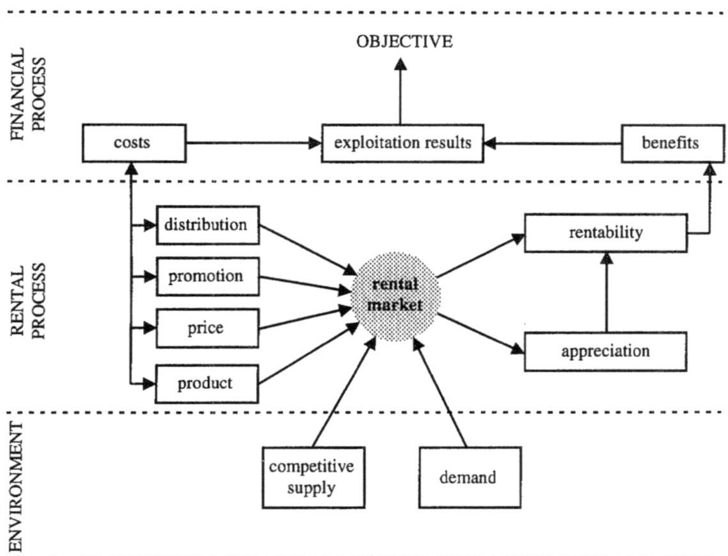

Figure 1. The effort, its results and the environmental factors
Source: Boekhorst and Smeets (1990)

The *input* efforts are distinguished as product, price, distribution and promotion; these are also called the marketing mix. The product is the dwelling and the bundle of housing services it offers. The price consists of the rental price and additional service costs for the amenities supplied with the dwelling. The distribution concerns the channels through which the product will be put onto the market (centrally, decentralized, or by third parties). The promotion, finally, is related to the way in which the product will become well-known (information, advertising, word-of-mouth publicity).

The *output* or outcome of the investor's efforts can be expressed in terms of appreciation, rentability and exploitation results. Appreciation refers to how the product meets the user's expectations. In a general sense, a high appreciation will lead to a good rentability. The rentability determines the estate's benefits. The efforts with regard to product, price, promotion and distribution together determine the estate's costs. Benefits and costs together form the exploitation result. This way, a good rentability does not necessarily lead to a positive exploitation result.

The *fit* of activities to the preferred results takes place during the tenancy process. This process has three phases: (i) the orientation and offer phase in which the (potential) occupant is orientating himself. He looks around, collects information and facts, makes a comparison and so on. Finally the occupant will express his preference in favour of one or more estates. The tenant's specific interest will lead to a direct reaction by the latter: an offer; (ii) the occupation phase: after the tenant has made his choice, the dwelling will be occupied for an indefinite period, and (iii) the exit phase when the tenant has decided to orient himself on the housing market again with the aim to find another dwelling.

The *environment* which has the following aspects. First of all, to housing managers, (potential) tenants are regarded as an internal environment. Here they form a number of primary and secondary target groups. Of course, competing housing suppliers are also relevant. In addition, a number of external environmental factors might influence an estate's rentability and exploitation results. The estate's manager has no influence on these. For instance, the interest rate could directly affect the price of a house without changing its quality. Also the consumer's confidence in the economy may strongly influence decisions with regard to a housing situation, as well as the (regularly changing) public housing policy of the city council and the central government.

In the framework of strategic management, a system of quality control can be considered as an instrument. It can be considered as a systematic procedure with regard to registration, signalization, assessment and strategy determination. With this procedure, the appreciation and rentability of individual estates can be controlled and information may become available for the benefit of quality policy and supply stock and, in the long run, new construction programming for the housing estate. Such systems do have different functions (Dochy 1990). First, they have an early warning function, which concerns locating and identifying the strong as well as the weak points (measuring) and the developments therein (monitoring). Then, they have an evaluation function which involves visualizing to what extent the chosen objectives have actually been reached and in what regard intensified intervention is preferred. Moreover, they have a planning function, which is aimed at stimulating the rationality of the method and the management. The methodology allows a more targeted effort of the managing instruments. Finally, they have a dialogue function which is directed at the promotion of communication between the various parties concerned. Such systems could help objectify the information exchange and thus offer a basis for dialogue between parties with different interests.

4. Method of Approach

In the system proposed here, a difference is made between product and process data. Process data give insight into the question whether or not the housing services fit the dwelling consumer's preferences. They indicate the efforts in the sphere of promotion and distribution (see Section 4.1.). Product data give insight into the question why this fitting turns out successfully or not. These critical success factors also offer guidelines for an adjustment of the quality policy (a given product at a given price).

4.1. PROCESS DATA AND THE 'RENTABILITY INDEX'

In a market-oriented approach the consumer's decision process can be taken as a basis for the evaluation system. In the offer, occupation and exit phases, important quantitative information can be collected. The information is already available to the organization (Figure 2). In the orientation and offer phases, amount and kind of interest can be measured, as can the degrees of acceptance and refusal. The duration of occupation of new houses is also relevant. In the occupation phase, duration of tenancy and kind of occupation behaviour can be measured. Indicators such as occupation duration, structural vacancy, paying behaviour, complaints and requests for comfort improvement can be registered and converted into signals. In the exit phase, matters such as mutation rate and friction vacancy are of importance.

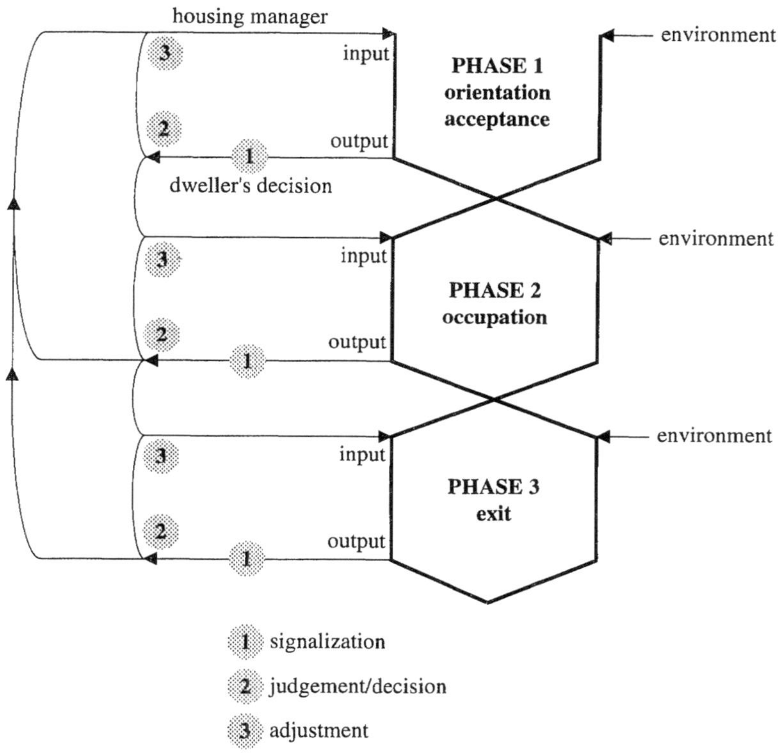

Figure 2. The phases in the rental process
Source: Boekhorst and Smeets (1990)

The tenancy process can be interpreted as a measurement path. In each phase, measuring points exist, making it possible to trace data in regard to the rentability. Data-collecting is taking place in an object-oriented fashion; one does record data per 'administrative' estate. At large high-rise estates, the task may be split up in storeys as well. A condition for reliable information, however, is a careful measurement procedure. Also important is that these data are collected over several years. As we know, important signals can be extracted from trends more than from the comparison of estates at one given time. To get good insight into the meaning of the indicators, it is recommended to develop an aggregated quality index (De Kaper and Knaapen 1991). Such an index, where indicators from the diverse phases are compounded into one index, can function as an early warning system.

For a Dutch housing association in Amersfoort, Nauta (1991) composed a so-called rentability index on the basis of an indicator from each phase of the rental process (offer, vacancy and mutation rate). In this index, it is supposed that the offer rate is the most important one; it tells most about the difference between the offered and the preferred quality (weighting factor 2). Moreover, the vacancy rate gets more importance than the mutation rate. After all, in case of a high mutation rate, the vacancy level remains low, and the estate is thus well rentable. So the vacancy rate gets a weighting factor of 1.5. At the same time a number of standards are set: the offer rate = 250%; the mutation rate = 10%, and the vacancy rate = 2%. These standard figures correspond with the means of the rates found in the city of Amersfoort.

The rentability index (*RI*) is calculated as follows:

$$RI \quad = \quad \frac{M + 1.5V + 2O}{4.5} \ \times 100 \qquad (1)$$

where,
M = mutation rate / 10;
V = vacancy rate / 2;
O = offer rate / 250.

If the index is 100, then it is said that there is a normal situation. The higher the RI, the worse the rentability and the greater the efforts to be made. The index can be calculated at three levels: per estate, per dwelling type and per neighbourhood. Table 1 shows a calculation on the dwelling level. The influence of the several indicators is different in various types.

The relatively high rentability index on single-family houses is a result of the offer rate's heavy weight. The offer rate on single-family dwellings is considerably higher than on multi-family houses. The very high rentability index on rooms, studio houses and the like is a result of the high mutation rate. In the background of this refusal behaviour lies the difference between 'starters' and 'movers'. A great number of the latter consist of 'preference movers' with a comparatively favourable negotiation position in the distribution process of housing accommodation. On the basis of this preliminary review we could go into greater detail on a number of estates. Such an index would be able to function as an 'early warning system'.

4.2. PRODUCT DATA AND QUALITY PROFILES

Further analysis is possible if, for the various phases of the rental process, in addition to quantitative data, more qualitative data are collected by means of intake interviews, satisfaction studies and exit interviews. To explain the reviewed processes, Boekhorst and Smeets (1990) identify five important factors: dwelling, dwelling type, physical environment and location, social environment and image (Figure 3). The dwelling concerns the household's private area, with everything belonging to it, such as the dimensions, number of rooms, rent level, house type, the view from the house. The dwelling type concerns all aspects of the dwelling situation which a certain number of neighbours have in shared use, such as the building's entrance, the hall, the stairwell, elevators and walkways. In short, it concerns the semi-public part of the dwelling situation. The physical environment and the location concerns all physical aspects of the dwelling's situation in regard to the public part, such as the access roads, shops and other services, green space, parking facilities etc. The social environment concerns how occupants maintain contact with each other, with their environment and with the manager, to what extent they associate with each other, the degree of social coherence, etc. The area's image in the public opinion is the 'condensation' of all aspects that may influence the quality, leading to the qualification of 'bad neighbourhood', 'nice area' etc. This opinion could be based on one's own observation or experience, on others' opinions, reports in newspapers and so on. Because this concerns opinion, it does not necessarily need to be consistent with facts.

TABLE 1. Rentability index by housing type (SCW, Amersfoort)
Source: Nauta (1991)

Dwelling type	Number VHE	Mutation rate	Vacancy rate	Offer rate	RI 4th quarter	RI 3rd quarter
Walk-up housing with elevator	1354	11.2	0.3	216	68	74
Walk-up housing without elevator	585	15.9	0.3	255	86	80
Shared doorway housing	4046	14.2	0.4	232	80	82
One family housing	6310	5.4	0.2	314	71	72
Senior apartments	547	7.3	0.1	315	74	67
Senior groundfloor dwellings	321	5.9	0.1	355	78	85
Group housing (seniors)	16	12.5	0.1	100	47	18
Group housing	39	15.4	1.5	100	77	69
Rooms	290	34.8	1.2	112	118	124
Studio houses	61	26.2	2.9	100	124	113
All dwellings	13020	9.4	0.3	247	71	72

In various phases of the rental process, these quality factors and aspects will be registered (Figure 3). A systematic analysis of acceptance and refusal behaviour on the estate level might give insight into the role of various quality aspects at play in the various estates. The analysis in the occupation phase has to be directed to the relation between quality factors and matters such as duration of occupation, pattern of complaints and desired improvements differentiated according to target group. After refusal motives, especially exit motives are an important source of qualitative information. Finally, the results of such an analysis can be registered in an estate profile, an overall picture of an estate's strong and weak points. In addition, it is possible to aggregate the data from the target groups' point of view. In the long term, more insight into the appreciation by the various target groups can be obtained suggesting which factors are very important and which ones less important for the appreciation of a dwelling. The target groups might possibly be identified by age of the main occupant, composition of the household, or household income. If we take into account the factor of life-style, we can also differentiate between 'modern' or 'traditional' households and those of an 'urban' and 'suburban' life-style (VROM 1992).

On the basis of data from intake interviews and exit interviews, in the long term insight can be obtained on such questions as the following: What factors are target group-dependent and independent? What factors lead to acceptance or refusal of a dwelling among various target groups? What factors are important but do not lead to refusal? What weak characteristics are compensated by strong quality aspects? When this information is available, it may be used for product development directed specifically to certain target groups. The information, however, will have to be translated into 'standards' for housing programmes. The quality profile of our example estate (Figure 4) shows that 'dwelling' and 'price' form the weak points. The stronger points are 'social environment' and 'dwelling type'. With increased competition, however, 'dwelling' and 'price' might form a threat. Corrections to dwelling and/or rent would then be necessary (see Section 4.3.).

Before turning to the strategic choices in the framework of the strategic planning, an exploration of environmental factors is of importance. These factors can be subdivided into 'opportunities' to attain goals and 'threats' which might be a danger to existing or prospective measures of management. Several examples have been suggested (Sweringa 1989; Nauta 1991): development in the consumer's legal position as an individual and as a group (rent protection, consumers' organizations); developments on the demand side of the housing market (for instance increase in small households); developments on the supply side of the housing market (competition behaviour); developments in the residential environment (changes in the supply of services, situations of inconvenience); developments in technology (increasing home automationand information networks); and macro developments. Environmental factors are developments which are hardly manageable. However, the manager can try to head them off.

THE DWELLING
ground floor/multi-storey
number of rooms
size living room
size other rooms
layout
closed/open kitchen
kitchen equipment
sanitation/2nd toilet
finishing
garden orientation/size
size and situation balcony
privacy
view
insulation level
rent
service charges
management
storage
separated shower/bath
quietness
own choice

THE HOUSE TYPE
house type
- single-family house
- multi-storey house
quality of entrance
- prestige
- comfort
- privacy
kind of entrance
- gallery
- shared doorway
recreation space
elevator

THE PHYSICAL ENVIRONMENT
accessibility by public/private transport
parking facilities
distance to (central) facilities
greenery
suitability for children
maintenance of the environment
recreational possibilities
layout
quietness

THE SOCIAL ENVIRONMENT
educational level
household composition
demographic structure
opinions about values and norms
social control
security
vandalism
public order
local authority allocation
suitability for children

IMAGE
the public opinion about:
- neighbourhood and location
- housing type and management
- the dwellings
- the manager
- the social environment
- layout of the neighbourhood
- design of dwellings and buildings (identity, style, prestige, diversity)

Figure 3. Quality factors and aspects influencing the appreciation of a dwelling
Source: Boekhorst and Smeets (1990)

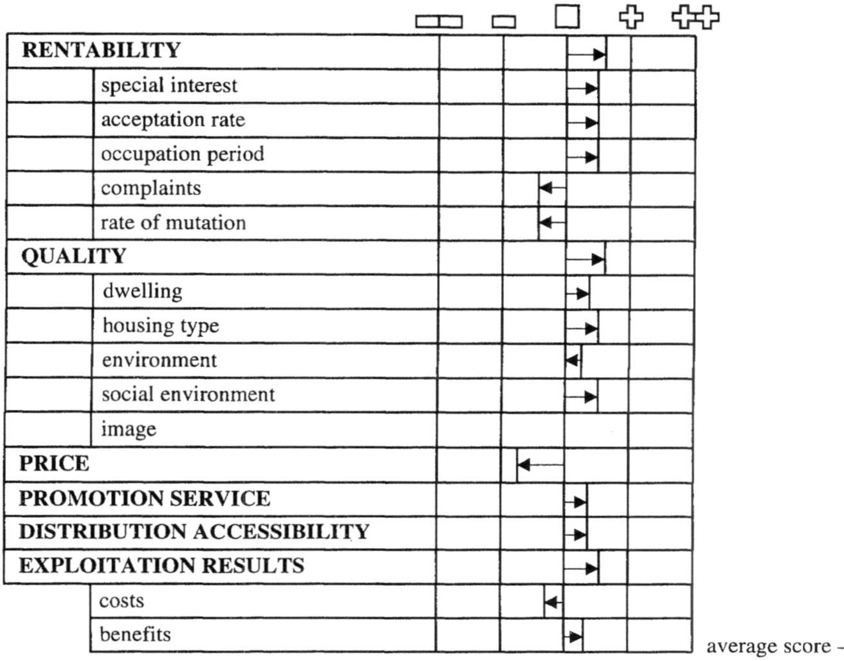

Figure 4. Quality profile of a housing estate (fictive example)
Source: Boekhorst and Smeets (1990)

4.3. DETERMINATION OF THE STRATEGY

On the basis of these analyses, a strategy can be formulated on an intervention planned. Such a strategy or planning intervention is based on identifying the state of affairs of a housing estate. The level of reference is the administrative complex toward which the system is directed. On the basis of the insights obtained, interventions can be determined in time (see Vijverberg 1989). What is necessary or preferred will be determined per project/estate in terms of the kind of strong and weak points, the opportunities and threats, and the possibilities to make adjustments.

In regard to the possible solutions, the extent of adjustability is of importance (Boekhorst and Smeets 1990). The manipulation of certain aspects of the dwelling's quality, such as rendering management service, price, equipment, etc. is relatively easy for a manager. It is more difficult to adjust the dwelling sizes and the bearing structure. Often this is only possible in the long term (upon major renovation). With regard to the dwelling form, the difference between single-family and multi-family housing is important. The extent of prestige, comfort, safety and privacy are of great influence here. The adjustment of these aspects depends strongly on the type of dwelling.

As far as the physical environment is concerned, the estate's direct environment can be upgraded by means of maintenance and small interventions. The urban design of the development, on the other hand, is difficult to correct. For that, managers have to depend on the local authority's policy. With the help of correct data, one may be able to influence the policy of the local authority. The social environment is relatively difficult to adjust. The housing allocation policy forms the most important instrument of a stable social climate. The image is the most difficult to adjust, in the sense that it is adjustable in the long term only. As soon as an estate has been stigmatized, it will be difficult to let the dwellings. Even when improvements have been made, they only influence the estate's image in the longer run.

The degree of adjustability also determines the gravity of the problem. If a very important characteristic is perceived very negatively and is difficult to correct, then there is a structural weak spot in the quality causing a continuing risk. If an important characteristic is appreciated neutrally and is adjustable, then there is a potentially strong characteristic that may turn structural as soon as it is necessary or advisable. To make a choice to tackle the problem at the source (eliminating the weak spot) or to choose another approach (weakening or avoiding the consequences by compensation with another plus factor) financial, organizatorial and human-oriented preconditions play a role. These include the deliberations on costs, the acceptance by the organization, and the consequences for individual persons.

5. Bottlenecks

One has to be conscious of the problems that may accompany the development of such systems. The quality, reliability, availability and accessibility of information may cause problems. Following Nijkamp (1984), Nieboer and Voogd (1990) distinguish several types of problems: outdated information, that is, the necessary information is not available in time; over-information, there is too much information to enable one to arrive at a structured decision; under-information, indicating there is too little information to be used in policy-making; incoherent information which reflects a lack of coordination between information related to the multiple policy sectors or levels, and misinformation, indicating a qualitative discrepancy between demand and supply information.

Some of these problems may be overcome by an adequate design of the information system. The distinction we made between process and product data works out to have a structuring effect. The process data have to be collected with a high frequency, for instance every three months. Because of this, the problem of outdated information can be prevented. The potential threat of over-information can be limited if we operate with the right indicators and if we determine the correct indexes. If we use data from different managers, we should employ a uniform definition of abstractions such as offer rate. Product data may be collected with a lower frequency (every one or two years). With this data type, other problems actually do occur. In the event of cooperation between managers, the collection of qualitative information must be standardized and structured. The checklist presented in Figure 3 may serve as a guideline (for example in exit interviews by various managers). It is also a great problem that data on the physical and social environment are insufficiently available (misinformation). Often information exchange between managers and local authorities is problematic. The harmonization of policy is necessary, for instance in the framework of an information plan for neighbourhood management.

6. Conclusion

The proposed system includes an integration of object-related process and product data. These data describe, through a combination of variables, to what extent the object (in this case the housing estate) is related to the demands of the dwelling consumer and the degree to which it provides points of departure for the managing measures that are permissable in the object's context. Process data indicate which tasks are to be accomplished in the sphere of promotion and distribution. Moreover, in each phase it is the manager's effort which brings about certain decisions on the part of the occupants. The interaction between both parties is registered, whereby an evaluation of the effort is possible and measures can be taken. The product data also give pointers for adjustment of the quality policy with regard to product and price (the product characteristics - dwelling, dwelling type, physical and social environment, area's image - and the price that has to be paid).

In the future, the above-noted bottlenecks will have to be solved. In addition to the quality of the product data, their relevance and importance will have to be checked as well. The system could then, whether it is automatized/computerized or not, be used as a management information system. It also offers building blocks for a decision-support system in the field of housing management.

References

Boekhorst, F. and Smeets, J. (1990) Naar een Kwaliteitsbesturingssysteem in het Kader van Strategisch Woningbeheer, Stichting Instituut Interface, Eindhoven.

Born, J., Nollen, M., and Krogt, T. van der (1991) "Kwaliteit van Dienstverlening: Het Ontwikkelen van een Meetinstrument", *Bedrijfskunde* **63**, 218-226.

Dochy, F., Segers, M., and Wijnen, W. (1990) *Management Information and Performance Indicators in Higher Education*, Van Gorcum, Assen.

Elffers, J. and Vulperhorst, L. (1991) "Klantgericht Werken Woningcorporaties Perspectiefrijk", *Cobouw* **3**, 5.

Kaper, A. de and Knaapen, R. (1990) *Praktische Kwaliteitsanalyse*, Kluwer, Deventer.

Morgan, G. (1986) *Images of Organisation*, Sage, London.

Nauta, A. (1991) *De Verhuurbaarheid Gemeten*, SCW, Amersfoort.

Nauta, A. and Smeets, J. (1989) "Informatievoorziening en Beheerprocessen", *Stedebouw en Volkshuisvesting* ***70***, 23-30.

Nieboer, N. and Voogd, H. (1990) "Housing Vacancy and Early Warning Systems", *The Netherlands Journal of Housing and Environmental Research* **5**, 237-249.

Nijkamp, P. (1984) "Information Systems: A General Introduction", in P. Nijkamp and P. Rietveld (eds.), *Information Systems for Integrated Regional Planning*, North Holland Amsterdam, 9-33.

Prak, N. and Priemus, H. (1986) "A Model for the Analysis of the Decline of Postwar Housing", *International Journal of Urban and Regional Research* **10**, 1-7.

Smeets, J., Nauta, A., and Thijssen, R. (1989) Kwaliteit, Verhuurbaarheid en Strategisch Woningbeheer, Stichting Instituut Interface, Eindhoven.

Spaans, M. (1990) *Naar een Signaleringssysteem voor de Naoorlogse Woningvoorraad*, Delftse Universitaire Pers, Delft.

Spaans, M. and Hoenderdos, A. (1989) Ontwikkeling van een Signaleringssysteem voor het Naoorlogse Non-profit Woningbezit in Rotterdam, OTB, Delft.

Suurmond, J. and Fokkema, E. (1989) Kwaliteitsmeting Dienstverlening bij Woningcorporaties, NCIV Studieconferentie Corporaties in de Markt, De Bilt.

Sweringa, G.N. (1989) "Strategie als Richtsnoer voor Beleid", in *Woonecologie: Tussen Consumptie en Existentie*, Stichting Voorlichting Huishoudwetenschappen, Wageningen, pp. 122-132.

Vijverberg, G. (1989) *Op Weg naar Systematisch Beheer en Onderhoud in de Sociale Huursector*, Delftse Universitaire Pers, Delft.

VROM (1992) "Stedelijke Woonmilieus en Stedelijke Woonvoorkeuren", in *Stadsvernieuwing en Verstedelijking*, VROM, 's-Gravenhage.

A DECISION SUPPORT SYSTEM FOR HOUSING OF (PUBLIC) ORGANIZATIONS

W.R. PULLEN, C.L.G. WASSENAAR, I. VAN HETTEMA,
J. TOONEN DEKKERS and I. JANSZEN
Ministry of Housing, Physical Planning and the Environment,
Dutch Government Buildings Agency, Directorate Consultancy and Research
P.O. Box 20952, 2500 EZ The Hague, The Netherlands
C.G.E. BOENDER, A. TAS, H. STEGEMAN
ORTEC Consultants, Groningenweg 6, 2803 PV Gouda, The Netherlands
C.G.E. BOENDER
Erasmus University Rotterdam, Econometric Institute, P.O. Box 1738,
3000 DR Rotterdam

ABSTRACT. In this paper we present a hierarchical decision support system for the allocation of organisations to available buildings, and for the allocation of employees of an organisation to the work units of a building. For both allocation problems a mathematical model and optimisation algorithm is developed, taking into account the relevant criteria, such as the extent to which the allocated floorspace is in accordance with the standards, and the extent to which departments are housed in connecting zones of a building. The decision support system is illustrated by two practical applications.

1. Introduction

The Dutch Government Buildings Agency (GBA) is responsible for the provision of work space to 170,000 people in the service of the Dutch central government. The building stock consists of about 4,500 buildings with a gross floor area of 6.5 million square metres and an estimated replacement value of about 10 billion U.S. dollars. With the exception of computer and furniture facilities etc., the housing cost of a work unit is more than 6,000 U.S. dollars per year. Thus, the housing cost of the employees of the Dutch central government adds up to more than 1 billion U.S. dollars per year.

The main objective of the GBA is to provide adequate housing for the government organizations in a flexible, effective and efficient manner. In the realization of its main objective the GBA has to take into account many developments, requirements and objectives, including the following. During the eighties several retrenchment operations were put into effect by the Dutch government, and many others are still expected. A crucial aspect of these operations is the wish to realize a smaller (but more effective) government apparatus. However, the extent of the cut-back in number of government employees is only partly known. Nor is it known precisely when these reductions will be realized, but they are obviously of enormous importance for the GBA. As a result of such an economizing measure, the GBA was furthermore given the task to reduce the yearly housing costs by 10%. There is a growing number of part-timers and a growing rate of churn in the organizations, which have considerable repercussions on housing requirements. Housing is increasingly understood as a means of production. That is, analogous to human capital and material resources (such as computers), the quality and size of the housing stock and its allocation to organizations and employees has to be determined in such a way that the size and the quality of the output of the organizations is maximized. To assist the GBA in realizing its

H. Timmermans (ed.), Design and Decision Support Systems in Architecture, 39–47.

objectives we developed a method to analyze, evaluate and optimize the matching of the available accommodation of real estate with the housing requirements and needs of the government organizations, taking into account the housing standards and policies of the central government.

In the process of evaluating and improving such a matching, many criteria have to be taken into account. A first criterion is the degree to which the standards set by the central government deviate from the numbers of square metres which are actually assigned. These deviations not only have to be minimized on a national, regional and urban level, but also the space assigned to the individual organizations and employees has to be in accordance with the norms. The relevance of this criterion is evident: if too much space is assigned, the housing cost will be too high, and insufficient space will affect the production of the organizations. A second criterion is the extent to which each organization is concentrated in one building. This criterion is of obvious importance: if an organization is scattered over more than one building, this will clearly hamper its readiness and production. Other criteria concern the distances of the buildings to a station of public transport (central government policy), the extent to which departments of an organization are housed in connecting zones of a building, etc.

It is practically impossible to take all the relevant criteria into account simultaneously. For example, is a matching in which an organization is concentrated in one building with a 14% shortage of space, preferred to another matching in which the shortage is remedied at the expense of a scattering over two buildings? If not, is the new matching nevertheless preferable if it implies that the situation of other organizations can be significantly improved? In order to be able to take all these aspects into account, a multi-criteria approach has been developed. In this approach each criterion is assigned a weight which reflects the relative importance of the criterion. Next, a score is assigned to a matching with respect to each relevant criterion. These criteria scores are aggregated to a total score of a matching using the criteria weights. For example, assume that only the used space and the concentration of organizations would be relevant, with relative weights of 40% and 60%, where the criterion scores of a matching would be respectively 80% and 100% (i.e. each organization would be housed in one building). Then the total-score of the matching would be $0.4*80\% + 0.6*100\% = 92\%$.

This multi-criteria approach not only enables the GBA to evaluate and analyze a current matching, but it also provides the basis for a systematic comparison of the pros and cons of adaptations and movements. That is, a matching can be optimized by moving (parts of) organizations or departments form one building (or zone of a building) to another in such a way that the total score of the resulting matching is maximized, taking into account certain conditions which matchings have to satisfy and the cost of moving. To carry out these optimizations, algorithms have been developed which are based on methods from Operations Research and Management Science.

The multi-criteria approach and the optimization algorithms have been implemented in a decision-support system called BOAST: Buildings Optimization and Analysis SysTem. BOAST is an interactive decision-support system, not a decisive system, which is intended to assist the analysts of the GBA rather than to replace them. In such an interactive decision-support system, the analyst and the computer join forces to obtain solutions which neither of them would have been able to obtain separately. The computer is unbeatable in quickly evaluating matchings, in verifying the feasibility of a matching with respect to all constraints, and in efficiently determining the best improvements. The human analyst controls the evaluation and optimization by fixing initial parameters (including the weights of the criteria and the threshold values of prohibitions) and by adjusting and updating (parts of) the computed matchings.

BOAST can be used on two hierarchical levels: on an urban or regional level for evaluating and optimizing the housing of government organizations in the available buildings (in the following we will refer to this part of the system as BOAST I; in the GBA organization this part is referred to as BOSS), and on a building level for evaluating and optimizing the housing of departments of a government organization in the zones of the building, and of the housing of the employees of departments in the work-units of zones (this part of the system will be referred to as BOAST II, which in Dutch is called GIOS). BOAST is implemented in the C language on PCs which communicate with other systems of the GBA to obtain their information. The level I part of the system is implemented on a MAC system, but it runs under MS-DOS and UNIX as well. The level II part of BOAST runs under MS-DOS and UNIX.

The sequence of topics in this paper is as follows. In Section 2, the general approach of evaluating and improving matchings is described. In Section 3, the criteria which have to be taken into account are described in more detail. Section 4 contains a succinct description of a priority method which can be used to determine the weights of the criteria. The underlying ideas of the optimization algorithms are described in Section 5: this section may be skipped by (some of) the readers. Section 6 contains some practical results. The conclusions are presented in Section 7.

2. The General Approach

In using BOAST, the user proceeds as follows. First, the user of the system has to assign basic scores with respect to each criterion separately.

TABLE 1. A level I case

Organizations	
. Number	10
. Total required space	29537 m²
Buildings	
. Number	6
. Available space	32909 m²

Evaluation of current matching

Criterion	Weight	*	Criterion score		
Used space	16%	*	44.68%	=	7.15%
Concentration	52%	*	27.90%	=	14.51%
Location	0%	*	100.00%	=	0.00%
Housing cost/m²	16%	*	62.81%	=	10.05%
Ownership/renting	0%	*	87.85%	=	0.00%
Public transport	16%	*	49.25%	=	7.88%
					39.59%

Evaluation after optimization

Criterion	Weight	*	Criterion score		
Used space	16%	*	100.00%	=	16.00%
Concentration	52%	*	98.15%	=	51.04%
Location	0%	*	100.00%	=	0.00%
Housing cost/m²	16%	*	66.62%	=	10.66%
Ownership/renting	0%	*	87.85%	=	0.00%
Public transport	16%	*	53.25%	=	8.52%
					86.21%

On level I, a basic score has to be assigned, for example, to each organization with respect to the criterion 'used space', which measures the extent to which the used space of the organization is in accordance with the standard. For example, the basic score of an organization with respect to the criterion 'used space' might be 100% if the assigned space deviates less than 1% from the standard, or 80% if the deviation is between 1% and 5%, etc. Next, the basic scores are aggregated to criteria scores by addition of the basic scores (see Tables 1 and 2). If it is relevant, the basic scores are weighted, as in the aforementioned example, where the basic scores are weighted with the size of the organizations. Then, in step 3, these criteria scores are aggregated to a total score of a matching, where each criterion score is weighted by a criterion weight, which has to be specified by the user of the system. These weights, which measure the relative importance of the criteria, can be determined by a pairwise comparison technique from the priority theory (Boender et al. 1989; see also Saaty 1980). These total scores, in combination with their decomposition in criterion scores, provide the information to analyze and evaluate matchings (see Tables 1 and 2). Finally, matchings can be optimized by moving (parts of) organizations and/or departments from one building and/or zone to another, in such a way that the total score of the resulting matching is maximized, taking into account certain conditions which matchings have to satisfy and the cost of moving. To carry out the optimization, the system invokes optimization algorithms from Operations Research (see, for example Nemhauser et al. 1989; Heyman and Sobel 1990), as will be described later.

To illustrate the general approach, we refer to the Tables 1 and 2, which we will describe in more detail below. The first part of Table 1 describes a multi-criteria evaluation of a matching of a number of organizations in a large Dutch city. The second part of this table describes the evaluation of the new matching which would result if the movements would be carried out which are computed by the level I optimization algorithm. Table 2 describes the results of the application of the level II optimization algorithm to allocate the employees of an organization to the work units of a new building.

3. The Criteria

3.1. LEVEL I CRITERIA

On an urban and regional level more than 20 criteria are considered, distinguished as so-called matching and technical criteria. The matching criteria measure the extent to which the organizations are accommodated in correspondence with their various requests and demands. Some relevant examples of these criteria are 'used space', 'concentration', and 'location'. The criterion score with respect to 'used space' measures the extent to which the space assigned to organizations is in accordance with the standards, which are set on a strategic level. Thus, if a high weight is assigned to 'used space', the total score of a matching will to a large extent be determined by the quality of the matching with respect to this criterion. Also, a high weight will clearly force the optimization algorithm to search for a new allocation in which the spaces assigned to the organizations are in accordance with the standards as much as possible (taking into account the remaining criteria as well). However, 'used space' is one of the examples of the criteria which also serve as prohibitions, since too large deviations (to be specified by the user) from the standards are not allowed. These prohibitions typically become active in the optimization process, where they impose restrictions on allowable movements to improve the total score. For each organization the criterion score with respect to 'concentration' is positively rewarded if the organization is concentrated in only one building. In the optimization process a movement which results in a deterioration of the scattering of any organization over buildings is not allowed. The concentration criterion is analogously applicable to groups of organizations which have to be concentrated in one building. The criterion score with respect to location measures the extent to which the organizations are located in accordance with the city area (e.g. city centre or suburb) they prefer.

TABLE 2. A level II case

Organization	
. Number of departments	7
. Number of solitaries	50
. Number of non-solitaries	384
. Total required space	4444 m^2
Building	
. Number of zones	33
. Number of work units	481
. Available space	5197 m^2

Optimization results

. Number of generated matchings	400
. Computation time (80386/33 MHZ with mathematical coprocessor)	10 min.
. Estimated probability of a better matching	0.5 %

Criterion	Weight	*	Criterion score		
Used space	10%	*	99%	=	9.9%
Connectedness	30%	*	100%	=	30.0%
Number of floors	30%	*	30%	=	9.0%
Relations	30%	*	100%	=	30.0%
					78.9%

The technical criteria are used to take into account strategic housing goals. Some relevant examples include housing cost, ownership/renting, and public transport. Housing cost (per m^2) of a building is included because GBA's task is to reduce the housing cost by 10% per employee. If this criterion is given a high weight, the total score will largely be determined by the housing costs of the matching, and the optimization algorithm will enforce movements to the most economical buildings. Ownership/renting is included in order to ensure that sufficient buildings are rented rather than owned, which increases flexibility, and the 'public transport' criterion reflects the government policy on the advancement of public transport and the reduction of the use of cars by commuters.

3.2. LEVEL II CRITERIA

At the building level 'used space', 'connectedness', 'number of floors', and 'relations' are taken into account. Analogous to level I, 'used space' measures the extent to which the space assigned to employees is in accordance with the standards. Furthermore, on level II the space which is assigned to a department is not allowed to be less than its standard. The 'connectedness' criterion measures the degree to which departments are housed in connecting zones. The criterion is obviously introduced to prevent the scattering of departments over a building. Analogous to 'connectedness', 'number of floors' measures the extent to which departments are housed on a minimal number of floors, and 'relations' is included to take into account that for some departments it is relatively important to be positioned in the neighbourhood of other departments, or in the neighbourhood of special zones of a building. In the determination of optimal allocations of departments to zones, and of employees to the work units of a building, several other restrictions have to be taken into account as well. These may include a maximum number of work units of office rooms, and the obligation that solitaries (i.e., people working alone) have to be allocated to a single room.

4. A Pairwise-Comparison Priority-Theory Method

To determine the weights of criteria, the user can apply a priority-theory method. The method infers the weights of the criteria from pairwise comparison of the criteria. For example, an analyst might judge 'concentration' to be three times as important as 'used space', and he/she might judge 'housing cost per m^2' and 'public transport' to be equally important. The method uses these pairwise comparisons of (possibly more than one) analyst to determine the weights of the criteria by logarithmic regression. Thus, if r_{ijk} denotes the evaluation of criterion i relative to criterion j by the kth analyst ($k=1,...,m$), the weights $w_1,...,w_n$ of the criteria are determined by minimizing the regression function

$$\sum_{i}^{n}\sum_{j}^{n}\sum_{k}^{m} \ln\left(\frac{r_{ijk}}{w_i/w_j}\right)^2 = \sum_{i}^{n}\sum_{j}^{n}\sum_{k}^{m} \{ \ln(r_{ijk}) - \ln(w_i) + \ln(w_j) \}^2. \tag{1}$$

That is, the weights $w_i,\ldots,w_n$ are determined such that the sum of the (squared logarithmic) deviations of the pairwise comparisons r_{ijk} from the (unknown) true values w_i/w_j is minimal. (The logarithmic transformation is partly motivated by the desire to obtain a decision problem which is linear in the decision variables $\ln(w_i)$.) For details the reader is referred to Boender et al. (1989).

5. The Optimization Algorithms

In this section we give a short description of the ideas underlying the developed optimization algorithms. The level I optimization algorithm is designed to optimize a current matching. Starting from a current matching, a new matching is determined which maximizes the total score (see Step 3 of the general approach), taking into account the restrictions, such as the prohibition to reduce the concentration of organizations, and taking into account the cost of movement. The level II optimization algorithm determines an optimal matching of the employees of an organization with the work units of a building, not taking into account their current positions in the building (if any).

5.1. THE LEVEL I OPTIMIZATION ALGORITHM

Starting from a current situation, the level I algorithm applies a so-called two-exchange heuristic to construct improvements (Lin 1965). First, a pair of units is selected, whereby a unit is defined as the part of an organization which is housed in a certain building. Thus, in an ideal matching the number of units would obviously be equal to the number of organizations. For the considered pair of units, the optimal allowable allocation of the two (parts of) organizations to the two buildings of the pair is determined. That is, the number of square metres which have to be assigned to each of the two organizations in the two buildings is determined such that the total score of the new matching is maximized, and all restrictions are satisfied. Such an improvement can be realized by (combinations of) the following actions: *Adaptation* (a reduction or increase of the number of square metres of an organization in the building where it is already positioned), *Transfer* (a move of an organization from one building to the other building), and *Exchange* (a move of an organization from one building to the other building, where its vacancy is, possibly partly, filled up by the arrival of the organization from the other building. After the determination of the optimal repositioning of the two organizations in the two buildings, it is checked whether a further improvement can be realized by the adaptation of the used space of the other organizations in the two buildings. These adaptations might be possible due to resulting vacancies of the repositioning. This step is carried out for each possible pair of units. If no permitted repositioning turns out to exist, the algorithm is completed, otherwise from all the possible repositionings, the one with the best total score which is permitted with respect to all constraints is chosen. This repositioning is carried out to obtain a new matching, and the whole process is repeated until no permitted repositioning exists. Note that the algorithm in each iteration for each unit considers each possible move to another building, as well as each possible 'return move' to fill a resulting vacancy (which is practically impossible for a human analyst). Nevertheless, in an efficient implementation each iteration takes less than one minute on an 80386 PC, even for large practical situations.

5.2. THE LEVEL II OPTIMIZATION ALGORITHM

The level II algorithm does not start from a current matching. Instead it assumes an empty building in which the employees of an organization have to be positioned such that the total score is maximized, taking into account all the relevant constraints. In order to accomplish this, first in step 1 (the *global phase)* a 'general' matching of the departments of an organization with the available zones of the building is generated. Next, in step 2 (the *local phase*) a 'detailed' matching of the employees with the work units of the zones which have been allocated to the departments in Step 1 is determined. If no such allowable detailed matching can be constructed, another general matching is generated. Otherwise, in step 3 (*stopping rule)* the best detailed matching is saved. This process is repeated until the estimated probability that a better matching can be found is less than a pre-specified value (say 0.5%).

The idea of iteratively applying a global phase and a local phase is generally accepted as an efficient and effective approach to solve complex optimization problems. The approach of the global phase is as follows. Starting from a partly constructed general matching, the algorithm generates one of the departments which are not yet allocated, and positions the selected department in the zones of the building which are adjacent to the zones which are already occupied. The selection of the department to be added to the partial matching, as well as the selection of the zones adjacent to the partial matching, occurs at random, where the selection probabilities are proportional to the expected attribution to the total score. That is, an extension of the partial matching which leads to a higher expected improvement of the total score has a greater probability to be carried out. In the local phase, the employees are allocated to the work units of the rooms of the zones which have been assigned to the departments of the employees in the global phase. This is accomplished taking into account all the restrictions which are mentioned in Section 3, such as the maximal number of work units of office rooms, and the obligation that solitaries have to be allocated to single rooms. In this phase we apply exchange heuristics as described above, and an adaptation of the first fit-decreasing rule, which is frequently used to solve the classical bin-packing problem (see Coffman et al. 1984).

In practice, the level II algorithm may iterate hundreds or thousands of times. The corresponding computation times (on a 80386 or 80486 PC) vary from a few minutes to many hours, depending on the size of the problem. The problem may embrace thousands of work units and several other factors, such as the relation between available and required space. In contrast with the level I algorithm, one is never sure if the level II algorithm has obtained the best possible result; each new iteration may lead to a better matching than the one obtained so far. Therefore, we applied recent results from the extreme value theory (see Dekkers 1991 and Dekkers et al. 1989). These can be applied since at each iteration of the algorithm each possible matching has a certain probability of being generated. Applying these results, we compute at every iteration of the algorithm the estimated probability that there is still a better matching than the best one obtained so far. This information can be employed by the user of BOAST to decide if the algorithm should be terminated or not.

6. Practical Results

We applied the BOAST system to several cases. The results of two of these cases are depicted in the Tables 1 and 2. The first part of Table 1 describes a multi-criteria evaluation of a matching of a number of organizations in a large Dutch city. The criterion scores are defined on a scale from 0% to 100%. Thus, the total score is defined on a 0%-100% scale as well, from which it can be concluded that the initial matching scores poorly (40%). Note from the table that the low score is primarily due to the fact that the organizations are weakly concentrated (a criterion score of 28% and a weight of 52%).

From the second part of Table 1 it can be concluded that the optimization algorithm yields an enormous improvement of the total score from 40% to more than 85%. Not surprisingly, this result is mainly determined by the improvement of the 'concentration' of the matching. To obtain this new situation, three adaptations have to be carried out, as well as one transfer and one exchange: the total number of square metres which have to be moved is 3480. For reasons of comparison we also computed a new housing situation taking into account the cost of moving. Then, a total score of 84% results, with two adaptations and one transfer, and only 480 square metres to move.

Table 2 describes the results of the application of the level II optimization algorithm to allocate the employees of an organization to the work units of a new building. In about ten minutes the algorithm determines the optimal location for each of about 400 employees in a new building. The total score of 78% indicates that the organization and the building match reasonably. The deviation from a score of 100% is clearly a result of the fact that (some of) the departments of the organization are scattered over more than one floor. The user of BOAST can employ the system to improve this phenomenon, but only at the expense of deteriorating the scores with respect to the other criteria.

7. Conclusion

In our opinion, the multi-criteria method and optimization algorithms, in combination with their implementation in BOAST, improves and structures the process of analyzing and optimizing the matching of demand and supply of building accommodation. The possibility to evaluate a matching with respect to several criteria (whose establishment required long and intensive discussions) clearly increases insight in the current situation. Furthermore, by varying the weights of the criteria and/or by adding (possibly imaginative) new buildings, several new matchings can be generated, which are optimal from different angles of incidence. From these optimal new matchings, the user him/herself can select a best admissible one, possibly on the grounds of considerations which are not taken into account by the system.

However, the authors realize that much work remains to be done. One important example is taking into account explicitly the (uncertain) long-term demand for accommodation. Another is the possibility that the system itself determines a building (or a reconstruction of an existing building) which minimizes the deviation between the quantity and quality of supply and demand of housing accommodation.

Acknowledgements:
The authors would like to express their gratitude to M. Salomon (Erasmus University and ORTEC), F. Spieksma (University of Maastricht), J.S. de Wit (ORTEC) and K. Walter (ORTEC) for their assistance in developing the models and methods of BOAST.

References

Boender, C.G.E., Graan, J.G. de, and Lootsma, F.A. (1989) "Multi-criteria Decision Making with Fuzzy Pairwise Comparisons", *Fuzzy Sets and Systems* **29**, 133-143.

Coffman, E.G., Garey, M.R., and Johnson, D.S. (1984) "Approximation Algorithms for Bin-packing: An Updated Survey", in G. Ausiello, M. Lucertini and P. Serafini (eds.), *Algorithm Design for Computer System Design*, Springer Verlag, Berlin, pp. 49-106.

Dekkers, A.L.M. (1991) On Extreme Value Estimation, Ph.D. Dissertation, Erasmus University Rotterdam.

Dekkers, A.L.M., Einmahl, J.H.J., and Haan, L. de (1989) "A Moment Estimator for the Index of an Extreme-Value Distribution", *Annals of Statistics* **16**, 1012-1021.

Nemhauser, G.L., Rinnooy Kan, A.H.G., and Todd, M.J. (eds.) (1989) *Handbooks in Operations Research and Management Science, Vol. 1*, North-Holland, Amsterdam.

Heyman, D.P. and Sobel, M.J. (eds.) (1990) *Handbooks in Operations Research and Management Science, Vol. 2*, North-Holland, Amsterdam.

Lin, S. (1965) "Computer Solutions of the Travelling Salesman Problem", *Bell System Technical Journal* **44**, 2245-2269.

Saaty, T.L. (1980) *The Analytical Hierarchy Process*, McGraw Hill, New York.

TOWARDS AN INTEGRATED FACILITY MANAGEMENT SYSTEM FOR MANAGEMENT AND USE OF GOVERNMENT BUILDINGS

A. DE SCHEEMAKER
Government Buildings Agency (Rijksgebouwendienst)
Research and Development Department
P.O. Box 20952, 2500 EZ The Hague
The Netherlands

Research Institute for Policy Sciences and Technology (OTB)
Delft University of Technology
Thijsseweg 11, 2629 JA Delft
The Netherlands

ABSTRACT. The Government Building Agency in the Netherlands is developing an integrated facility management system for two of its departments. Applications are already developed to support a number of day-to-day facility management activities on an operational level. Research is now being carried out to develop a management control system to better plan and control housing and material resources.

1. Introduction

The management of organisations takes place on different levels with different planning horizons. In designing structures for adequate information supply and information exchange, one must be aware of the different information needs that exist between the management levels and their decision processes. This paper focuses on the relationships between planning, control, operating and information supply in a facility management organisation. A case study concerning the development of an integrated facility management system by the Government Building Agency (GBA) in the Netherlands will be discussed.

Commercial facility management systems that can be bought off the shelf nowadays only support the operational management level, where day-to-day facility management activities are being carried out (Joudah 1992). These facility management systems usually consist of large databases where amounts, types and location of facilities are stored. Research is now being carried out at the GBA to extend the functionality of such a 'standard' system to the tactical-strategic management level. For this purpose a management control system is being developed that monitors a set of well-defined indicators (the critical success factors), which are aggregated from the operational data stored, and predicts the housing consequences of policy measures that are proposed on the higher, strategic management level.

This paper will focus on the research that is undertaken to develop the management control part, on top of the existing FM system. Section 1 gives a brief overview of the tasks of the Government Building Agency and developments in the public sector that are changing patterns of work and housing conditions. Section 2 explains the strategic goals of a facility management system that integrates the information supply of both the GBA and its clients. In Section 3, the general approach of the research to extend the functionality of the existing FM system to a tactical management level is described. In Section 4, some preliminary results are presented. A proposal

H. Timmermans (ed.), Design and Decision Support Systems in Architecture, 49–59.

for a facility management control system is discussed in Section 5 and some conclusions are formulated in Section 6.

2. The Government Building Agency

The Government Building Agency (GBA) is responsible for the provision of work space to 160,000 people in the service of the central government in the Netherlands. The stock consists of some 4,000 buildings with a gross floor area of six million square metres and an estimated replacement value of ten billion U.S. dollars. Integration of housing and facilities information is needed for a better control of housing costs and to meet the higher requirements which constantly changing organisations pose on their buildings.

In the past, the GBA has focused mainly on building and technology. According to this approach, a building was largely seen as an end in itself: a technological product which had to be designed, constructed and maintained. The organisation using a building came in second place. Since the publication of the first Housing Policy Plan in 1988, this strategy has changed. A building is no longer looked upon solely as a technological accomplishment, but primarily as a means to support the primary production process of a public organisation. The focus in the management of real estate has shifted since then from the supply side (the building stock) to the demand side (the users of the buildings) (De Jonge 1990). Real estate decisions are now aimed at creating a work environment that improves the organisational performance outcome and job satisfaction in general (see also McLennan 1992; Drucker 1991; Becker et al. 1991). From the point of view of the GBA, more investments in housing can be made if they contribute to the overall results of the organisation.

2.1. FACILITIES AND HOUSING COSTS

While the management of real estate and the provision of housing is the primary function of the GBA, it also forms part of the facility process of a public organisation, a customer of the GBA. Facility management involves planning, providing and maintaining appropriate physical spaces, support and services for an organisation. Its main goal is to support the primary functions of a single organisation in a flexible, effective and efficient manner. The GBA takes part in this process by providing adequate housing.

The annual costs of providing people with facilities to operate in office environments are about NLG 23,000 per work unit (Figure 1). The costs for housing represent some 10% of the total costs of a public organisation: about NLG 12,000 per work unit yearly. The GBA pays about 40% of the facility costs of an organisation. These consist of capital costs, costs of furniture and fixed assets, energy, maintenance and part of the cleaning. Many other facility functions, however, which are not performed by the GBA and are not paid for by it either, are strongly affected by characteristics of the building. Examples of these functions are cleaning, security, communication, data services, postal service, space, inventory accounting etc.

Although facilities costs and housing costs are relatively small compared to the total costs of an organisation, the importance of facility and housing management should not only be stressed from the point of view of the direct costs related to it. The importance lies in the effects that efficient operation of facilities has on the primary process and the disturbance of it when facilities are not efficiently operated. These costs, however, are difficult to calculate.

Obviously, the management of public real estate and the facility process is a complex matter in which the GBA and its clients have shared responsibilities. Cooperation is vital, not only to provide an efficient, integrated operation of facilities, but also in order to respond to important developments that are now taking place in the public sector. These developments have major consequences for the quality of government buildings.

	cost of:	minimum	maximum	average	percentage
1	capital	4900	7700	6500	32.5 %
2	energy	400	1200	825	4.1 %
3	maintenance	500	1040	800	4.0 %
4	cleaning	420	560	500	2.5 %
5	security	750	1100	950	4.8 %
6	catering	1300	2150	1500	7.5 %
7	management	1800	3500	2600	13.0 %
8	reproduction + postal service	800	3000	1900	9.5 %
9	stationery	500	1250	800	4.0 %
10	fittings	400	1400	950	4.8 %
11	communications	400	1400	900	4.5 %
12	data service	1000	2500	1750	8.8 %
	Total	13170	26800	19975	100 %
	Total including 18.5 % Tax	**15606**	**31758**	**23670**	

Figure 1. Annual costs per work unit in NGL
Source: NEFMA survey of office buildings, 1989

2.2. DEVELOPMENTS IN THE PUBLIC SECTOR

The Dutch government wants a smaller but more effective public apparatus. This means a re-evaluation of tasks in the primary production process, but it also entails a more effective operation of secondary, supporting functions. During the eighties, several austerity measures were put into effect and are still in progress. These operations are meant to reduce expenses by decreasing the number of civil servants. The measures have a direct impact on the way facility management units and the GBA are working.

The facility management units in the public sector must respond to developments such as concentration on primary tasks (management by objectives, contracting out), a growing rate of churn in an organisation which requires a flexible, quick-witted organisation, teamwork and project structures, better qualified personnel and a deconcentration of responsibilities from staff to line management. These developments suggest that GBA and FM units have to operate under uncertain circumstances. Their main task is therefore to create deliberate flexibility in the facility management and housing process. For example, at stock level a mix of rented and owned space is used (see De Jonge 1991), at building level office systems to reduce space costs and improve flexibility are used (see Wassenaar 1990), and at service level short-term contracts where conditions might be altered are advocated.

In the first Housing Policy Plan (1988), the GBA was called upon to reduce housing costs by 10% per employee between 1988 and 1992. To realise this reduction in housing costs, the GBA has implemented two strategies. First, the effectiveness of the use of public housing was improved by better matching the available accommodation in the real estate stock with the housing demands of an organisation. For this purpose an active portfolio management is carried out (see Boender 1992). Secondly, spatial costs and quality standards were tightened up and these standards were enforced on new and, if possible, on existing housing. The actual space per work unit is still well above the spatial standards set for the public sector.

Housing costs and facilities costs in general still show a tendency to rise, however. This is caused by a growing numbers of part-time employees, who all have a right to their own work unit, facilities needed for telecommunication, and requirements for public housing which stem from general public policy on environment, indoor climate, accessibility of buildings for the handicapped. To find solutions for the growing tension between housing supply and housing demand and to reduce the housing costs per work unit, the GBA must have accurate, well-structured information about the total housing costs, irrespective of whether these costs are paid for by the GBA or its client, the technical and functional quality of the real estate stock, and existing and prospective housing situations. Unfortunately only costs which are paid for by the GBA are reasonably well known. Data of other housing and facility activities, performed and paid for by her clients, is hardly available or otherwise difficult to compare.

3. An Integrated Facility Management System

To meet the need for more integrated information on housing (demand) and housing-related facilities and to better support the primary functions of its clients, the GBA is developing a CAD-based facility management system, together with a Dutch software house. The FM system consists of a central relational database and 'loose-fit' modules for specific applications. An Intergraph system (Workstation) was chosen for graphic representations, which can be linked to several relational databases; e.g. Oracle and Informix. The FM system runs under UNIX and MS-DOS and operates on a client-server base (Figure 2). This means that data (alphanumerical and graphic) are stored separately from specific FM applications, where the actual data processing and calculation takes place. In this way several users can consult facility management data at the same time through different applications ('multi-user environment'). At this moment modules of the system are used by facility management units of the Department of Social Affairs and the Department of Housing, Physical Planning and Environment (VROM). The goals of the FM system for the GBA and its clients are threefold:

1. *Task control at the operational level*
   Task control in facility management assures that specific tasks are carried out effectively and efficiently. The FM system supports and integrates the day-to-day activities of the FM units. Applications have already been developed for space planning and optimization, inventory management, drafting coordination, move coordination, call handling and reservations. Although many FM systems are already available to deal with the above-mentioned tasks, integration of these processes is very often problematic.

2. *Management control at the tactical level*
   Management control concerns the implementation of strategies which are used to attain strategic business goals. Management control does not involve the detailed operating decisions and activities that are the focus of task control; it is the means by which management assures that the organisation carries out its strategies effectively and efficiently (Anthony and Young 1984). Management control in facility management is needed to link facility management to the basic goals of an organisation. These goals are decided upon by top management on a strategic planning level. Facility management and computer-aided facility management should support management control activities by supporting the implementation of business strategies, and

deriving management information from different facility management processes, in order to better plan the required facilities for the organisation.

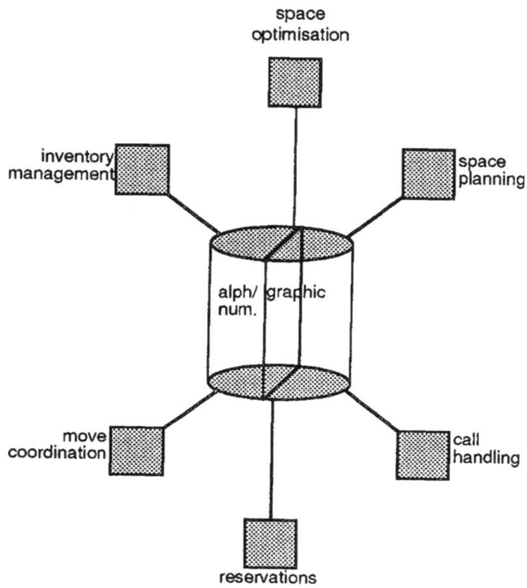

Figure 2. Model of the facility management system

3. *Information exchange*

   By developing an FM system, the GBA wants to establish a uniform and structured information exchange between the facility management process of its clients and the public housing process. The information exchanged will probably concern housing costs, maintenance, housing standards, housing demands and housing analyses on the sectoral, urban and building levels. These goals indicate that the FM system can play an important role on different managerial levels in the FM organisation and that the system will help in linking facility management to the overall goals of the organisation. Moreover the system can be used as an instrument to exchange housing (related) information between the GBA and its clients. The facility information supply in an organisation is depicted in Figure 3.

4. Approach

When a facility management concept is introduced in an organisation, often little attention is given to linking FM to the primary goals and activities of an organisation. Usually facility management only draws the attention of top management when it does *not* meet the needs of the organisation (people start complaining). Housing and material resources are still not generally regarded as strategic means of production for an organisation, which need as much careful planning and control as human and capital resources do. Existing FM systems reflect this attitude towards facility management. They are primarily used for the support and integration of day-to-day activities on an operational level; e.g. cleaning, moving, inventory management, security etc. Management reports for the top level are not generated on a regular basis.

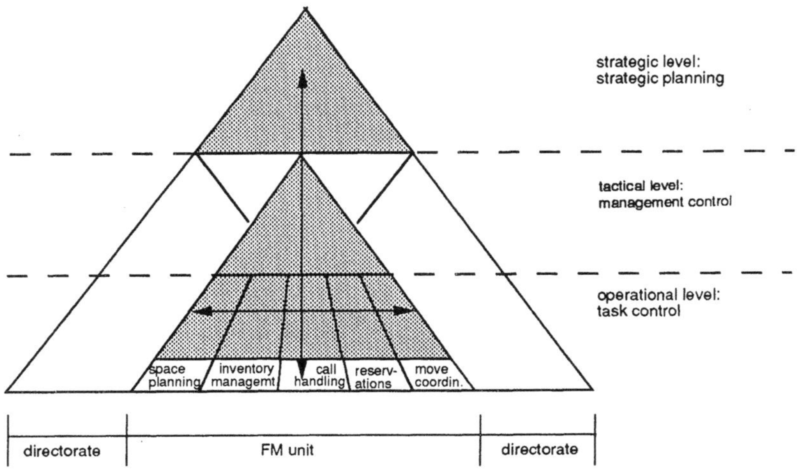

Figure 3. Facility management information supply in the organisation pyramid

When the GBA and the Department of Housing, Physical Planning and Environment (VROM) started to develop an FM system, they decided that this system should also offer tools to exercise some sort of management control over the facility management process. Research was needed to answer two questions: (i) to what extent are housing and facilities important for the primary activities of a public administrative organisation?, and (ii) what is the facility information need on a tactical and strategic management level?

4.1. RESEARCH STAGES

The research for the development of a facility management control system is conducted in three stages. The research is now at the beginning of the second stage. The three stages are:

1. *Describing the as-is situation*
   Information and organisation analysis of the way facilities are being managed at present in central public organisations. During the last quarter of 1991 and the first quarter of 1992, a series of interviews was held with directors of FM units of six departments and with line managers of the FM unit of the Department of VROM. These are the future users of the FM system. The interviews were held to get an impression of the way facilities are managed in a public administrative environment. Moreover, we tried to define the global information needs of facility managers and their staff and the problems they are dealing with in daily practice. The results of the interviews, the data flow diagrams and the procedure charts that were drawn up are used to define the functionality of a facility management control system.

2. *Describing the to-be situation*
   Development of a management model and an information model for a desired situation. These models form the basis for a facility management control (sub)system that will enable the user to derive management information from daily facility management processes and to forecast financial, housing and material consequences of housing plans, retrenchment strategies and reorganisation plans. These plans are made on a strategic level.

3. *Drawing up an implementation plan for a facility management control system*
The implementation plan will comprise the order in which automated tools have to be implemented, the necessary alterations of the FM system, which is now primarily used for task control (operational activities), and the organisational consequences of a management control system.

5. Management Control in Current Facility Management Processes: Preliminary Results

5.1. THE MANAGEMENT CONTROL PROCESS

The management control process takes place in the context of an organisation that has goals and that has decided on broad strategies for achieving these goals (Anthony and Young 1984). There are five main steps in a formal management control process (Figure 4). In the standardisation step, requirements to which the products of an organisation have to conform are derived from strategic business goals. These requirements are stated in the form of cost and quality standards. Then, in the programming step, decisions are made with respect to the major programmes the organisation plans to undertake during a coming period. These programmes are usually stated in the form of long-range and mid-term plans. Next, in budget formulation and allocation step, a budget which is a plan expressed in quantitative, usually monetary, terms covering a specific period of time (Anthony and Young 1984) is formulated and allocated to programmes for execution.

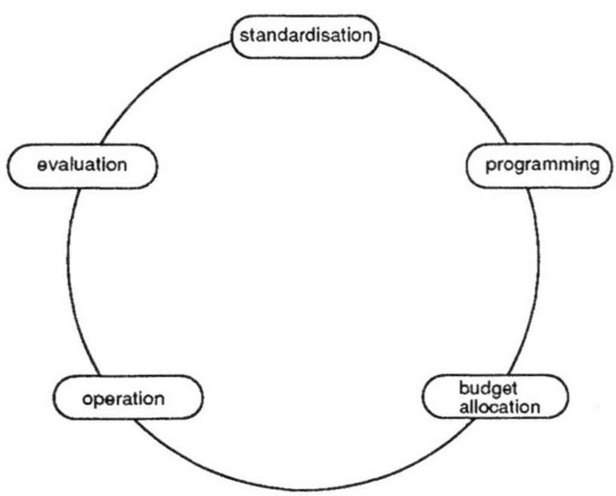

Figure 4. The management control cycle

During the period of programme execution, records are kept of resources actually consumed and outputs actually achieved. This is the step of operating and measurement. Finally, in the reporting and evaluation step, accounting information is summarised, checked against cost andquality standards and reported to top management on tactical and strategic levels (directors and secretary-general of a Department).

5.2. PROBLEM ANALYSIS

The following major problems exist in the facility management processes which have been studied. These problems may obstruct adequate management control in the facility management process.

Weak management control
The five steps of the management control process are already filled in and applied in the public housing process of the GBA and form a closed loop of planning and control. This cycle, however, is almost nonexistent in current facility management processes. Most directors of FM units who were interviewed do not have their own policy plan, nor do they use clearly defined standards to compare actual outputs and inputs of the facility management process to planned outputs and inputs. None of the FM units studies policy plans of other directorates (their clients) to make detailed quantitative and qualitative plans of the facilities which must be provided with housing services. Reporting to top management does not occur frequently or in a structured way. Developments in personnel growth and decrease are only taken into account in a limited sense. On the other hand, the directorates, clients of the FM units, do not specify their housing and facility needs in a structured way; for example, in the form of short-term housing plans. So it is very hard for an FM unit to foresee how many housing mutations can be expected in one year's time as a result of moves, projects or internal reorganisations. Obviously, most of the FM units do not have the power to 'force' their clients to make plans in which housing and facility needs are specified. One facility manager told us he was given a week's notice to move 500 employees.

The relationship between the primary production process and the facility management process is indistinct
FM units can be classified as service centres. Their output contributes to the work of other business units (directorates). Only a few of the questioned facility managers, however, could give a fairly good description of the primary process and the overall goals of their department. In most cases FM units only react to complaints and faults. They have little idea of how to give better support to their organisation, since they do not know exactly how good their level of support is at the moment. Perhaps this problem is caused by the nature of the primary process, which is rather 'vague' in public organisations. Their function is threefold: policy-making, legislation and realising income transfers (subsidies). Especially the output of the first two process functions is hardly quantifiable. Further research into this matter has yet to be done.

Low cost consciousness
There are insufficient incentives for the facility manager to work in a cost-conscious way. An important goal of the budget accounting system which is being used in public organisations is to pose limits on the expenses incurred by directorates. The relationship between expenses (called budgets) and the non-monetary output of the facility management process (in the form of housing, support and services for the organisation) is often not clear. Budgets of an FM unit are in most cases based on the expenses of the previous year. Usually a budgeting technique called 'input budgeting' is used to calculate a budget. This budgeting technique only measures the expenses incurred, but does not measure the monetary or non-monetary value of the (FM) unit's output.

Facility management information is scattered
Facility management information is usually scattered among several automated and hand-operated information systems. Separate applications exist to manage inventory, telephone and data network, support space planning etc. Integrated FM systems are not used yet. None of the departments visited exchanged information in a structured way between FM applications and the personnel information system or the budget accounting system. Cost classifications and registrations of employees in these systems are not sufficiently tuned to one another. It also

became clear that in some cases deconcentration of responsibilities from staff to line management hampered the information supply for facility management. Deconcentration in these cases has taken the form of a delegation of tasks, without an obligation to report frequently to the top level. The consequence is that information, e.g. about acquisitions of material supplies, repairs and maintenance, can only be obtained from the client himself or from the contractor. Information gathering in this way is very costly and time-consuming.

GBA's housing plans are not detailed enough to develop a strategic facility plan
The GBA produces housing plans on various managerial levels. On the strategic level, it generates the Housing Policy Plan (1988), in which housing goals and strategies for attaining these goals are stated for the next four years. On the tactical level, it generates housing analyses to design planning strategies on the sectoral (per department), regional and urban level. These strategies are worked out to meet future housing demands. Existing housing situations are compared to functional and technical criteria or standards, derived from strategic housing goals. Then several housing alternatives can be proposed which will improve the existing situation (see Boender 1992). The translation to tactical housing plans on a more pragmatic operational level is not yet conducted by the GBA. That would involve identifying the consequences, for example, of a specific housing plan for the number of workplaces, space costs, the nature of information technology and the equipment needed, possible types of space layout and of space planning, possible security regimes etc. GBA can help facility managers of public FM units to make their own strategic facility plans, which would enable them to make more informed choices.

6. Proposal for a Facility Management Control System

A management control system as part of an integrated facility management system should contain an accounting system which provides historical information. That is information on what has happened and what the costs were. In addition, the facility management control system should provide two types of information that are not found in the accounting system (Figure 5): (i) estimates of what will happen in the future, called forecasts, and (ii) estimates of what should happen, called standards or budgets.

6.1. ACCOUNTS

Accounts are devices for collecting data about what has happened (historical data) and what is planned to happen (future data). The interviews with facility managers have given an impression of the sort of accounting information needed. Much of this information will be provided by the applications for space planning, inventory management, move coordination and call handling, which form the task control part of the FM system at this moment. First, *process measurements* such as turnover data (acquisitions per product group), expenditure data (energy, water), cycle time (complaint and fault handling, post), churn rate of an organisation, and the use of equipment are required. In addition, *product measurements* such as faults, occupation rates, types of space layout and space layout losses are needed. Then, there is a need for *cost information* : personnel costs of FM employees, material costs, contract costs, facility and housing costs per workplace, per employee, and *client information* such as complaints and employee satisfaction. *Standards* are required to compare process and production measurements, involving spatial standards, cost standards, average workplace costs and checklists for maintenance, cleaning etc. Finally, there is a need for *planning information* for stock management and contract management, and *product and contractor information.* Cost information, standards, planning information and product and contractor information is not yet provided by the FM system of VROM.

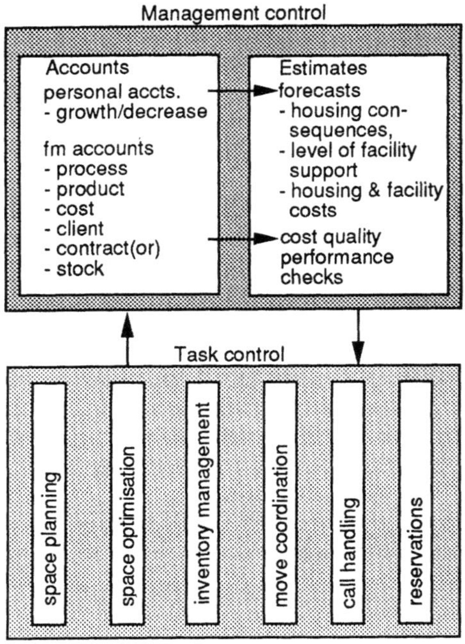

Figure 5. Proposed model for a facility management system, comprising a task and management control part

6.2. ESTIMATES

In order to perform the estimation function well, the facility management control system should be linked to the personnel system of the organisation. In this way, forecasts can be made of housing consequences and the desired level of facility support, based on future personnel strengths. Also housing and facility costs should be calculated, based on both the characteristics of the building and the organisation. Moreover, an FM control system should be built around a financial structure, to which process and production information can be related. For this purpose the facility cost classification structure probably has to be tuned with two other cost classification systems used in public organisations: the official budget accounting system of a department and the housing cost classification of the GBA. The latter is needed to establish an adequate information exchange on housing and facility costs between the GBA and its clients. Last but not least, the GBA can help set up a sound planning and control cycle for facility management byworking out housing plans and housing analyses per department on the operational working level of the facility manager. Good cooperation is vital.

7. Conclusions

Facility management must be linked to the primary activities of the organisation. Otherwise an FM unit runs the risk of becoming a department with no other mission than to solve complaints and faults. The positive effects of adequate housing and facility supply will only be noticed by the organisation if an FM unit is able to work in a pro-active rather than a reactive way. This means that planning and control instruments have to be developed that can help the facility manager make more informed decisions that affect the organisation's ability to attain its basic goals.

However, more research still has to be done on the relationship between the primary process and facility management.

The GBA can offer instruments for standardisation, programming, operating and evaluation, as far as it concerns the management of housing. In cooperation with its clients, the GBA can extend its own management control system for housing to comprise other housing-related facilities as well. An integrated FM system, consisting of a task and management control part, is an opportunity for a facility manager to be ahead of developments now taking place in the public sector. These developments have major consequences for the way public organisations are being managed, housed and serviced.

References

Anthony, R. and Young, D. (1984) *Management Control in Nonprofit Organizations*, Irwin, Homewood, Illinois.

Becker, F., Sims, W., and Davis, B. (1991) Managing Space Efficiently, Final Summary Report of Cornell University International Facility Management Program, Ithaca, New York.

Boender, G. (1992) "A Decision Support System for Housing of Public Organizations", paper for the Design and Decision Support Conference, Mierlo, The Netherlands, July 6-10.

Drucker, P.F. (1991) "The New Productivity Challenge", *Harvard Business Review* **52**, 69-79.

Housing Policy Plan I (1988) Rijkshuisvestingsplan: De Vierkante Meter van Morgen.

Jonge, H. de (1990) "Changes in FM", paper presented at the MIT Conference Facilities Management Thriving on Chaos, Boston.

Jonge, H. de (1991) "Delivering the Buildings we need", paper presented at the AFM International Conference The Changing Workplace, London.

Joudah, A. (1992) "Developing Facilities Information Management Systems", paper presented at the EuroFM Conference, Rotterdam.

McLennan, P. (1992) "Facility Planning: Understanding the User and the Cycles of Change", paper presented at the the Euro FM Conference, Rotterdam.

Scheemaker, A. de (1991) *Prestatiebeheersing in het Rijkshuisvestingsproces*, Delft University Press, Delft.

Scheemaker, A. de (1992a) Analyse van Interviews met Directeuren en Hoofden van Facilitaire Diensten, GBA Internal Report, GBA, The Hague.

Scheemaker, A. de (1992b) Bestudering van het Facilitaire Proces in de Bestaande c.q. Toekomstige Situatie van de Facilitaire Dienst (VROM), GBA Internal Report, GBA, The Hague.

Wassenaar, K. (1990) Kantoor-inrichtingssystemen, GBA Report KO 12013, GBA, The Hague.

WORKING WITH PROTOTYPES: FROM CAD TO FLEXIBLE TOOLS FOR INTEGRATED BUILDING DESIGN

A. DRACH, M. LANGENEGGER and S. HEITZ
Institute for Industrialized Building Production
Department of Architecture
University of Karlsruhe, Englerstraße 7, 7500 Karlsruhe
Germany

ABSTRACT. The formulation of design knowledge as concepts, goals and rules cannot be captured in fixed and valid statements. The dynamic modelling of concepts and goals is, on the contrary, part of the design process itself. Tools that effectively support architects in their design should therefore never use predefined mechanisms, but must be definable interactively according to design specifications. We propose the concept of prototypes as a cognitive model to represent and structure design knowledge. Prototypes incorporate an individual view of design in a synthetic and organizational model for a defined area of interest. They actively control and guide design processes in supporting the organizational concepts for solutions. The a+Tool implements these concepts on the basis of a modelling language. It provides a dynamic toolkit and user interface to support design as well as knowledge modelling.

1. Introduction

"The most successful designs are not those that try to fully model the domain in which theyoperate, but those that are in alignment with the fundamental structure of that domain and hat allow for modification and evolution to generate new structural coupling" (Winograd and Flores 1986, p. 53).

CAD systems mainly support geometric modelling and drawing. Concepts to represent adequate product models and thus to enable integration and communication in a design process are still major issues in current research, and not only for building design. Our approach goes beyond product modelling in emphasizing the process of design supported by tools. To reflect and support the evolutionary aspects in this process, these tools have to combine fundamental structures for product modelling with an idea of design that is dependent on a designer's knowledge and interpretation.

The usual approach to provide decision support in design systems is to provide a set of knowledge bases that analyze and solve standard problems. This was the first approach of the authors when they participated in the development of the "Intelligent Design Tool Armilla" (Drach, Gauchel and Hovestadt 1990), a system that supported designers in the layout of complex service systems in buildings. Experience showed, however, that the problems of knowledge acquisition, extension and maintenance became more and more unsolvable with the development of larger design applications. Because of these problems, the systems remained limited in their scope and did not support complex design relationships.

The new approach has been fostered by a long period of experience in the development of expert systems and knowledge-based applications for building design. Our research involved investigations with a variety of tools and many activities in user-interface design. This practice and understanding of the problem provides the background to our proposal. The present concept of "a+" was developed in continuing research to provide a toolbox for applications in building

H. Timmermans (ed.), Design and Decision Support Systems in Architecture, 61–82.

design. On that common platform, cooperating planners can configure their own specific tool to support different tasks like floorplan, structural or service-systems design.

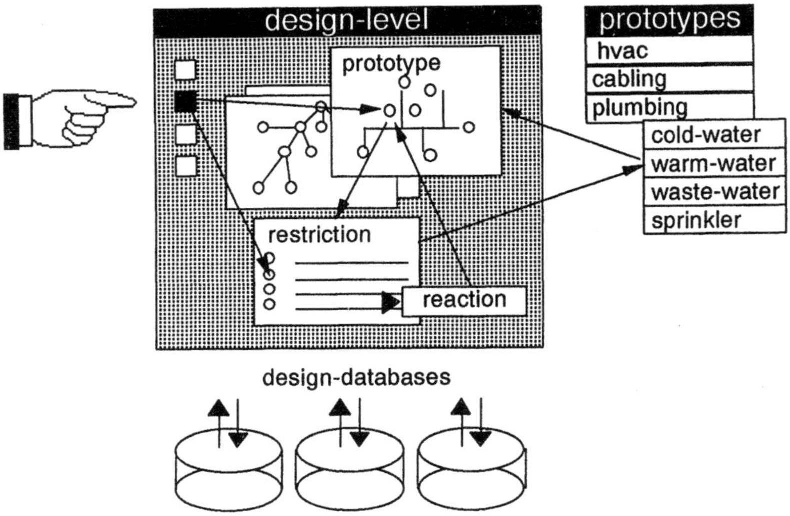

Figure 1. Scenario a+Tool

The central idea of prototype configuration is based on a modelling language using semantic expressions, enabling designers to flexibly extend and change system behaviour. Interactive visual representations and editors support the parallel modelling of design solutions, specifications and control. The language incorporates new schemes that naturally describe semantic knowledge in different levels of abstraction as structural, geometric and functional aspects. The basic elements to define prototype knowledge are design elements, associations, shapes and constraints. Actions, events and queries supply the verbs and prepositions for design control in the formulation of active constraints. Facts and knowledge are not isolated, but treated as explicit and descriptive entities. Therefore, new environments can easily be built upon others, using and specifying existing versions. The a+ system is developed as an object-oriented, extensible model and implemented in CLOS on workstations. Applications demonstrate design examples for conceptual service-system layouts.

2. The Problem of Design Integration

The growing complexity of building design, particularly through mechanical systems and building services, the shorter cycles of technical innovations and the overwhelming amount of different codes and regulations in the field of construction have created a situation where not only architects, but all parties involved tend to lose control of the design process if they do not use computer technology. These problems become vital with intelligent buildings and automated production facilities. The traditional separation of architectural and engineering tasks in building design, on the other hand, has brought on isolated software solutions and applications in the field. The different CAD systems and software applications are not able to communicate. This is not a technical problem of software engineering, but the problem of a common language. Product models could provide this common basis for communication but despite many efforts that have been made (Björk and Penttilä 1989), there is still disagreement about the capacity for expression and standardization of building representations, not only during the many stages of design and construction, but also for the different engineering contexts. This situation demands integration models and tools that incorporate the architect's generalized view of design.

2.1. ARMILLA INTEGRATION TECHNIQUES

The ARMILLA research project started at the Institute for Industrialized Building Production in the early 80s, to deal with the integration problem in building design. It was a common belief that solving the problem required the combination of architectural knowledge and new developments in computer technology. The project's formulation and results are based on the special architectural experience of Fritz Haller, who is best known for his designs of construction systems (USM-Haller, MINI, MIDI, MAXI).

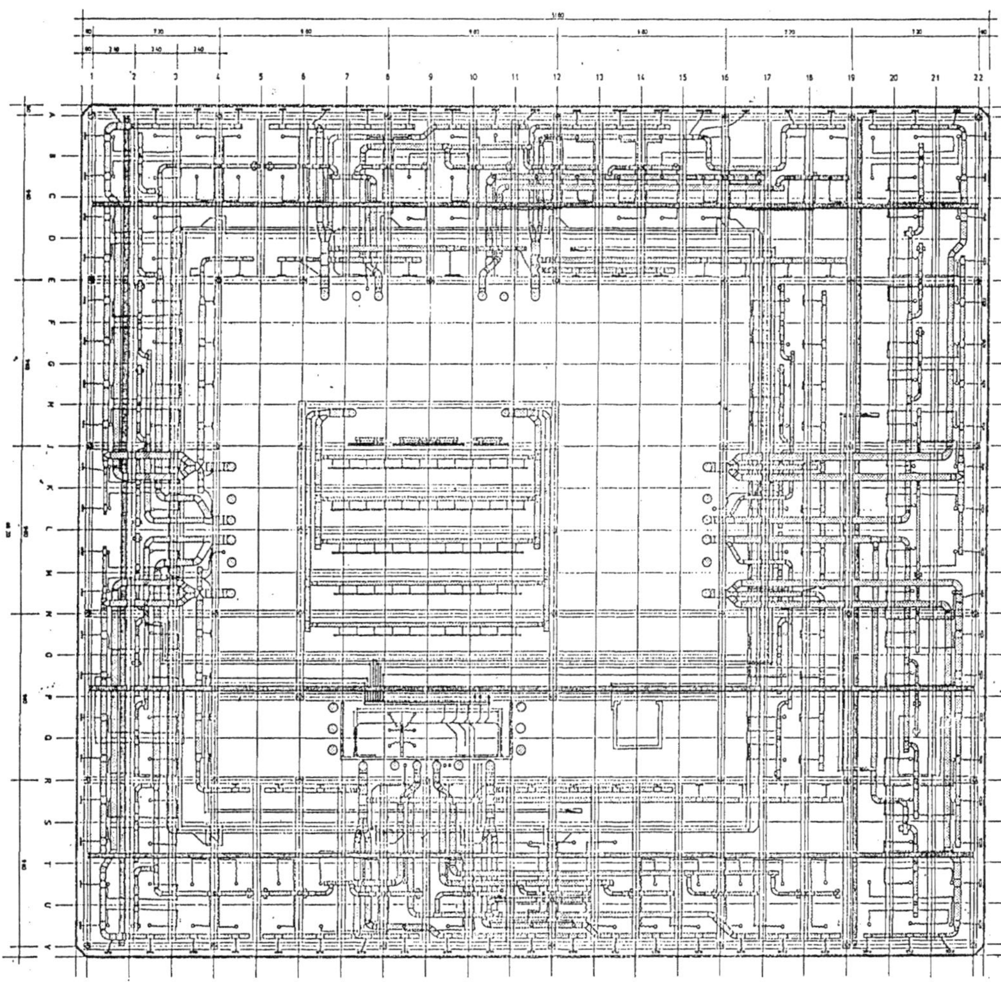

Figure 2. Installation ARMILLA, SBB Trainingcenter Löwenberg/ Murten CH

The ARMILLA installation model (Haller 1985) defines an integration technique based on geometric coordination and control, using spatial templates and different levels of abstraction. It had been developed as an abstract methodology to organize service systems in space and is applicable to any building design without the need for computers. The basic concept of the installation model is the structuring of space through a three-dimensional modular grid. Possible layouts and locations of pipes and ducts are restricted to certain routes or templates defined in the spatial grid. Different templates exist for different types and sizes of pipes. The modular and integrated organization of templates reduces possible conflicts between interfering systems to a minimum.

Since 1985 there have been efforts to apply the ARMILLA methodology to knowledge-based design environments. The results include several prototype expert systems to deal with the problem of automating service-system design (Mathis 1988; Raetz 1989). These systems, however, failed to represent and automate design solutions for problems of acceptable complexity. ARMILLA 2 (Gauchel et al. 1990; Drach, Gauchel and Hovestadt 1990) was the first approach to combine expert systems and CAD. It proved that the interactive approach was a promising concept to overcome the problems of brittleness and rule-explosion in complex and unstructured design tasks like building design. Such tasks not only require reasoning at different levels of abstraction, but also the integration of innumerable concepts from various domains and a great amount of common sense.

The work presented in this paper extends and combines earlier results in introducing a new level of flexibility and interaction in a model for supporting design tools. User control is suggested not only at the level of product modelling in finding appropriate design solutions, but also at the knowledge level, through interactive definitions for design consistency and support. Product modelling is only one part of the problem. Providing macro-languages to incorporate knowledge as part of a design is a major goal of our work. In this sense it is explorative in narrowing the gap between cognitive representations and software technology.

Corresponding research at the institute focuses on aspects of communication, cooperation and database technology in models for building design (Hovestadt 1991). Projects are funded by the government of Baden-Württemberg (Germany), CAD/CAM- Schwerpunkt, BMFT (Minister for Research and Technology, Germany) and DFG (German Research Association).

3. Expert Systems versus Flexible Tools

Despite the successful application of the ARMILLA model in expert systems, major questions of integration and representation of design knowledge remained unsolved. These problems concerned large and unstructured knowledge bases, knowledge acquisition and maintenance. Our experience and recent research in cognitive science supported various theories and goals for flexible knowledge environments. First, design knowledge cannot be captured in constant and valid statements. It is the individual property of a designer and must therefore be modifiable by designers. Knowledge is a lifetime process directly dependent on the person of expertise. The knowledge of an expert does not represent a defined optimized status of knowledge. Rather, it is subject to permanent changes, based on individual experience and learning (Becker 1992, p.47). Secondly, complex and contradictory tasks require permanent interactive design decisions, supported by informational and assisting tools. Expert systems and knowledge-acquisition models tend to treat design knowledge as logically consistent expressions. This contradicts the architect's view, in which goals and strategies are evolutionary concepts that evolve and change during the process of design. Architects have to take responsibility for their designs. Tools that support design must therefore be understood by the users. Thirdly, the permanent process of technological innovation in the building industry involves frequent changes in design methods and specifications. Easy adaptations and extensions of knowledge modules are therefore essential qualities for design knowledge bases. Finally, a modular representation of different aspects of design in the form of macro-languages could facilitate communication between distributed environments and users.

Expert systems have proved to support designers in solving clearly-defined standard problems (Kröplin et al. 1990; Lenart and Ketteler 1990). To date, however, applications have failed to effectively deal with complex and unstructured problem spaces. The question of structuring building design into standard problems that match standard solutions remains the main focus of present research and becomes even more complex for highly interactive design environments. Our proposal suggests a bottom-up development of decision-support systems to overcome the problem of blindness in expert-system applications. Design support in interactive environments may provide a basic level of control to inform users, if and how their solutions comply with defined goals and constraints. A user can then proceed with the design interactively while solutions are evaluated and checked for conflicts. On a second level, active knowledge can be added to constraints, proposing fixes for violations (Marcus and McDermott 1986).

Declarative and intuitive structuring of the supporting concepts and mechanisms is essential for users to be able to configure their own tool. According to classic expert system theory, applications are designed and maintained by knowledge engineers as domain-specific sets. Representation and formalism depend on the individual designs of knowledge engineers and are not visible to users, except through explanation control, which does not allow any rule modifications. The advantages of inference strategies in the separation of facts, rules and control tend to change into weaknesses with growing knowledge bases. Representations in blackboard systems lose their inherent relationships, which are evident and essential at the time of their formulation. Even experienced knowledge engineers admit that this problem makes large knowledge bases hard to maintain. On the other hand, advances in software technology have improved modularization-techniques for software applications. Object-oriented representations provide a more cognitive way to represent problems. Declarative representations of composite chunks of knowledge are more expressive and therefore easier to change and adapt to similar problems.

4. a+ Prototypes

The key concept and basic cognitive image for knowledge representations in a+ is the concept of a prototype. Prototypes are classification schemes for types of solutions, but their representation differs in some aspects from comparable concepts and terms used in software engineering. Prototypes aim at expressing information about different types of solutions, in the sense of giving an actual description of the solution spaces and interdependencies involved. In architecture and also in engineering, the term type or standard type refers to its synthetical nature. The pieces of knowledge it incorporates are already organized in relation to each other and coordinated in such a way that they do not conflict but have an integrating impact on the practical results. The type, as operational knowledge, may therefore guarantee coherence between an action and its goal (Muratori 1960).

An example of an architectural type is shown in Figure 3. All three floorplans of residential buildings by Frank Lloyd Wright follow the same organizational principle of a floorplan layout, despite their obvious variety in size and form. Their common prototypical concept could also be represented by a set of constraints that confine a solution to certain standards but do not restrict variations outside its range. We consider this concept as a basic cognitive model for design, which could be translated to all relevant areas and aspects. We therefore introduce it asthebasicstructure of our knowledge representation. We will show later how we use this concept to express design dependencies for service-system layouts.

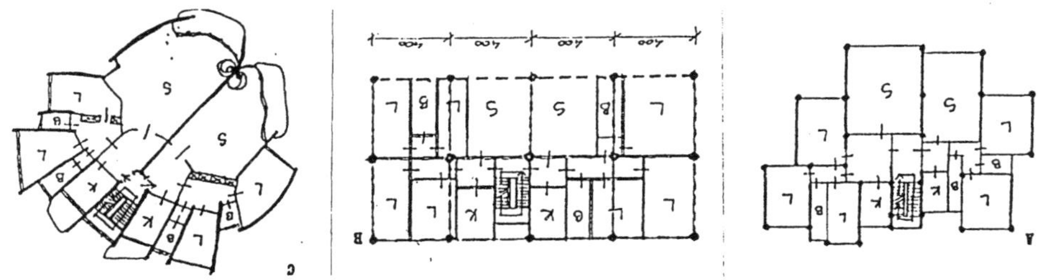

Figure 3. F.L. Wright, Variations of a Type, in: Canniggia/Maffei, Il progretto nell'edifizia di base, Venice 1984

4.1. WHAT IS A PROTOTYPE?

The following aspects may give a summarized definition of what is meant by prototype:
(i) *Classification scheme for types of solutions.* The language of architects and engineers already provides a variety of generally used types, to denote specific solution types for configurations in building or construction. This representation focuses on the descriptive power of prototypes in order to facilitate knowledge modelling. Prototypes may be organized in hierarchies, specializing certain compatibility and restriction rules for specialized systems.
(ii) *Cognitive model of design.* Prototypes represent individual categories for knowledge, incorporating aspects of context, time and experience in a single representational scheme. This notion follows the concept of "Gestalt" theory, where cognition is determined by the special structure and organization of information, assuming that the whole is more than the sum of its parts (Becker 1992, p.53). In representing information as a relevant section of the whole design, a prototype structures and reduces complexity. In this goal it corresponds to other concepts based on analogy like case-based reasoning (Hua et al. 1992) or shape grammars (Flemming 1987).
(iii) *Synthetic, not operational model*. We run into problems when we try to express our cognitive model in sequential operational activities. The information stored in prototypes is rather synthetical in nature. When designers think of prototypes, there is no operational model of how to generate a type of solution from a given situation. Nor is there a ready analogy for a situation, which only has to be mapped and adapted from a similar situation (case-based reasoning, routine design). In our representation, prototypes define the boundaries for their valid solution spaces, through specification of valid elements and configurations of elements, which are checked and organized when applied to a design situation.

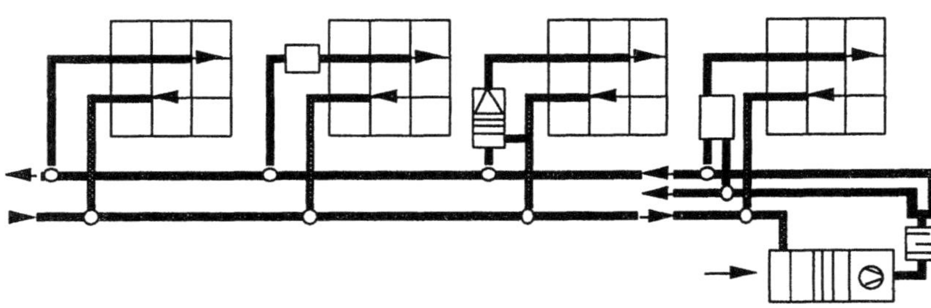

Figure 4. Different prototypes for air-conditioning systems in service zones of a building

4.2. THE MODELLING LANGUAGE

Before we can illustrate how prototypes may be represented and integrated in an interactive design-support environment, we have to define a terminology for the problem space. To provide a maximum of modularity in specifications, the modelling language is based on object-oriented modelling techniques (Booch 1991). It should therefore be no problem for users to specify their own objects from taxonomies and add control rules, instead of configuring and defining macros in a CAD environment. In order to describe prototypes in their area of interest and relevance for design solutions, we have to introduce an underlying level of description which we will refer to as the *building-model* level. This is the basic level of information and communication between prototypes. It could also be referred to as a product model of a building that provides an interface between cooperating applications in the process of building design, construction and maintenance, such as engineering, simulation, construction and facility-management tasks.

The following categories form our basic concepts with which to describe a design model of a building in all stages of design and in adequate semantics to serve as an interpretation background where prototypes can apply their knowledge in a useful way. *Projects* specify the actual building models for specific buildings. Design is not a linear process of refinement from the abstract to the more detailed. Rather, it progresses through recursive steps of revising, changing and refining decisions. A building model has to represent all levels of abstraction that were necessary to generate it. This is important to allow users and prototypes to step into and view through any level of abstraction at any time. Design levels form a structuring concept similar to views in filtering information in respect to certain abstraction levels. ARMILLA (Drach et al. 1990) proposes five major *design levels*: 1. spatial organization of the building structure, 2. requirement definition for service systems, 3. strategic layout design of service systems, 4. spatial coordination of piping routes, and 5. component specification and allocation. *Design elements* are the basic architectural concepts to describe structural, functional and geometric semantics in building design are systems, components and zones. *Shapes* constitute a separate language to model geometry provides representational formalism for three-dimensional orthogonal structures and functionalities. These include expressing geometric hierarchies and relations, layout preferences, grid and functions, as well as queries about geometric relationships. A final category is *associations*: interpretations and conclusions about design states can rarely be deduced from the states of the elements alone, but mostly involve relations between elements. We therefore provide schemes to model user-defined associations to express semantic networks. More details concerning the modelling language are given in Sections 5, 6 and 7.

4.3. WORKING WITH PROTOTYPES

We now turn to an example of how our theoretical concept of a prototype might guide a design and what kind of functionalities are provided when integrated in our interactive design system. Basically, a+ prototypes do represent nothing else than the cognitive concept or mental image a designer would use in exploring design problems. They represent a specific solution concept during a design session, through managing, controlling and organizing a defined space of interest (as a section of the whole state of design). Through managing elements in their relation to each other, it also defines the rules of consistency for associations between elements. We will assume that we have already worked out a layout for an air-supply system with our tool (see Figure 5) and would now like to be supported by prototype knowledge in the subsequent steps.

Installation. a+ prototypes can be installed and used in a design situation interactively through selection menus and may interactively be removed or replaced by other prototypes. In order to install the knowledge only in the relevant section of the design, prototypes use their special installation knowledge. This means that they first check what kind of prerequisites have to exist, what kind of states and situations are relevant, which are the elements of interest, that should be controlled and organized, which elements can form a solution, and when is it appropriate to warn, criticize, repair or make proposals. This type of knowledge is formulated in active constraints (see Section 7), which directly link prototypes to design elements.

Instantiation methods take care of the proper mapping of prototypes and elements to design situations defined in the building model, whereas attached rules and reactions supply assisting knowledge and procedures. They also incorporate knowledge about their appropriate design level and how to propagate themselves to other levels.

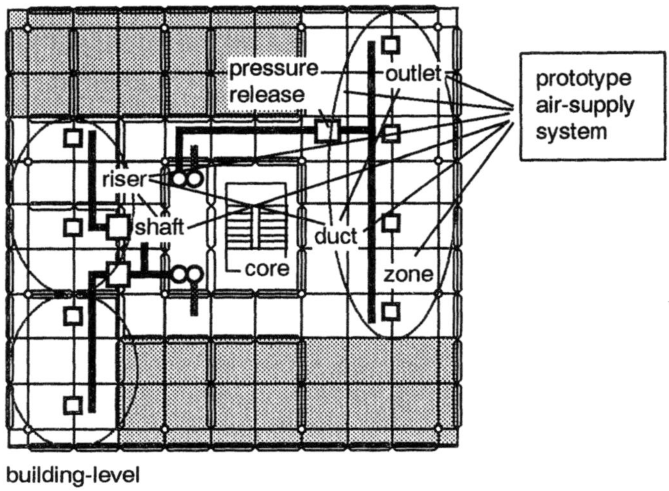

Figure 5. Design example air supply system

Interactive design. As long as a prototype is installed, its rules remain active. This means that they react to certain manipulations (events) in a protecting way, to keep the design consistent to a defined solution space. In case of violations, prototypes may either warn or criticize a user or make proposals and try to repair. In our example, the prototype for an air-supply system manages the relationships between a supply zone, the outlets and the ducts that connect an outlet to the riser for supply air. When an outlet or riser is interactively moved to another location in the building, the prototype will take care that the connecting ducts are adjusted. When a new zone is added or an existing one is resized, the proper layout for outlets may be generated or altered. If the size of a zone and its necessary air-supply volume/h extends the capacity of the connected riser, this information is brought up to the user, proposing several reactions like dividing the zone and introducing a new riser, or changing the riser's diameter if possible, which also increases its capacity.

Design conflicts. Constraints or goals in a prototype definition or between concurrent prototypes may contradict each other in trying to organize the same elements to meet their antagonistic goals. This is a very common case in design. You may, for example, state the rule, that risers are only allowed inside of shafts, but your HVAC prototype defines a maximum distance between outlet and riser. In some cases there will be no way to comply to both constraints. Then only an experienced designer can decide, depending on the overall situation, which rule can be relaxed. We therefore do not provide any automatism to resolve these inconsistencies, like other approaches with truth maintenance systems. On the contrary; conflicting goals are inherent in the design process. The advantage of our approach is that they are made explicit and in that way can be resolved interactively. The difference between an acceptable and an excellent design solution often lies in the decision which criteria should be relaxed in a certain situation.

4.4. CONFIGURATION OF PROTOTYPES

Instead of providing users with a predefined set of prototypes that represent a limited selection of solution types for a problem, a+ enforces the interactive redefinition and configuration of prototypes in order to incorporate individual requirements and knowledge. Any time during design modelling, the level of prototype configuration can be entered and existing prototypes can be modified in their functional behaviour or "semantic" interpretation of a problem. Prototypes are either defined from scratch or redefined and extended by inheriting elements from other prototypes.

User definitions. When using prototypes it often happens that their standard behaviour may be improper. Certain goals or technical constraints may be either obsolete or just inconvenient in the actual situation. Experiences have shown that even simple mechanisms can provide excellent design support when accompanied by a proper model for semantic control. A tracing mechanism for piping nets, for example, might determine that connected parts should be dragged along in order to keep the net together. This rule might be applicable only under certain conditions, or not at all, depending on the prototype's definition. The procedure can vary with different prototypes defined by different designers and their individual style of design.

Relaxation. Prototypes provide additional switches to enable relaxations. When turned off, interactive design can go on without warning. As soon as the switch is turned on again, the new situation is evaluated for restrictions. This applies to whole prototypes as well as to selected constraints.

Layers of information. A great problem, but also a challenge, for our approach is providing adequate user interfaces that allow informational transparency about a prototype's scope and its semantic and behavioural details as well as meaningful tools for its modification. Prototype configuration is structurally organized around layers of information (Figure 6). A complete view of the actual state of knowledge becomes visible through navigation along knowledge paths. Starting from any selected design element of the generated building model, a user can request and browse through information about its actual state (design view), including geometric and semantic properties, associations (association view) and related prototypes and constraints (prototype view).

Constraint rules. The best way to identify constraints, which are supposed to handle new conditions and behaviour for a prototype, is to concentrate on each class of elements involved and check for all kinds of roles and parts it may take in the solution. Forms facilitate the formulation of conditions. Queries support definitions in supplying a language to express complex search patterns. Constraints are triggered by events. In the conditional parts, design situations can be described by combining queries. Queries may not only relate to geometric information like distance or relative location but also concern type and association-specific states of design elements. In the reaction part, the results of queries can be linked with actions. A detailed description of constraints and queries is given in Section 7.

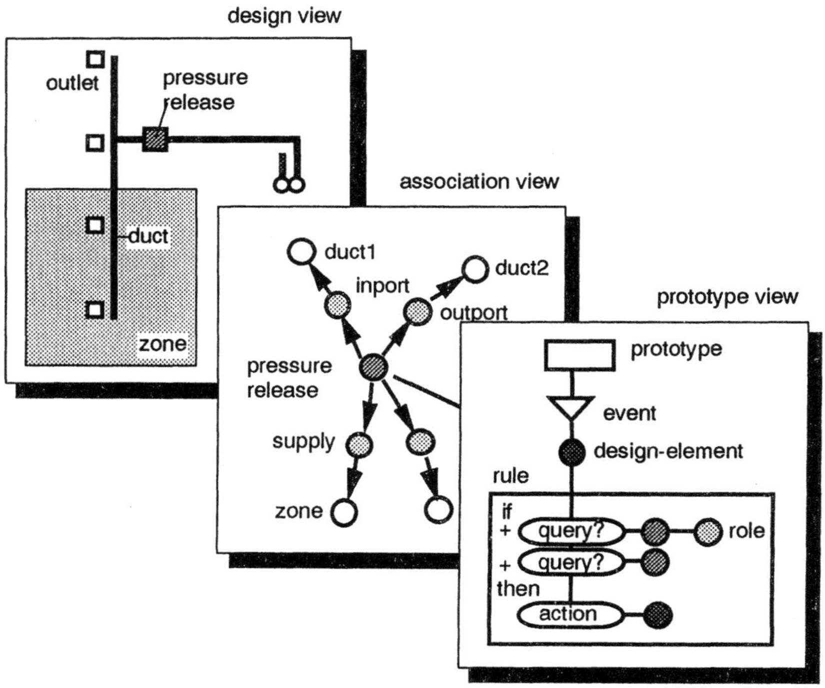

Figure 6. Layers of information

5. Entities and Associations

Marvin Minsky's theory of mind (Minsky 1985) captures many cognitive aspects of design. Most striking is his theory of k-lines and societies of k-lines. He distinguishes organizations of higher-level and lower-level agents. Each higher level of description must add to knowledge about lower levels rather than replace it. Our approach uses entities organized in object hierarchies and association networks to represent the notion of agents and agencies which are organized through k-lines. In the representation of entities and associations, we relate to languages for product modelling like STEP/Express (Anderl 1992) and NIAM (Nijssen and Halpin 1989).

Design elements serve as the base class for entities. Design-specific entities may be built upon more general concepts that are provided with our modelling language. General concepts in building design include: (i) systems to represent structural entities in capturing hierarchies of functional decompositions into subsystems or components, which semantically relate to certain functional tasks in a building and have no geometric representation (i.e., structural system, circulation system, air-conditioning, cold-water supply), (ii) components as parts of systems, which form the end nodes in a design solution and, depending on the level of abstraction, may represent actual physical components and require geometric representations (i.e., partition, riser, duct, outlet), and (iii) zones that describe spatial organizations in buildings. They are important concepts to organize building hierarchies as semantic and spatial containers for components. Systems need zones to define spatial requirements for their functional tasks. Zones may overlap. Several zones representing different aspects of the design may even occupy the same space.

Examples include room, core, shaft, service zone, circulation zone, installation zone.

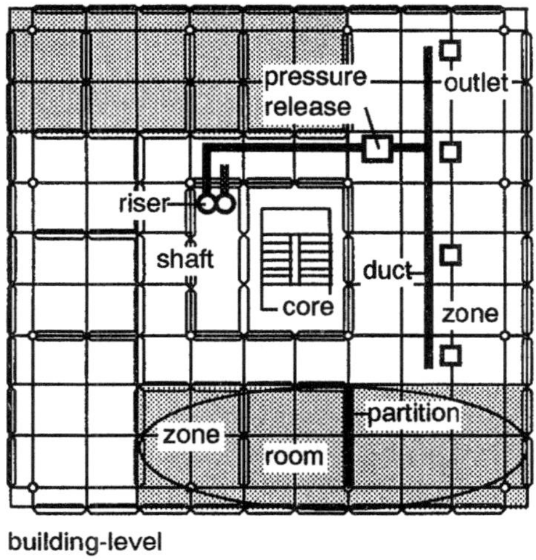

Figure 7. Design elements

Associations are complementary elements to entities in forming the k-line connections between design entities. They are important features in representing design, not only as structural solutions, but in describing intermediate steps or direct goals between corresponding entities. A similar approach has been used in ACCORD, a system for arrangement and assembly tasks that uses roles to denote valid states and goals between objects (Hayes-Roth et al. 1986). This holistic or coherent view reflects much better mental images in design than the operational view in rule-based representations, where agents form an unstructured mass of independent views.

Our representation uses the following definitions based on concepts introduced by Rumbaugh et al. (1991): roles define the direction of links between two or more entities. Each role has its name tag, denoting a proper identification of the semantic type of corresponding association. The language provides forms to build user-defined associations from basic concepts. Links create association networks during design (product modelling) through instantiating direct paths between objects. That way they provide an additional level of information in forming search patterns for problems. Because associations are often interrelated (as shown in Figure 8), the problem of keeping association networks consistent with defined semantics is an important issue.

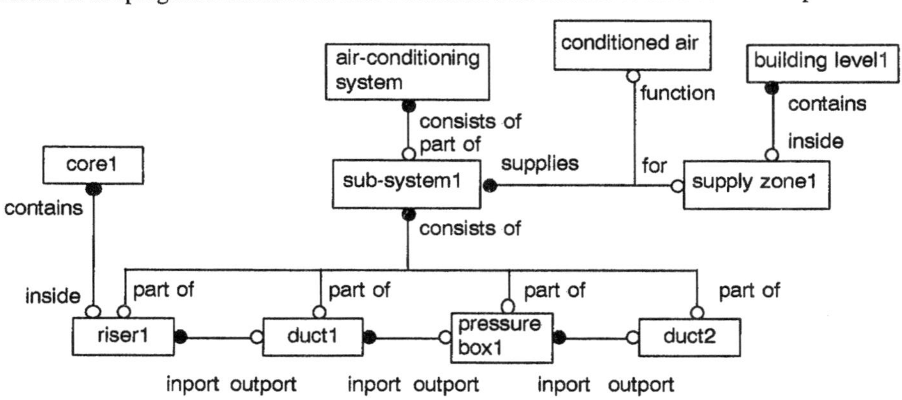

Figure 8. Associations and roles

In our approach this is managed at knowledge level (prototype specification) through user-defined constraints on associations, that watch for inconsistencies and use applied methods to restore semantic coherence.

6. Shapes

Geometric representations in a+ use shapes as a uniform description model to represent geometry in buildings (Langenegger 1992). In many CAD approaches, but also in shape grammars (Flemming 1987), geometry is the main focus for representation; geometric and semantic aspects of design are represented in a single concept. The shape model introduced with a+ proposes a more refined solution. Unlike traditional CAD systems, where geometry and visualization, as inseparable concepts, are the only representations for design, we propose the deliberate separation of semantic and geometric aspects. Shapes do not represent design semantics and therefore can never exist independent of design elements. They only form geometric properties of design elements which may be represented through one or more shape objects. These properties provide the design element with powerful representational schemes and functionalities. Shapes are defined by an enclosing box and a direction vector, and are designed to manage their own coordinates in a three-dimensional, orthogonal structure of space. They support interactive geometric manipulations as well as geometric queries on shapes and relations between shapes. The major representational schemes and functionalities of shapes are described below.

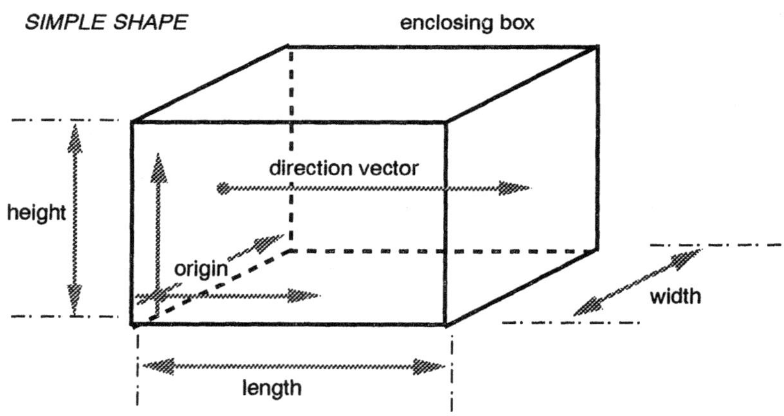

Figure 9. Three dimensinal representation of shapes

Shapes are associated with design elements. Shapes represent geometric properties of design elements. They provide functionalities to query their associated elements and receive or propagate information about changes from the geometric (shapes) to the semantic (design elements) model.

Geometric hierarchies. Container shapes may include other shapes. This enables the structuring of shapes into purposeful hierarchies to reduce geometric search for a given problem. Relevant subsets of design elements may be grouped in meaningful containers to decompose a design in relevant sections. Building hierarchies and zones, for example, may geometrically be represented by corresponding shape hierarchies. Included shapes define themselves relative to the boundaries and coordination system of their parent. Methods are provided to query groups of shapes along hierarchies for defined criteria.

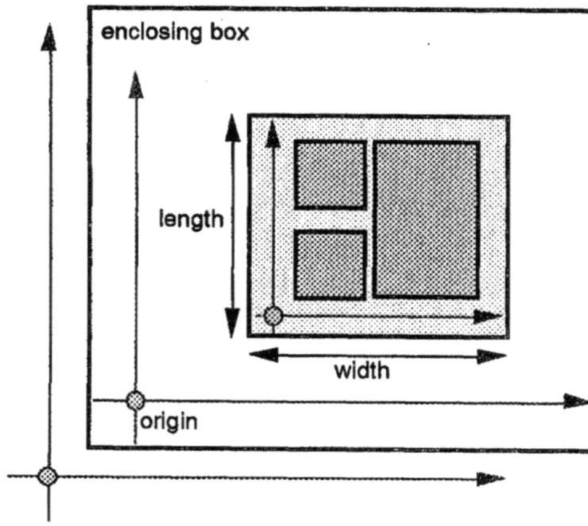

Figure 10. Shape hierarchies

Geometric coordination is supported by three-dimensional coordination systems in the form of a complex grid. Specifications define the proper alignment of shapes in position, size and direction relative to the grid. This functionality allows the specification of abstract coordination models that manage possible locations of components and elements in a building. Higher-level shapes determine coordination for the next lower level. Shapes may align themselves in all three dimensions to rulers or zoning templates of their parent shape. The individual shape defines the details of alignment through methods that specify if and how to adjust to the grid, using additional information about adjacent shapes that might effect the proper layout. After modifications, all affected shapes are informed automatically to check and readjust their alignment to their surrounding coordination systems.

Layout preferences describe the proper size of shapes in all three dimensions. Shapes may be constrained in their length, width and height to minimal and maximal sizes, or dimensions may be defined in relation to each other. Values are automatically adjusted to follow these constraints after modifications of shapes. Layout preferences define the range of size and form of shapes; they do not specify possible positions and locations.

Geometric manipulation. An interface to model the distinct geometry of shapes provides methods to move, place, rotate and scale shapes or to change their location in geometric hierarchies. Changes are automatically propagated to associated shapes through hierarchies to enable their adjustment according to the changes.

Geometric queries. Shapes supply a query language to test geometric relationships between shapes. It includes basic functionalities to test the relative locations (north-of, adjacent, inside etc.) as well as comparisons (smaller, larger, distance, etc.). Primitive queries may be combined in expressions to test complex relationships.

7. Active Constraints

The concept of active constraints as the constituent elements of prototypes is introduced as the central structure to define and provide dynamic control in an interactive design environment

(Heitz 1992). Constraints comprise flexible and powerful representational schemes for semantic specifications of prototypes and are designed to assist users during design in several ways.

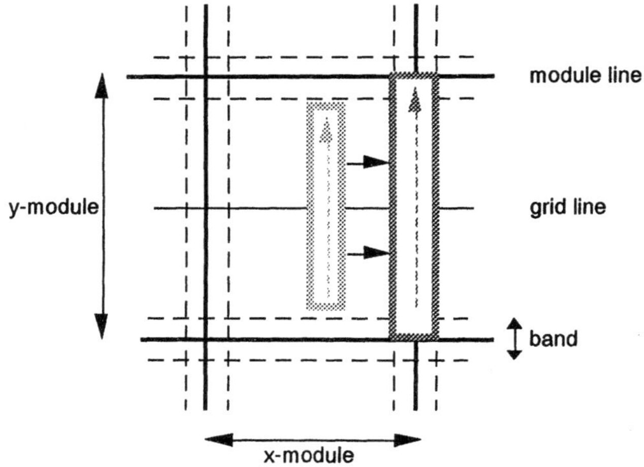

Figure 11. Geometric coordination with shapes

First, user-defined constraint rules check for semantic consistency of associations between systems, components or zones in a design; they supply methods for correction. For instance, the consistency rule for an inport-outport association between two components of a piping system may determine that both have to be members of the same subsystem. If one component is attached to a different subsystem during design, the improper links have to be removed. Secondly, user-defined or library constraints check if a design situation meets certain specifications. In case of a violation the same constraints can then supply users with detailed information about the problem. For instance, a restriction for waste-water systems may state that outlets can be placed only within a certain distance of a riser. If a user should move a riser or outlet outside these limits, the constraint may then pop up the information about the restriction. Thirdly, constraints actively assist users in certain tasks through adding operational information about methods of possible fixes of the problem. For instance, in the previous example, the constraint could be extended by a modified reaction method, indicating several possibilities to fix the problem. This might include undo previous command, move riser, move outlet. Or the reaction might suggest a standard fix method, which could involve installing a new riser at the appropriate distance. Finally, constraint-directed search and optimization strategies can assist users in solving tedious configuration or allocation tasks. Reaction methods of rules are used to formulate knowledge-based correction rules following a propose and correct strategy (Marcus, Stout and McDermott 1988; Puppe 1990, p.143). For instance, a typical problem to which this strategy could be applied is the allocation of competing piping systems in their optimal spatial templates in a building, where many geometric and functional constraints may interact.

Active constraints and prototypes. Active constraints are always associated with prototypes. They may be linked to one or more user-defined prototype, which enables the easy definition of prototype variations. As soon as a prototype is installed in a user environment, all associated constraint rules are installed on their defined design elements. When a user interactively deinstalls the prototype again, all constraints are removed as well, unless they are not used by a second prototype which remains active. The association of constraints with prototypes defines prototype-specific sets of constraints. This provides a powerful instrument to structure constraints in problem-specific units. Any time during design, a user may follow up which constraints are active with installed prototypes.

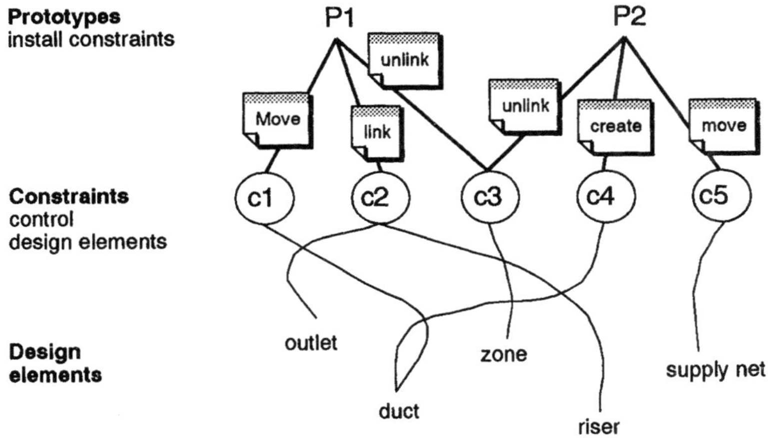

Figure 12. Prototypes and constraints

Constraint formulation. Constraints consist of a trigger part, a conditional part and an active part. Control is supplied through a kind of event trigger mechanism, which directly connects design elements with constraints. This way inference can be provided without the overhead of a global matching algorithm. Another advantage is a more declarative structure of the knowledge base in the direct connections between design elements and constraints. The conditional parts of constraints represent situation patterns describing the valid solution space for the defined constraint. Queries are intermediate representations to describe conditional parts in constraints.

```
Constraint
trigger    design element <system>
           event: link <link-obj>
condition
   if and
           <system> = of-type zoning air supply system
           <components> = <system> is-linked-with-role has-components
           <risers> = get-all <components> of-type riser
           number-of <risers> greater-than 1
   end and
reaction
   divide-system <system> = <new-system>
   unlink <link-obj> with-role component-of
   link <link-obj> with-role component-of <new-system>
```

The goal is to provide reusable templates for an easier description of situation patterns. Elements of queries are set definitions, predicates, type checking, logical operations, variables and the access to design elements, associations and geometric relations. A user may furthermore combine queries to define reusable macros for repetitive conditional expressions. The reaction part provides the same interface for actions on design elements that designers use through graphic manipulation tools. For example, a zoning air supply system may only be linked to one riser. The constraint states that if a second riser has been linked (either interactively or by reactions), the system has to be divided.

Event trigger mechanism. Reactions of constraints invoke the same event trigger mechanism as top-level user commands. Design elements incorporate knowledge of how to transform actions into object-specific operations. Actions are linked to corresponding types of events. Events are created after the execution of an action and sent to a constraint handler. For each action-event cycle, the appropriate constraints are instantiated and sorted into a queue, which is processed while the conditional parts of the constraints are tested. In case of a violation, their active parts are called, which may invoke additional actions and events until the queue is completely processed. Despite the limits of this inference strategy, the idea of flexible constraints can be supported by the present implementation.

8. a+Tool

The a+Tool provides an interface and mechanism to use the concepts of prototypes, design elements, shapes and constraints in a system and demonstrates design support and assistance according to the ARMILLA model. Graphic editors and views are the major means of information and system interaction. Design control and assistance is enabled using event-driven inference based on object specifications. We provide a *toolbox* with a basic set of prototypes, typical design elements, associations and constraints, which can be used as is, or extended by the user. All relevant functionalities that form the interfaces to shapes, queries, actions, user-commands and visualizations are encapsulated and inherited from the basic concepts. Public protocols are supplied for these interfaces, based on the idea of a meta-object protocol (Kiczales et al. 1991), which allows users to adjust the design and implementation to suit their particular needs. In other words, users are encouraged to participate in the language design process. Applications are built as sets of additional domain-specific prototypes, design elements and associations. These sets can also be combined to easily reuse existing knowledge for new areas.

8.1. VISUALIZATION AND INTERACTION

The interactive graphic user interface reflects the different layers of information already introduced as design view, association view and prototype view. Basic concepts for visualization and interaction are view objects, editors and commands. The design was strongly motivated and supported by the following goals: (i) direct interactive user control (any time during processing a user may control system behaviour according to his or her will), (ii) design elements are visualized as graphic symbols (a user may directly manipulate these symbols by mouse commands, similar to drawing tools like MacDraw), (iii) editors view and structure the building model of a project under certain aspects and relations, and (iv) editor tools support hypermedia connections and browsing along design views of a building model.

Design elements in a+ are not able to visualize themselves in the user interface; they use *view objects* as independent graphic-symbolic representations. View objects may represent different appearances, characteristics and information of design elements under different criteria like geometric, functional or semantic aspects. In providing concepts for symbolic, geometric or textual representations, they are specifically designed to handle the different kinds of visualizations of design elements. For each representation in an editor window, a design element uses a separate view object. Visualization, user interaction and communication of and with design elements is entirely managed by view objects. This is done by a communication protocol between both objects that allows view objects to query a design element for the relevant information it needs for representation. Visualizations are constantly updated after changes on design elements to reflect the proper and actual state of the design.

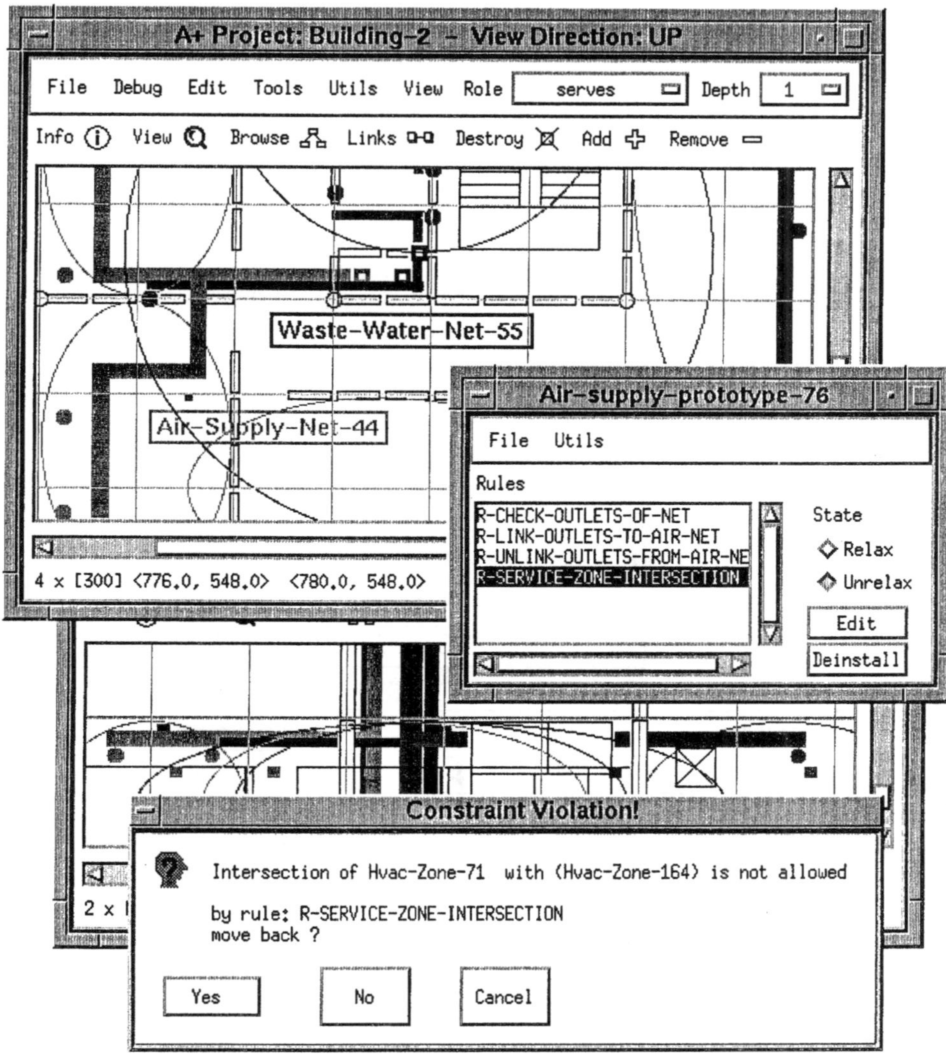

Figure 13. a+Tool- Geometric editor, prototype viewer and constraint violation

Editors provide a view on a defined portion of a project. Editors may, for example, visualize the geometries of design elements of a building in plan or section the way we are used to view a design on a CAD screen. The same situation could be represented as an association network in another window. For a+ we developed the abstract concept of an editor description to describe the various functionalities of different view concepts and to support the generation and redesign of editors. Editor descriptions include definitions for the scope of the editor as the relevant portion of a building model, the form of visualizations and the supporting tools for manipulation. Filters define criteria for representation like type, location, relationships or other properties of design elements. For the filtered elements, an associated view object is generated and exposed in the editor. This way a building model can be viewed, checked and manipulated differently according to the editor's definitions.

Design elements in editors may be manipulated through manipulation tools. The underlying concept of any interactive manipulation is the a+ *command*, which is performed by a user event in the form of a button or mouse command on a view object. The command may represent a series of actions, which are instantly propagated to the related design element where they invoke the appropriate operations. Design elements represent at any time the actual state of a project; their visual representations are only updated as a the result of the operations.

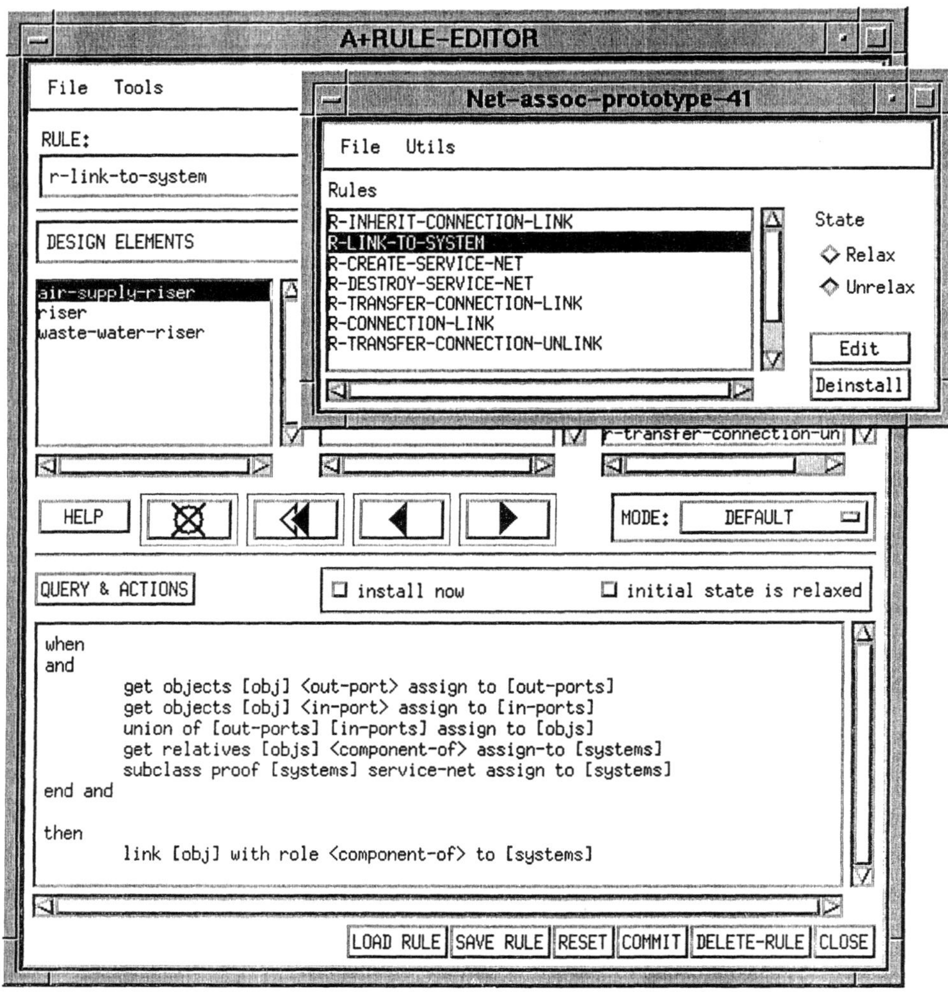

Figure 14. a+Rule editor

8.2. USER INTERFACE

The a+Tool includes the following editors and tools which are modifiable to support interactive design in different environments: *geometric editors* represent geometric properties of a building model which can be viewed in plan, section or elevation at any scale. Tools are provided for three-dimensional geometric browsing and manipulation. *Association viewers* explore the semantic networks of associations. Associations and design elements may be viewed or manipulated graphically. *Instance viewers* display the actual state of design elements during design. The information is structured in appropriate textual windows and panels that enable browsing through all information and instances accessible in a project. *Prototype tools* show the state of installed prototypes in a design environment. Additional functionalities allow the user to view and relax or unrelax associated constraints. *Prototype editors* support designers in configuring or modifying prototypes and constraint rules. *Design palettes* display libraries of predefined design elements and prototypes. They may interactively be selected and copied (instantiated) to a project.

The current implementation of the a+Tool and language is in CLOS on DEC-3100 workstations using GINA (Spenke and Beilken 1990) as an interface toolkit. The implementation is designed to be portable to other object-oriented languages, or object-oriented data base management systems and implemented as an extensible model.

9. Summary and Conclusion

The impetus for the a+model was formed by experiences in the field of building design integration. To date, architectural practice does not offer standard models to structure building design, nor does it provide any integrated design methodologies. The process of building design still appears as unstructured and orderless and is to a large degree dependent on human creativity and control. In this situation, we do not believe that tools should replace human control and communication, but tools could be quite efficient in facilitating it. Our proposal is centred on a user's perception of design, which means he or she should be enabled to structure his or her own environment within the limits of an integrating language framework.

The concepts developed in our model concentrated on the idea of providing public interfaces for design knowledge. Questions of data management, CAD integration, product modelling, concurrent design and multi-user environments, which are important issues in this field of research, were only touched upon by our approach.

The model has proved to be quite promising during implementation. In structuring and combining exchangeable knowledge modules with semantic and geometric representations, we could provide a useful model for the integration of user-defined environments and data bases in building design. The aim of our flexible tool is to point out perspectives. This aim might conform well with the next generation of CAD systems based on object-oriented data base management systems. In some respects, this vision is not so different from using style formats in a word-processing system, instead of using prototypes in a CAD system. We think that knowledge modelling in design-support systems could become as easy as programming Hypercard or Excel.

References

Anderl, R. (1992) "STEP - Grundlagen, Entwurfsprinzipien and Aufbau", in Krause, F.L., Ruland, D. and Jansen, H. (eds.) *CAD'92, Neue Konzepte zur Realisierung Anwendungs-orientierter CAD-Systeme*, Springer Verlag, Berlin, pp. 361-381.

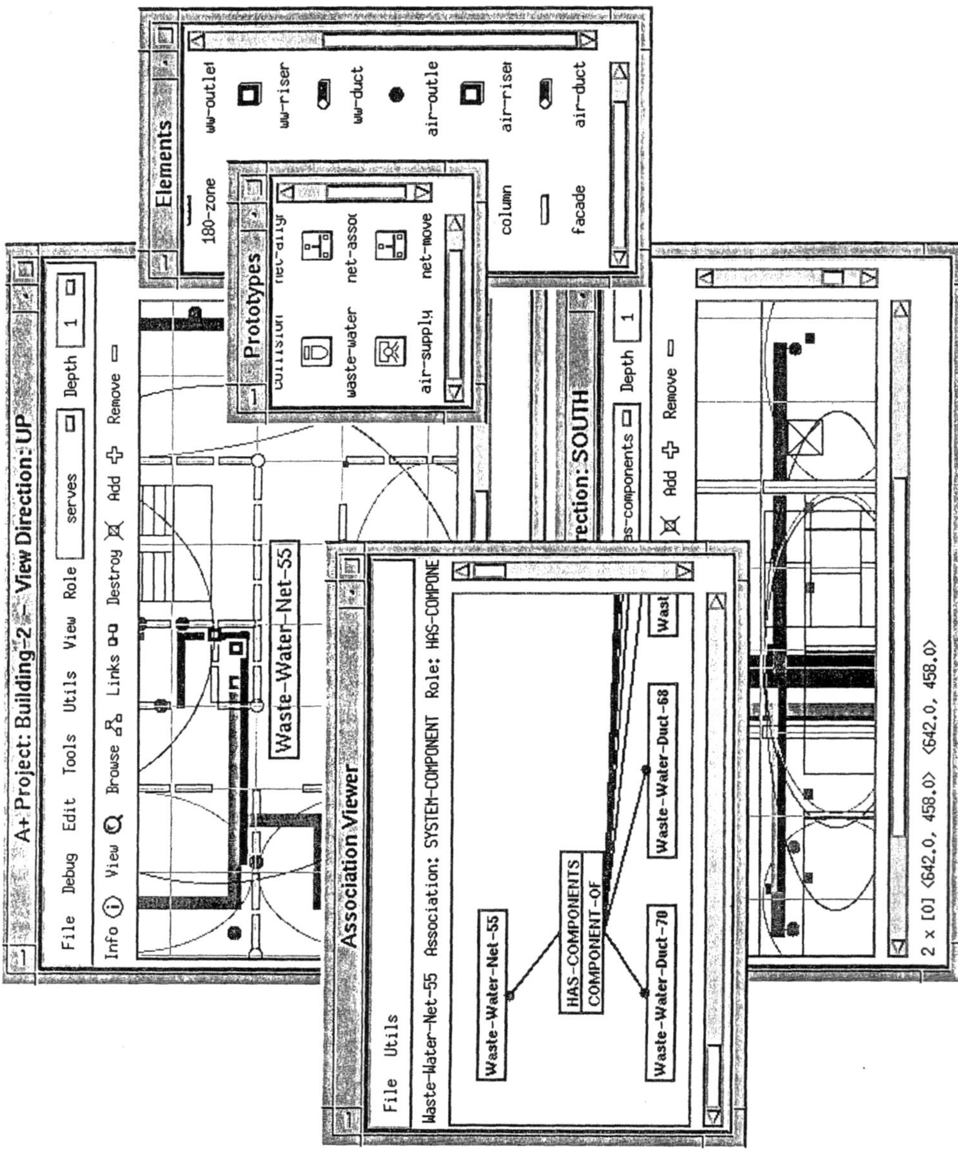

Figure 15. The user interface of a+

Becker, B. (1992) *Künstliche Intelligenz, Konzepte, Systeme, Verheißungen*, Campus Verlag, Frankfurt/Main.

Björk, B. and Penttilä, H. (1989) "A Scenario for the Development and Implementation of a Building Product Model Standard", paper presented at Current Research and Development in Integrated Design, Construction and Facility Management, CIFE, Stanford University, March 28/29, 1989.

Booch, G. (1991) *Object Oriented Design with Applications*, Benjamin/Cummings Publishers, Redwood City, California.

Drach, A., Gauchel, J., and Hovestadt, L. (1990) "ARMILLA - Ein Intelligentes Designwerkzeug für die Layoutplanung von Leitungsnetzen in Hochinstallierten Gebäuden", in J. Gauchel (ed.), *KI-Forschung im Baubereich*, Ernst & Sohn, Berlin, pp. 85-100.

Flemming, U. (1987) "More than the Sum of the Parts: The Grammar of Queen Anne Houses", *Environment and Planning B* **14**, 323-350.

Flemming, U. (1989) "More on the Representation and Generation of Loosely Packed Arrangements of Rectangles", *Environment and Planning B* ***16***, 327-359.

Gauchel, J., Drach, A., and Hovestadt, L. (1990) "Ein Intelligentes Planungswerkzeug für das Verlegen Haustechnischer Leitungen in Hochinstallierten Gebäuden", *VDI-Berichte* **861.5,** VDI-Verlag, Düsseldorf.

Haller, F. et al. (1985) Armilla- Ein Installationsmodell, Institut für Baugestaltung, Universität Karlsruhe.

Hayes-Roth, B., Johnson, M.V., Garvey, A., and Hewett, M. (1986) "Application of the BB1 Blackboard Control Architecture to Arrangement Assembly Tasks", in R. Engelmore and T. Morgan (eds.), *Blackboard Systems*, Addison-Wesley, Wokingham.

Heitz, S. (1992) Entwurf und Prototypische Implementierung eines Regelinterpreters, Institut für Industrielle Bauproduktion, Universität Karlsruhe.

Hovestadt, L. (1991) "A4 - A Model for an Extensive Use of Computers in Architecture", paper presented at the Second International Workshop on Computer Building Representation, Aix-les-Bains.

Hua, K., Smith, B., Faltings, B., Shih, S., and Schmitt, G. (1992) "Adaptation of Spatial Design Cases", J.S. Gero (ed.), *Artificial Intelligence in Design '92*, Kluwer Academic Publishers, Dordrecht, pp. 559-578.

Kiczales, G., Rivière, J. des, Bobrow, D.G. (1991) *The Art of the Metaobject Protocol*, MIT Press, Cambridge.

Kröplin, B. et al. (1990) "Ein Konzept für die Wissensbasierte Konstruktion von Standardbauteilen im Stahlbetonbau", in J. Gauchel (ed.), *KI-Forschung im Baubereich*, Ernst & Sohn, Berlin, pp. 183-195.

Langenegger, M. (1992) Shapes - Geometrische Strukturen und ihre Beziehungen in dem Flexiblen EDV Werkzeug a+ zur Integrierten Gebäudeplanung, Institut für Industrielle Bauproduktion, Universität Karlsruhe.

Lenart, M. and Ketteler, G. (1990) "Expertensysteme für den Entwurf von Baukonstruktionen", in J. Gauchel (ed.), *KI-Forschung im Baubereich*, Ernst & Sohn, Berlin, pp. 120-130.

Marcus, S. and McDermott, J. (1986) SALT: A Knowledge Acquisition Tool for Propose- and-Revise Systems, Technical Report, CMU-CS-170.

Marcus, S., Stout, J., and McDermott, J. (1988) "VT: An Expert Elevator Designer that uses Knowledge-Based Backtracking", *AI Magazine* **9**, 95-111.

Mathis, C. (1988) *Installationsplanung mit einem Wissensbasierten System*, VDI-Verlag, Düsseldorf.

Minsky, M. (1985) *The Society of Mind*, Touchstone, New York.

Muratori, S. (1960) "Studi per una Operante Storia Urbana di Venezia", Instituto Poligrafico dello Stato, Rome.

Nijssen, G.M. and Halpin, T.A. (1989) *Conceptual Schema and Relational Database Design (A Fact Oriented Approach)*, Prentice-Hall, Sydney.

Puppe, F. (1990) *Problemlösungsmethoden in Expertensystemen*, Springer Verlag, Berlin.

Raetz, P. (1989) *XNET: Ein Intelligentes CAD-System für die Planung von Lokalen Netzwerken in Gebäuden*, VDI-Verlag, Düsseldorf.

Rumbaugh, J., Blaha, M., Premerlani, W., Eddy, F., and Lorensen W. (1991) *Object Oriented Modelling and Design*, Prentice Hall, Englewood Cliffs, New Jersey.

Spenke, M. and Beilken, C. (1990) "An Overview of GINA, the Generic Interactive Application", paper presented at the Esprit/Eurograpics Workshop on User-Interface Management Systems and Environments, Lisbon.

Winograd, T. and Flores, F. (1986) *Understanding Computers and Cognition*, Addison-Wesley, Reading, Masachusetts.

MODULAR BUILDING MODELS

J. GAUCHEL, L. HOVESTADT *, S. VAN WYK and R.R. BHAT
Center for Building Performance and Diagnostics
Department of Architecture, Carnegie Mellon University
Pittsburgh, Pennsylvania 15213-3890 USA

* *Institut für Industrielle Bauproduktion (IFIB)*
Universität Karlsruhe, Karlsruhe, Germany

ABSTRACT. The development and implementation of a modular building model appropriate for computer aided design is described. The limitations of a unified building model with regard to concurrence and complexity in design is discussed. Current research suggests that to model real-world complexity, one must trade centralized control for autonomy. In this paper we develop a modular approach to building modelling that is based on object-oriented autonomy and makes it possible to define these models in a distributed concurrent manner. Such a modular and autonomous implementation brings inherent uncertainty and conflict which cannot be determined a priori.

1. Building Models

The design, construction, management and maintenance activities of buildings involve various experts, each having their own mental models of the work involved. These models are partially described by the regular communication tools of the professions, for example drawings and specifications. Any implementation of these models as software tools for the building profession must consider building design as a life-cycle activity distributed among different organizations. This activity involves many concurrent partial design decisions that are often not synchronized for organizational or practical reasons. The implementation of a unified building model with traditional database management systems has proven to be difficult, due to the following constraints: (i) a building model should serve the information interests of different experts with unique views of the data, (ii) given the potential size of a building model, large-scale information management will become a performance problem for all related design activities in a centralized data model, and (iii) all possible conflicts among the design decisions, as well as design data, cannot be resolved by an a priori implementation of design knowledge or rules. In this paper we propose an alternative to the unified building model. We develop a modular concept of building models that can be shared and customized by individual experts according to their specific needs.

2. Previous Work - The ARMILLA Project

The present approach draws on the experiences of the 'ARMILLA Project' (Haller 1988) at the Institut für Industrielle Bauproduktion (IFIB), Universität Karlsruhe, Germany. The on-going effort of the team resulted in three prototype computer-aided design (CAD) tools (Mathis 1988; Raetz 1989; Hovestadt 1989; Gauchel 1990; Gauchel 1991). These prototypes implement knowledge-based tools derived from design strategies for the spatial layout and coordination of mechanical systems in buildings.

For each mechanical subsystem, the design task of ARMILLA can be viewed as a problem of

H. Timmermans (ed.), Design and Decision Support Systems in Architecture, 83–95.

routing services from a delivery point to several supply points. The design constraints of the routing problem are the cross-sections of the services (pipes, ducts, or cables), technical concerns (required slope for wastewater drainage, etc.), and the geometry of the building design. The constraints are dynamic, since routing alternatives for one subsystem depend on the layout of other subsystems, and the resulting cross-section of the service depends on the length of run and number of service points. The design goals can also be expressed as constraints: for example, minimize the length of the service; minimize the number of changes in service direction; or make any changes in direction as 'smooth' as possible.

2.1. A2 PROTOTYPE

The A2 prototype of ARMILLA works on the basis of predefined layout strategies that are defined for each mechanical subsystem - e.g. pipe, duct, or cable network. Each layout strategy uses a system of three-dimensional orthogonal grids and a system of productions, which checks for conflicts with other strategies, to resolve the spatial routing between the mechanical system's supply and delivery points as specified. The result is an integrated and systematic design solution (Haller 1974; Haller 1985; Hesse 1987). The strategies used in the A2 prototype were successfully tested in the design of several recent buildings by Professor Haller, most notably, the education center of the National Railway Company of Switzerland in Murthen. The system architecture of the A2 prototype had the following features.

The design world is represented as a semantic network whose nodes are object-oriented descriptions of design objects. Design objects can be *building objects,* such as spaces, building components and assemblies; *modelling objects,* such as design tasks, intentions, production rules, constraints; and *interface objects,* such as graphic primitives, display primitives and interactive tools. The design process is represented as tasks or design intentions that can be expressed at any level of abstraction, from conceptual layout sketches to the complete parts list. All tasks can be accessed either non-sequentially or in stages. The implementation is based on a system kernel that supports the interactive definition and manipulation of building objects. The kernel can be made more intelligent, or extended, with domain-specific, user-supplied modules of knowledge-based generative and diagnostic tools.

The user interface supports three levels of user interaction or design modes. In the traditional CAD sense, the user interface provides graphic interface and modelling objects to define and manipulate building objects interactively. In the expert diagnostic design sense, the interface provides background consistency and plausibility checks on actions initiated by the user or by knowledge-based generative tools. In the expert generative design sense, the interface provides knowledge-based generative tools that react to actions initiated by the user. The latter two levels provide the user with knowledgeable design assistants whose actions can be overridden at any time. The graphic interface, like the design process, is based on the concept of tasks or design intentions. Each task is configured to provide a graphic window and a specific view or representation of involved design objects, a set of graphic and interactive tools to define and change the state of design objects, and a set of user-activated generative and diagnostic tools that correspond to the design modes introduced earlier. The system kernel is supported with utilities to import and export the building model and its data so that a standard drawing tool such as AutoCAD can be used to define the geometry of building objects and to visualize the building objects defined by A2.

2.2. LIMITATIONS OF A2

While A2 was specifically designed for the spatial layout of mechanical systems, it was viewed as a prototype for a generic and dynamic building model that could be applied to general building design problems. When all the mechanical systems required in a single plenum of a laboratory building were implemented, the building model resulted in several thousand instances of more than a hundred building objects. We describe below some of the shortcomings of this approach. The three major shortcomings were increasing complexity, inability to generalize to a multi-user scenario, and unpredictability in the behaviour of the model.

The design world can be understood as a set of interacting design objects, as illustrated in Figure 1. An obvious implementation strategy of this design world is a semantic network, where design objects are described as nodes and the edges as interactions. The edges are pre-defined bonds between design objects and serve as channels for passing messages.

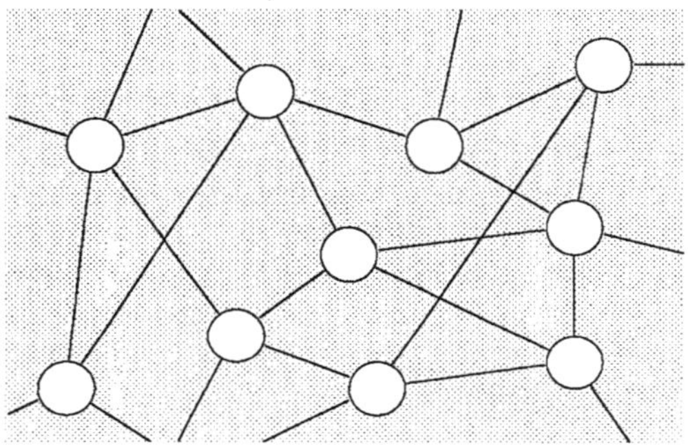

Figure 1. Interacting design objects

In a dense network, the number of connections approach the square of the number of nodes. The introduction of new nodes leads to increased density of connections. It became increasingly difficult to model, implement, predict, and effectively test the behaviour of the A2 prototype. This was because the *complexity of* the model increased and there was a sharp degradation in performance. Though the A2 prototype demonstrates important features of how different experts can work together in one environment, it was a single-user system. For this prototype to function as a true *multi-user system,* different users would have to work concurrently on the entire network, or on separate partial semantic networks defined for specific purposes. For both these scenarios, A2 offered no promising solution. Working concurrently on the entire model led to unpredictable results due to the connectivity. Defining partial networks led to problems in handling interactions between interconnected nodes as well as the problem of overlapping partial networks (Figure 2).

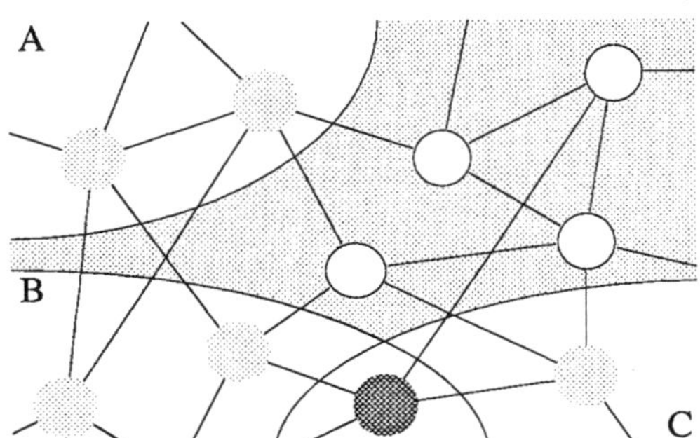

Figure 2. Interconnected and overlapping partial semantic networks A, B and C

Another problem we encountered concerned modifications to a tightly connected network.

Adding or deleting nodes or edges led to erratic and *unpredictable* behaviour in other parts of the system, which in turn led to the unpredictable design state of the system. Maintaining the integrity of a semantic network and the inheritance properties of the various objects in such a dense network proved to be extremely difficult.

These limitations led some members of the ARMILLA group to conclude that most of the problems of the A2 prototype arose from the fundamental decision to implement the design world as a semantic network. This sparked a search for other approaches that would allow one to easily describe and modify separate sub-models while retaining the useful features of semantic networks, class structure, and inheritance mechanisms.

3. Current Developments

It is useful at this point to consider some of the emerging issues and paradigms in computing that deal with issues of representing and modelling in large-scale distributed databases and in concurrent design. Such issues have emerged largely because traditional methods of data management and representation have been unable to cope with arbitrarily complex real-world problems. In addition, there is a global effort to formalize conceptual data models for buildings from the product model viewpoint. Since these issues are fundamental to any software implementation, we will briefly introduce them.

3.1. DATABASES

The traditional database management systems (DBMS) have been subjected to criticism for several reasons. It is difficult to represent data at a logical level appropriate to the expert users. It is difficult to model the complex behaviour of interacting objects using data or state variables alone. They do not provide an adequate framework for a multi-user environment. Object-oriented DBMS have been suggested (Kim 1989) to incorporate state data as well as behaviour. OODBMS for distributed worlds (Purdy 1987) has been proposed to address issues of distributed environments and shared information models through cooperation and communication.

3.2. AUTONOMY

The premise of ubiquity, whereby all items in the database are potentially accessible to any user at any site, necessarily leads to a large degree of centralized control. Experience suggests that this premise cannot resolve the largely concurrent and asynchronous nature of the design world. Object autonomy at the database level has been suggested in a new paradigm to replace ubiquity (Kemper 1990) and would have the following five properties: (i) an object-oriented data model that can model structure and behaviour of the design object, (ii) object dependencies that can be resolved by cooperation without hierarchic control, (iii) objects that reside in different physical sites, (iv) structure and behaviour of the object that can change according to the current stage in the object life-cycle, and (v) an object that can know about its acquaintances and can influence their behaviour.

Similar approaches based on autonomy have been proposed in concurrent programming. The actor model (Agha 1986) provides for modelling shared objects with changing local states, dynamic reconfigurability and inherent concurrence. In this model, communication is achieved through message passing. Such a data model of autonomous objects, with built-in control over cooperation with other objects, with distributed locations at different physical sites, and with modifiable behaviour over the life-cycle of the objects, would ideally meet the requirements of the design world.

3.3. DATA MODELS

The design decision process involves a large number of objects at different levels of abstraction. When viewed as a product model, the abstractions can be the product as a whole, as a composition of its functional systems or its subsystems, or as a collection of low-level individual parts of the design. A large-scale international effort by the International Standards Organization is under way to standardize the product model. The STEP/PDES effort seeks to define a conceptual model as well as a physical data model for actual transfer of data in product design (Owen 1987). Even a partial standardization of the underlying data structures would result in great savings in data communication and data exchange between applications and also ease human interaction with the data (Bjork 1991).

In contrast to the logical view of the model, there is a graphic view of the data model, as represented by different subsets of structural and behavioural attributes needed by the individual expert user. Graphic views would at least partially depend on the design intentions and interests and would present graphic representation and modelling methods common to the expert's experience. The traditional view of data objects has been based on fixed types for the various design objects. From a design point of view, an interesting alternative to fixed data types is the idea of type evolution (Skarra 1987), where design objects evolve from an abstract idea to a recognizable entity with clearly defined properties throughout the design process.

4. Present Work - The A4 Prototype

Modular building models are parts of building design environments that allow all experts involved in the life-cycle of the building to communicate and work together. In principle, the environments include two types of components: decentralized data communication modules; and applications that synthesize, analyze, evaluate, visualize, etc. Data created by applications in the design environment is stored and communicated by the data communication module (Figure 3). The background of the A4 prototype is elaborated elsewhere (Gauchel 1992; Hartkopf 1991; Hovestadt 1990; Hovestadt 1991a; Van Wyk 1991).

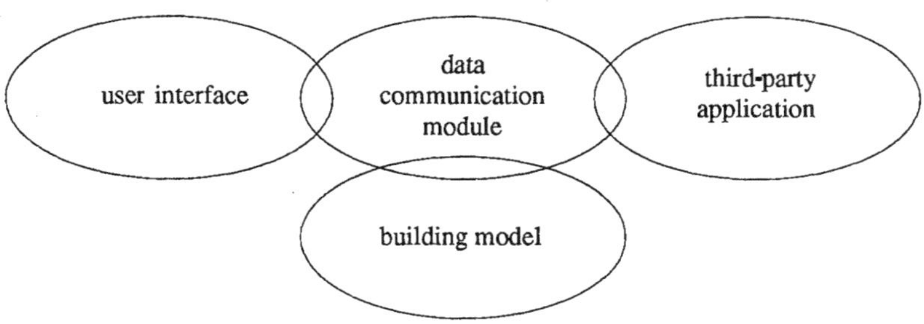

Figure 3. The A4 Prototype - an integrated building design environment

Since building models serve special purposes, they are always paired with at least one application. Though the applications are separate design tools, building models can overlap because they are modular and can share modules.

5. The Concept of Modular Building Models

In developing a concept of modular building models that avoids the problems mentioned earlier in regard to the A2 prototype, the following design and implementation questions arise: how can highly connected building descriptions be split into separate modules to support design decisions? How can the modules interact with one another to achieve the quality of a highly connective description? Can there be a system of architecture that is common to all modules? Can there be a unified interaction principle between the modules? To address these questions, it is necessary to reduce the complexity of the highly integrated design world into *building models,* as determined by descriptions of building objects, and *user interfaces,* as determined by process-oriented descriptions, such as modifying and editing building descriptions.

5.1. BUILDING MODULES AND BUILDING OBJECTS

The basic approach is to split building descriptions into building modules, each of which describes a simple interaction pattern or behaviour of one building object. Let us consider Figure 4, which shows a network of building objects. A and B represent some space or building component and are sources of interaction with other building objects, for example object C.

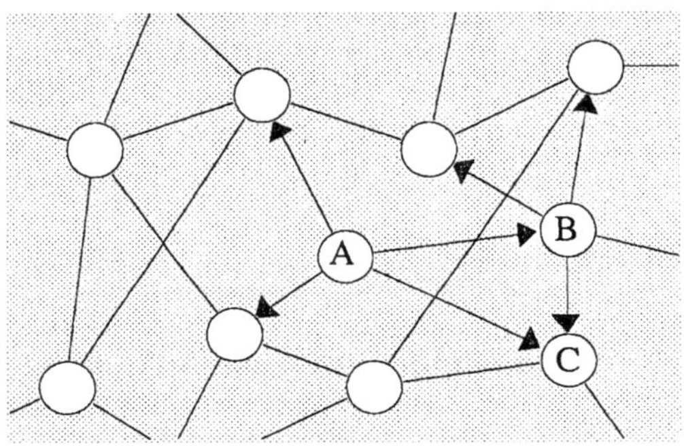

Figure 4. Three building objects as sources of interaction with other building objects

Figure 5 shows how these interactions can be separated into two modules: module A represents the interaction pattern from object A to B, C and others, while module B represents the interaction pattern of object B to C and others. Encapsulating both patterns into separate modules is possible only by introducing redundant representations of the objects B and C that they share. The approach is general and allows all building modules to be structured in the same manner. Modules interact if they share redundant representations of building objects. Based on this simple principle, it is possible to develop a single general interaction mechanism for all modules.

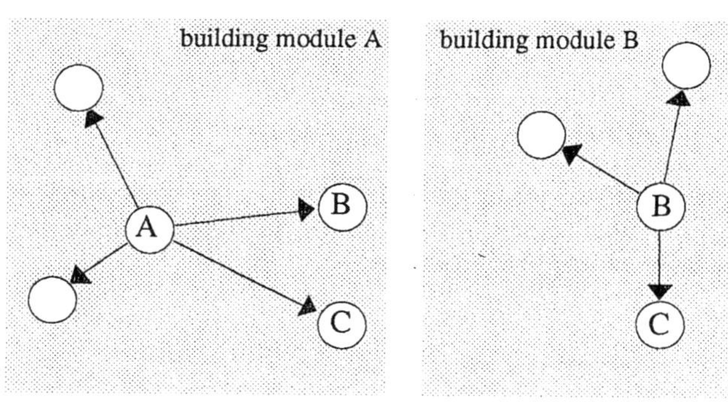

Figure 5. Building modules encapsulating interaction patterns

Building objects are sets of interacting structural and behavioural attributes inside a building module. They can be visualized as nodes, as in Figure 5 above. The structural attributes are descriptions of some real-world design object such as spaces or building components, while behavioural attributes can be characterized as design or decision supports which normally take the form of functions. Building objects can be thought of as systems that seek stable, internal states or structure. Both structural and behavioural attributes may be linked to form chains of internal interactions that respond to changes initiated outside the building object. In principle, changes are directed at one structural attribute of a building object that, if linked to a behavioural attribute, will trigger a change in the state of other structural attributes, until the changed structural attribute has no link to a behavioural attribute. Such structures can be implemented as 'frames and slots'; the internal links between structural and behavioural attributes may be implemented as 'demons' (Figure 6).

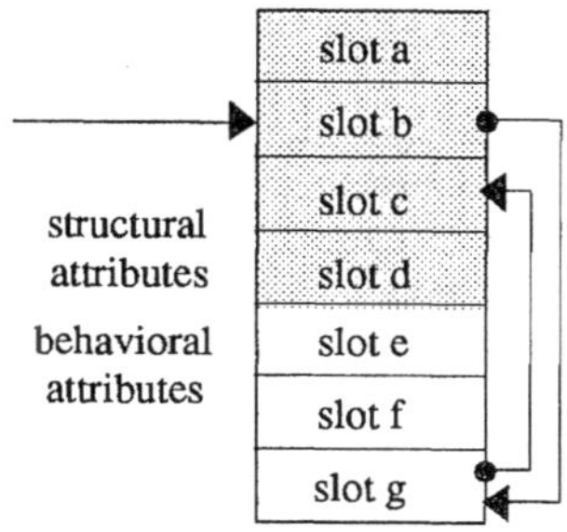

Figure 6. Structural and behavioural attributes of a building object

Building objects inside a building module have a two-level hierarchy. There is one central object and one or more surrounding objects. Only the central object can receive changes initiated outside the module. By its own changes, it can interact with surrounding objects, which can then respond with changes of their own. Thus, a building object is a central object of its own building module and can be a surrounding object in other building modules.

The interactions of the central object with its surrounding objects are modeled as production rules. Changes in structural attributes of the central object can automatically trigger these rules and result in changes of structural attributes in surrounding objects. Such changes may include creation, deletion or modification of the surrounding objects (Figure 7). Thus, a building module can be characterized as a concept or class-type. It is then possible to create instances of building objects, as well as instances of building modules. The instances of a building module inherit the characteristics of the concept.

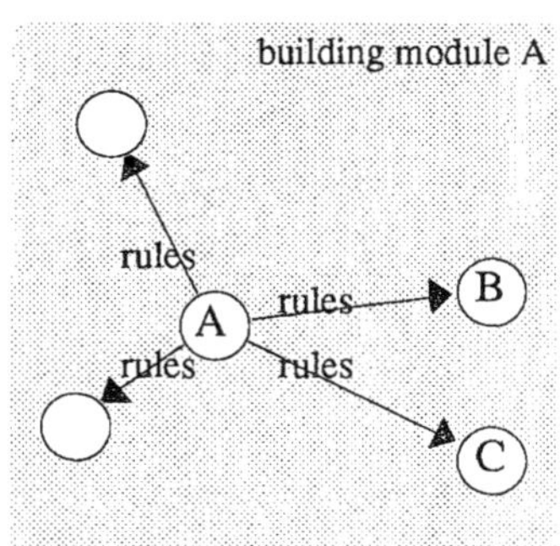

Figure 7. The architecture of a building module is based on the interactions of a central object with its surrounding objects

5.2. BUILDING MODELS

Building models are named collections or sets of building modules that are selected and configured for special purposes. The building model, when combined with a desired view of the expert and an area of interest, together support some task or design application. Building modules can be considered as design or domain-specific knowledge; building models can be perceived as design applications.

5.2.1. *Expert Views.* Views are, in principle, representations of one individual's expertise and role in the design process. In our approach, views are initially related to the roles of experts (architects, civil engineers, HVAC engineers, facility managers). These are standard views that, through a built-in facility, can be customized further by individual experts. Expert views are declared and always associated with a design application; different compositions of views are also possible. Expert views depend on the specific nature of building objects and are linked to their structural attributes. Views declare which structural attributes of a building object are sensitive to messages or inputs from outside. They thus determine the behaviour of the object, as well as the range of possible interactions between a central object and its surrounding objects. In principle, the description of a building object is understood to be a canonical one. That is, all structural attributes relevant to an expert are described within the building object and are then declared to be relevant to the specific expert view.

5.2.2. *Areas of Interest.* Areas of interest may be spaces in a building, the union of spaces occupied by certain objects, or a specific bounded portion of the design world. Given a building model and an area of interest, only those instances of the selected building modules that are located within the declared area can become members of the actual design application. This selection process is accomplished automatically by a loading mechanism. With this approach, it is not possible to manipulate instances of building objects that lie outside of the active, or current, area of interest. Instances can be loaded into separate areas or into one single area of interest. Instances of the building modules A and B loaded into separate areas of interest result in redundant representations of the shared building objects (Figure 8). Even if areas of interest overlap or if they are identical, there will always be redundant representations of shared instances of modules or building objects.

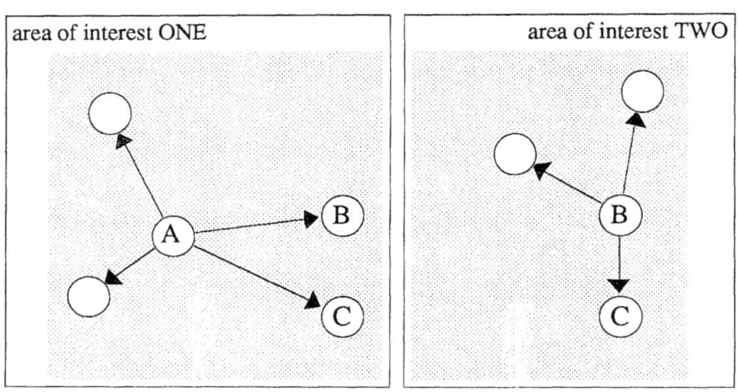

Figure 8. Modules A and B loaded into separate areas of interest

Instances of the building modules A and B loaded into one area of interest result in a union of the instances of building objects that is automatically performed by the loading mechanism

(Figure 9). Using an import mechanism, the design objects in a single area of interest can be updated to reflect changes made to the design world by other modules in the design world. In such case, only those instances in the *current* area of interest that have been changed concurrently in *other active* areas of interest are actually imported and updated. Areas of interest can be dynamically changed during the course of design to allow the users to navigate through the design world. While we start with a spatial basis for defining an area of interest, we realize that other valid criteria exist which may also be used. Other areas of interest, for example, could be defined for time, version, morphology, level of detail, set membership, etc. (Hovestadt 1991b).

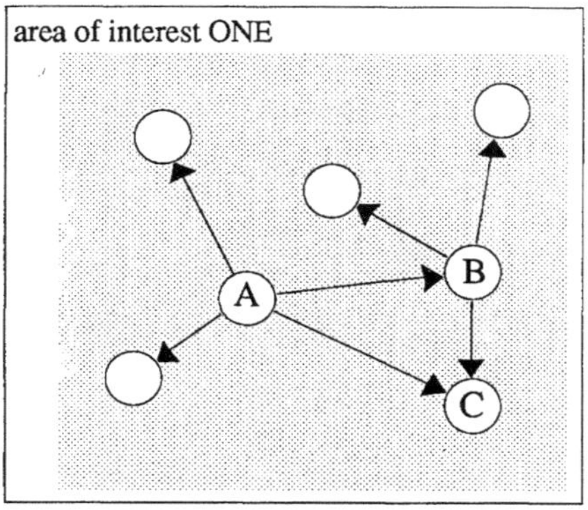

Figure 9. Building modules A and B loaded into the same area of interest

5.3. LEVELS OF AUTONOMY

There are several levels of autonomy in the approach we have described for modular building models. First, building objects can be designed as independent systems. Instances of building objects can only receive 'inputs' from outside. The internal or behavioural activities triggered by changes to its structural attributes have no predefined linkages to objects or activities outside of the building object. Secondly, within a building module, the activities of the central object are modeled using production rules. These are software tools with no predefined linkages to objects outside the building module. Thirdly, building modules are small environments of domain-specific knowledge. They can be described without reference to other building modules. Moreover, building models are independent collections of building modules. Finally, design applications use areas of interest and expert views to describe encapsulated scenarios of instances of building objects. These scenarios are only connected with the outer world by a single, unified import-export interface.

6. Object Communication

6.1. IMPORT-EXPORT

The interaction between areas of interest and the outside design world depends on explicit calls from an area of interest to import data from or export data to the external world. Importing data serves to instantiate a new design state. Exporting data communicates the internal design state to the external world. The advantage of this import-export mechanism is that an area of interest always allows exploratory design and validation before communicating the results outside. The objects of import and export are always instances of building objects. The import mechanism assures that all instances inside an area of interest are non-redundant (Figure 10).

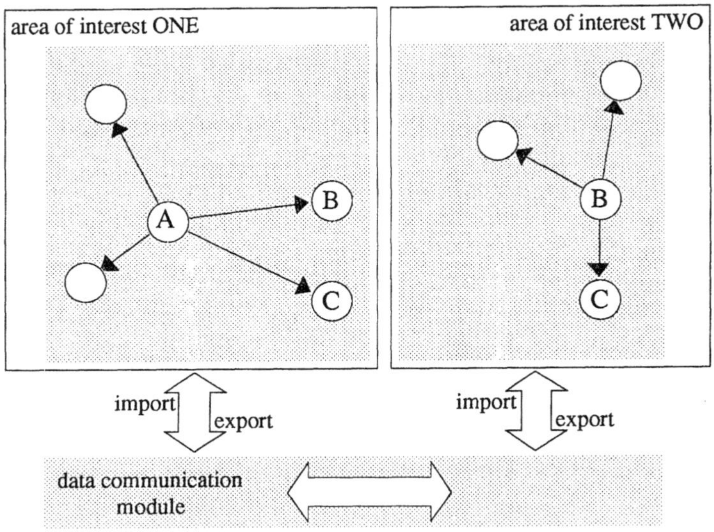

Figure 10. The interaction among areas of interest

6.2. COMMUNICATION MODULE

Since working sessions often include several areas of interest that can be distributed over different machines and physical sites, the storage of any data has to be distributed as well. We are currently attempting to formalize a data communication module that will meet this need. We propose to implement the communication between objects in a distributed environment by drawing on principles of electronic mail systems across networks using a client/ server model. A similar implementation for autonomous object-oriented systems is described in the literature (Casais 1988).

6.3. CONFLICT RESOLUTION

The modularity of the building model, in terms of conceptual as well as underlying data, offers great advantages when one considers the distributed needs of the design profession. The design objects are distributed over various physical sites. Yet the stability and success of the entire system depends both on the independence of the individual objects and their cooperation. A system constructed from distributed objects is inherently uncertain. Spatial uncertainty exists because at any instant the exact location of all constituent objects cannot be known. And temporal uncertainty exists because none of the objects can detect a single global state. The coordination and mobilization of the objects and their processes is a difficult problem and is an active area of research (Raynal 1988).

Of all the uncertainties, spatial conflicts are perhaps the easiest to isolate. Other possible conflicts can arise due to concurrent access, data storage and inconsistent states of building objects. There are many potential conflicts and they cannot all be isolated or dealt with in advance. Conflicts which are not a part of the system knowledge can be resolved only through mutual recognition and resolution by interested users and involved objects. Communication modules have no domain knowledge to resolve such conflicts. Automatic conflict detection, such as concurrent read/write, concurrent modification, and inconsistent object states, are detected. Such conflicts can only be indicated through some signalling and mail mechanisms and must be resolved by the users through dialogue.

7. Exploring Design Spaces

The approach we have presented should not be taken as a complete pre-definition of all possible objects and their possible interactions for a building design problem. Software is necessarily predefined in terms of data structures and domain-specific knowledge. In developing and using building models, it would be useful for design objects and design knowledge to be defined interactively during the course of design.

The concept of exploring design space must include provisions to introduce new design objects, modify and delete them, and to alter their behaviour or influence it on a exploratory basis. Thus, initial descriptions of objects can be tentative, and knowledge and resolution can be introduced as the design progresses. Our approach to building models as characterized by the several levels of autonomy provides opportunities for exploratory design. It does so by allowing users to dynamically create objects and to encapsulate their contexts. Though at present interaction rules are predefined, advanced interfaces for describing knowledge in a dynamic manner are needed to support exploration in design. Such advanced interfaces should allow exploratory design objects to be created, manipulated, and modified in real time during the design process, provide powerful graphics to communicate simulations or design space visualizations and their integration with hypermedia and multimedia tools (Hovestadt 1991a, 1992), and provide knowledge-based interfaces that can capture design knowledge, add behavioural attributes to building objects, define and modify the knowledge-based interaction in a building module, and modify given standard views. This level of functionality is necessary if the exploratory nature of design is to be supported by a computing environment. It is also necessary if the development of building models is supposed to be part of the design process.

8. Conclusions

The general approach described in this paper is the basis of three prototypes under development at this time. One prototype that focuses on the user interface, as described earlier, is being developed by Ludger Hovestadt at the University of Karlsruhe, Germany, and is part of his on-going doctoral research. Another prototype is being developed by the other authors at Carnegie Mellon University (CMU) in Pittsburgh, Pennsylvania. The latter focuses on the development and integration of building models with existing applications, such as building performance simulation tools. A third prototype is being developed by students at the Software Engineering Institute (SEI) at CMU under direction of the authors. This last one focuses on the development of a data communication module based on these principles. All three prototypes are in early stages of development. We hope that the on-going work will prove to be helpful in our effort to develop an integrated building design environment.

Acknowledgements:
The work reported in this paper has been supported by the Center for Building Performance and Diagnostics (CBPD) at CMU; by the Institut für Industrielle Bauproduktion der Universität Karlsruhe; and by a research grant from the RETEX project, financed by the Bundes Ministerium

für Forschung und Technologie (BMFT), Germany and the company ROM Hamburg, Germany. We thank our colleagues, especially the SEI group and their advisors, for their contributions.

References

Agha, G.A. (1986) *ACTORS: A Model of Concurrent Computation in Distributed Systems,* MIT Press, Cambridge, Massachusetts.

Bjork, B.C. (1991) "Intelligent Front Ends and Product Models", *AI in Engineering* **6**, 46-56.

Casais, E. (1988) *An Object Oriented System Implementing KNOS in ACM-COIS,* Palo Alto.

Gauchel, J. (ed.) (1990) *KI-Forschung im Baubereich,* Ernst & Sohn, Berlin.

Gauchel, J. (1991) ARMILLA - A Knowledge-Based Design Tool for An Integrated Layout of Complex Mechanical Systems in Buildings, Carnegie Mellon University, Pittsburgh.

Gauchel, J., Skip Van Wyk, Jr., C.G., and Bhat, R.R. (1992) "Building Modelling Based on Concepts of Autonomy", paper presented at the Artificial Intelligence in Design '92 Conference, Pittsburgh.

Haller, F. (1974) *MIDI - Ein Offenes System für Mehrgeschossige Bauten mit Integrierter Medieninstallation*, USM Bausysteme Haller, Munsingen, Switzerland.

Haller, F. (1985) ARMILLA - Ein Installationsmodell, Institut für Baugestaltung der Universität Karlsruhe, Germany.

Hartkopf, V. and Gauchel, J. (1991) "A Computer-Aided Design Approach for Total Building Performance", paper presented at Building Systems Automation-Integration Conference, Madison, Wisconsin.

Hesse, R. (1987) ARMILLA - Fallstudie, Bürogebäude IBM-Nürnberg, Institut für Baugestaltung, Universität Karlsruhe.

Hovestadt, L. (1989) "Ein 'Intelligentes CAD-System' für die Planung und Verwaltung von Leitungsnetzen in Hochinstallierten Gebäuden", *VDI-Berichte* **775**.

Hovestadt, L. (1990) The Armilla Film, MacroMind Animation for Macintosh, University of Karlsruhe, Germany.

Hovestadt, L. (1991a) "A4 - A Model for an Extensive Use of Computers in Architecture", paper presented at the 3rd International Symposium on Systems Research, Informatics and Cybernetics, August 13-14, Baden-Baden, Germany.

Hovestadt, L. (1991b) BO - A Generative Model for an International, Integrated and Distributed Computer Environment for the Building Industry; ESPRIT III Proposal, R&D Task IV.2.3.

Hovestadt, L. (1992), A4 - Digitales Bauen, Draft Document, Ph.D. Dissertation, University of Karlsruhe, Germany.

Kemper, A., Lockemann, P.C., Moerkotte, G., Walter, H.D., and Lang, S.M. (1990), "Autonomy over Ubiquity: Coping with the Complexity of a Distributed World", paper presented at International Conference on Entity-Relationship Approach, Lausanne.

Kim, W. and Lochovsky, F.H. (eds.) (1989) *Object-Oriented Concepts, Databases, and Applications,* Addison Wesley, Reading, Massachusetts.

Mathis, C. (1988) *Installationsplanung mit einem Wissensbasierten System,* VDI-Verlag, Düsseldorf.

Raetz, P. (1989) *XNET: Ein Intelligentes CAD-System für die Planung von Lokalen Netzwerken in Gebäuden,* VDI-Verlag, Düsseldorf.

Owen, J. and Bloor, M.S. (1987) "Neutral Formats for Product Data Exchange: The Current Situation", *Computer Aided Design* **19**, 436-443.

Purdy, A., Schuchardt, B., and Maier, D. (1987) "Integrating an Object Server with other Worlds", *ACM Transactions* **5**, 2747.

Raynal, M. (1988) *Networks and Distributed Computation,* MIT Press, Cambridge, Massachusetts.

Skarra A. H. and Zdonik S.B. (1987) "Type Evolution in an Object-Oriented Database", in B. Shriver and P. Wegner (eds.), *Research Directions in Object-Oriented Programming,* MIT Press, Cambridge, Massachusetts, pp. 392-415.

Tsichritzis, D.C., and Nierstrasz, O.M. (1989) "Directions in Object-Oriented Research", in W. Kim and F.H. Lochovsky (eds.), *Object Oriented Concepts, Databases and Applications,* Addison Wesley, Reading, Massachusetts, pp. 523-536.

Van Wyk, Jr., C.G. Skip, Gauchel, J., and Bhat, R.R. (1991) "A Knowledge-Based Approach to Building Design and Performance Evaluation", paper presented at the ACADIA '91 Reality and Virtual Reality Conference, Los Angeles.

A FUNCTIONAL APPROACH TO REALIZING DECISION SUPPORT SYSTEMS IN TECHNICAL REGULATION MANAGEMENT FOR DESIGN AND CONSTRUCTION

G.L. LUCARDIE
TNO Building and Construction Research
P.O. Box 49
2600 AA Delft
The Netherlands

ABSTRACT. Technical building standards defining the quality of buildings, building products, building materials and building processes aim to provide acceptable levels of safety, health, usefulness and energy consumption. However, the logical consistency between these goals and the set of regulations produced to achieve them is often hard to identify. Not only the large quantities of highly complex and frequently changing building regulations to be met, but also the variety of user demands and the steadily increasing technical information on (new) materials, products and buildings have produced a very complex set of knowledge and data that should be taken into account when handling technical building regulations. Integrating knowledge technology and database technology is an important step towards managing the complexity of technical regulations. Generally, two strategies can be followed to integrate knowledge and database technology. The main emphasis of the first strategy is on transferring data structures and processing techniques from one field of research to another. The second approach is concerned exclusively with the semantic structure of what is contained in the data-based or knowledge-based system. The aim of this paper is to show that the second or knowledge-level approach, in particular the theory of functional classifications, is more fundamental and more fruitful. It permits a goal-directed rationalized strategy towards analysis, use and application of regulations. Therefore, it enables the reconstruction of (deep) models of regulations, objects and of users accounting for the flexibility and dynamics that are responsible for the complexity of technical regulations. Finally, at the systems level, the theory supports an effective development of a new class of rational Decision Support Systems (DSS), which should reduce the complexity of technical regulations and restore the logical consistency between the goals of technical regulations and the technical regulations themselves.

1. Introduction

Technical building regulations range from housing law to national building codes and standards issued by professional bodies. These regulations aim at securing acceptable levels of safety, health, usefulness and energy consumption for future occupants. But the logical consistency between these goals and the means to achieve them is hardly tractable. One problem is the large quantity of highly complex and frequently changing building regulations. Another is the variety of user demands; and yet another the steadily increasing body of technical information on (new) materials, products and buildings. Together, these developments have produced a very complex set of knowledge and data, which must be taken into account when dealing with technical building regulations.

As building regulations grow in complexity, scope and importance, the need for sophisticated regulation management systems becomes urgent. One of the entry points to realize such systems is the use of insights developed in information technology (IT). It is widely recognized that knowledge and database technology are important components of IT for improving the management of knowledge and data in regulations. In particular, the *integration* of knowledge

H. Timmermans (ed.), Design and Decision Support Systems in Architecture, 97–109.

and database technology seems to provide a wide range of possibilities (see for instance McCullough 1991; Koster and Stott Parker 1990). In general, two strategies of integrating knowledge and database technology can be distinguished. The first emphasizes the transfer of data structures and processing techniques from one field of research to another. In contrast, the second approach is concerned exclusively with the semantic structure of what is contained in the knowledge-based or data-based system (Newell 1981; Levesque 1984; Brachman and Levesque 1986; Twine 1989).

The aim of this paper is to show that the second approach, the *knowledge-level strategy*, in particular the theory of functional classifications, is more fundamental and more fruitful than the first *symbol-level strategy*. The fire-safety code, one of the most complex parts of the building regulations in the Netherlands, will serve to illustrate the main issues. To state them clearly, Section 2 first reviews the notions of the knowledge level and the symbol level. Section 3 explains the two strategies of integrating knowledge and database technology. Next, the knowledge-level strategy is discussed following a functional view. Finally, some implications of following a functional approach when developing Decision Support Systems (DSS) in the field of building regulations will be dealt with.

2. The Knowledge Level versus the Symbol Level

In his presidential address to the American Association of Artificial Intelligence (AAAI), Newell proposed a new computer-systems level, which he called the knowledge level (Newell 1981). The introduction of the knowledge level is primarily intended as a separate level within which knowledge can be defined. The global definition of knowledge that is often employed views knowledge as competence for selecting actions to realize goals. This competence is accomplished through knowledge of goals, knowledge of actions and knowledge relating goals to actions. From this notion of knowledge, it follows that relations between goals and actions are the basic components of knowledge. The *principle of rationality* structures these relations. More precisely, according to the principle of rationality, if the knowledge is present that an action will lead to a goal, that action will be selected. Under certain simple conditions the principle of rationality will be effective, but in many situations it is not suited for determining behaviour. Some of these situations can be covered by auxiliary principles. In case multiple actions are involved, the principle of rationality is extended with the auxiliary *principle of equipotence of acceptable actions*. The latter asserts that every action leading to a goal is equally acceptable from the viewpoint of the goal itself. A second auxiliary principle, covering situations in which actions are connected to multiple goals, is the *preference of joint goal satisfaction.* "For given knowledge, if goal G1 has the set of selected actions {A1,i} and goal G2 has the set of selected actions {A2,j}, then the effective set of selected actions is the intersection of {A1,i} and {A2,j}" (Newell 1981, p. 9).

Notwithstanding the utility of these extensions, on many occasions, they fail to provide effective guidelines for predicting the system's behaviour. Even after the introduction of other auxiliary principles accounting for goal preferences, risk and uncertainty, the elementary extensions of the central principle of rationality are not sufficient to cover all situations. Knowledge-level models describing the environment do not contain encompassing principles which state that multiple goals need to be compatible or which solve incompatibility. The failure to determine behaviour uniquely, the probabilistic elements and the incapability to describe the entire range of behaviour all indicate that knowledge-level models are approximations of reality.

Knowledge used in the service of goals includes semantic issues (about the functionality of a system). But it completely excludes user-interface aspects (how to present the functionality of a system to users) as well as implementation aspects (how to encode the functionality). At the knowledge level, attention is explicitly focused upon what knowledge is present. An essential part of Newell's proposal is the existence of a symbol level. Knowledge - from a knowledge-level perspective viewed as competence to select goal-related actions - is reduced at the symbol level to structures and processes. At the symbol level, data structures with particular properties and associated processes carry out problem solving to realize a goal-oriented functionality. The data structures contain knowledge and the processes provide access to this captured knowledge. This is represented by the following well-known equation:

Representation = Knowledge + Access

Newell (1981, p.14) explains this equation as follows: "The representation consists of a system for providing access to a body of knowledge, i.e. to the knowledge in a form that can be used to make selections of actions in the service of goals.

This paper adopts the presumption that a clear conception of knowledge should logically occur prior to that of representation. This implies that work should be done at the knowledge level. But what is to be gained from this knowledge-level perspective? First, a major advantage of knowledge-level research is the provision of an adequate entry point for the analysis of the nature of knowledge. Such an analysis permits one to understand or predict behaviour without the need to construct an operational model of the implied process. A second advantage is the spin-off of knowledge-level research to conceptual modelling.[1] A knowledge-level analysis explores the fundamental principles of conceptual models which will lead to conceptual models, free of inadvertent implementation biases (Brachman and Levesque 1986, p.77; Twine 1989, p.125). Levesque (1984) formulates this as follows: "In terms of system design, the main reason for distinguishing between the knowledge level and the symbol level, is to allow the functionality of a system to be treated independently of its symbolic implementation. In particular, it allows us to consider new operations on a knowledge base (that can be explained in terms of existing ones) without necessarily committing ourselves to any particular implementation style" (Levesque 1984, p.206). Furthermore, knowledge-level models can be effectively used for validation and design purposes. According to Steels (1990), a knowledge-level perspective permits the development of deep conceptual models enabling the realization of robust systems. These systems are capable of providing - when compared with explanation facilities in traditional systems, which are simple replays of used data structures - more subtle and sophisticated justifications of conclusions. Thirdly, the knowledge level can serve as a vantage point to compare and examine critically the properties of representation formalisms. Those properties will continue to hold, no matter what symbol-level decisions are made. These comparisons and examinations should relieve information-systems developers of the burden of having to mould each problem to suit their implementation tools.

Though Newell's ideas have an ever-increasing impact on IT, there is still great confusion between the knowledge level and the symbol level. Brachman (1985), for instance, points to the problems encountered by Fahlman, Touretzky and Van Roggen (1981) when they tried to ascertain the meaning of their inheritance mechanism and could not find a consistent interpretation. Conversely, semantic issues are often mixed up with symbol-level or implementation concerns. When discussing the functionality of a system, notions of inheritance mechanisms, frames and the like often crop up.

---

[1] Conceptual modelling refers to a fundamental phase in the design of information systems in which knowledge and data about the application domain are collected, modelled and documented. It is also used to denote the methodology applied in this phase for collecting, modelling and documenting knowledge and data (see Di Battista et al. 1989, p. 245). The results of the conceptual modelling phase are documented in conceptual models or conceptual schemata consisting of concepts.

3. Integrating Knowledge Technology and Database Technology: Two Strategies

Relationships between knowledge and database technology are central issues in computer science. Some researchers stress the similarities of knowledge technology and database technology, such as the common logic basis and the appearance of triggers. Other researchers point to differences such as the deductive proof-theoretic inferencing of knowledge-based systems versus model-theoretic query evaluation in data-based systems (Brodie and Jarke 1986) or emphasize the complementary nature of knowledge and data-based systems (Risch et al. 1988).

In spite of these efforts, the state of affairs with respect to the distinction between the knowledge level and the symbol level is no different from the general situation in IT. The main thrust of the research directed at the integration of knowledge and database technology is concentrated on transferring data structures and processing techniques from knowledge technology to database technology or vice versa. The use of records to store production rules (Herwijnen, Van Houten, Houtsma and Romkema 1990), the use of frames to store relational data (Chow 1987), or the addition of rule-processing algorithms to databases (Stonebraker 1984) are all examples of research conducted at what we have called the symbol level.

It cannot be denied that knowledge and database technology can help each other at the symbol level. But integration at this level without understanding fundamental issues at the knowledge level leads to serious difficulties. Too much concern with representation mechanisms at the expense of knowing what function a system is computing has led to disadvantageous implementation biases. All too often, discussions about the utility of representation techniques like inheritance taxonomies, production rules and connection graphs lack correct understanding of what, if anything, these symbol structures indicate (Brachman 1983; Etherington and Reiter 1983; Berg-Cross and Price 1989).

A more fundamental and fruitful view of the relationship between knowledge and database technology is concerned only with the semantic structure of what is contained in the knowledge or data-based system. At the knowledge level, the distinction between these systems is non-existent. In terms of what knowledge is represented, as opposed to the data structures and processing techniques to encode the knowledge, both systems are simply collections of facts serving representational ends (Brachman and Levesque 1986, p.72; Twine 1989, p.125). The similarity between knowledge-based and data-based systems can be illustrated by applying some basic mathematical logic to describe the components of knowledge. At the knowledge level, a component of knowledge can be viewed as a set of functions. A function is defined as follows (De Brock 1989, p.6):

Definition 1

F is a function $\overset{D}{\Leftrightarrow}$ F is a set of ordered pairs and

$$\forall\ (x;y) \in \mathrm{F}{:}\ \forall\ (x';y') \in \mathrm{F}{:}\ \text{if } x = x' \text{ then } y = y'$$

If F is a set of ordered pairs (i.e. the ranking of the ordered pairs is not relevant and there are no duplicates), $\mathrm{dom(F)} = \{x \mid (x;y) \in \mathrm{F}\}$ and $\forall\ x \in \mathrm{dom(F)}$ *exactly* one y exists (i.e. the same x cannot have different y's, but the same y could be connected to several x's), then F is a function. To illustrate the definition of a function, we revert to an example taken from the fire-safety code in the Netherlands. More precisely, we will look at the fire-safety regulations which attempt to limit the spread of fire by imposing demands on the fire-resistance of walls. For the sake of simplicity, we assume that two properties are important for determining the fire-resistance of walls: *thermal insulation* and *irradiance*. The following sets of ordered pairs contain the knowledge describing the values of these properties of two walls:

T1 = {(*identifier*;*we123*), (*thermal insulation*;*25*), (*irradiance*;*35*)}
T2 = {(*identifier*;*wi214*), (*thermal insulation*;*35*), (*irradiance*;*45*)}

T1 en T2 are functions consisting of three ordered pairs. Every domain element of a set is linked to exactly one value. Such functions are grouped into components, to make them more amenable to survey techniques. The definition of a component (C) is as follows:

Definition 2

C is a component over the set A $\overset{D}{\Leftrightarrow}$ C is a set and
$\forall\ c \in C: c$ is a function over A

taking into account the following definition:

Definition 3

F is a function over the set A $\overset{D}{\Leftrightarrow}$ F is a function and
dom(F) = A

If C = {T1,T2} then C complies with the definition of a component. C is an example of a component occurring in regulations for construction and industry. Those regulations comprise a set of functions with the explicit purpose selecting fire-resistant walls to limit the spread of fire. C is an element of an allowed set of components. The specification of this set is as follows (see De Brock 1989, p.26):

Definition 4

$$\text{WW} = \{C \mid C \subseteq \prod(\text{FW}) \text{ and } \forall\, t \in \text{WW}: \forall\, t' \in \text{WW}: \text{if } t \neq t' \text{ then } t(\textit{identifier}) \neq t'(\textit{identifier})\}$$

taking into account that:

Definition 5

FW = {(*identifier* ;*{we120,we121...}*),
(*thermal insulation* ;*N*),
(*irradiance* ;*N*)}[2]

Definition 6

$$\prod(\text{FW}) = \{f \mid f \text{ is a function over dom(FW) and } \forall\, x \in \text{dom}(f): f(x) \in \text{FW}(X)\}$$

[2] N is the set of integers (including 0).

From a symbol-level perspective, component *C* could be represented as a table of a database over *A* with exactly four elements. Each element of *C* corresponds to a tuple of a table. As every element of *C* is a function, a tuple can be viewed as a function over the relevant set of field names (Figure 1 and Definition 3). A second option is the implementation of the elements of *C* as production rules. Then *C* forms a set of production rules (Figure 1). The two types of representation structures, namely records and production rules, employ search in quite distinct ways. The underlying search mechanism for extracting knowledge from tables is a form of query evaluation. The fundamental access mechanism of a knowledge base consisting of production rules boils down to some sort of deductive inferencing (Brodie and Jarke 1986, p.191; Smith 1986, pp. 8-12).

Walls: a record-based database

Identifier	Thermal Insulation (in minutes)	Irradiance (in minutes)
we123	25	35
wi214	35	45

Walls: a rule-based knowledge base

Domain = {identifier, thermal insulation, irradiance}
IF: identifier = we123 THEN: thermal insulation = 25 and irradiance = 35
IF: identifier = wi214 THEN: thermal insulation = 35 and irradiance = 45

Figure 1. Two symbol-level representation structures

The example shows that the distinction between both kinds of systems is non-existent at the knowledge level. A component described at the knowledge level has several representation options at the symbol level. But this does not imply that choice of a representation structure should take place at the symbol level. On the contrary, as we will argue, only at the knowledge level can well-founded choices of representation structures be made. One of the advantages of working at the knowledge level, mentioned in the previous section, is the availability of a separate, implementation-independent level from which representation structures and access mechanisms can be analyzed. An example of this is the view of databases from the knowledge level as described by Brachman and Levesque (1986). They state that it is possible to assess class membership after some theorem proving that one of two conditions is true without saying which one (using disjunction) or to state that an object[3] satisfies a certain condition without saying what that phenomenon is (using existential quantifiers). A solution to this intractability problem is to limit the uncertainty expressible in an implementation language. Because record-based data bases are restricted in precisely this way, they can be considered as knowledge-based systems of a limited form. A data base is comparable to the knowledge representation component of a knowledge-based system.[4] To illustrate a view of databases from the knowledge level we increase the number of functions of the previous example to four.

---

[3] Not only physical objects are meant by the term object. It also refers to phenomena, processes etc. that are of interest.

[4] "A knowledge-based system is any system that uses an explicit knowledge base in some capacity. The knowledge representation component is the part of the overall system that manages the knowledge base" (Brachman and Levesque 1986, p.72).

T1= {(*identifier*;*we123*), (*thermal insulation*;*25*), (*irradiance*;*35*)}
T2= {(*identifier*;*wi214*), (*thermal insulation*;*35*), (*irradiance*;*45*)}
T3= {(*identifier*;*wi329*), (*thermal insulation*;*20*), (*irradiance*;*25*)}
T4= {(*identifier*;*wi491*), (*thermal insulation*;*25*), (*irradiance*;*35*)}

This produces the following table:

Walls

Identifier	Thermal insulation (in minutes)	Irradiance (in minutes)
we123	25	35
wi214	35	45
wi329	20	25
wi491	25	35

Figure 2. A table on the fire resistance of walls

The range of uncertainty in these functions is quite limited. Disjunctions, negations or existential quantifiers are not present. A consequence of these limitations in expressiveness is a lack of knowledge. This becomes clear when we rephrase the question as: *How many walls meet the fire-safety requirements for limiting the spread of fire?* It appears that the knowledge expressed by the logic functions is not sufficient to answer this question. For example, we also need to know in what way the two heat-insulation attributes of walls are related to fire resistance.

Assume that we possess the knowledge that a wall meets the minimal fire resistance requirements if the heat-insulation properties both exceed 30 minutes. Then, the formal query to successfully answer the question should be:

$\lambda C \in$ WW: $|\{t \in C|$ t(thermal insulation) >=30 and t(irradiance)>=30$\}|$[5]

But it is still not certain that the query yields the right answer. For it is possible that the list of functions is incomplete or that we123 and wi491 are different identifiers of the same wall. Toreinterpret the symbol-level question as the question originally posed, additional knowledge is needed:

$c_i \neq c_j$

(each constant represents a unique individual)

$\forall$ [wall(x) $\supset x = we123 \vee x = wi214 \vee x = wi329 \vee x = wi491$]

(universal quantifier naming explicitly the positive instances)

Because of the extra knowledge, no reasoning is needed to find out how many walls are present. Inference is simply reduced to calculation. All the system has to do is count how many

5 If A is a set, than $|A|$ is the number of elements of A. The notation λ $C \in$ WW: U_x in which U_x is an expression in x, is a shorthand for $\{(C;U_x)|$ $C \in$ WW$\}$.

appropriate tuples appear in the wall relation. It does not have to reason by cases or contradiction. But suppose that the following knowledge about the relations between various types of walls and the minimal required fire resistance is captured in the system:

{(*wall-type*;*exterior*),(*minimal required irradiance*;*30 minutes*)}
{(*identifier*;*we123*),(*wall-type*;*exterior*)}

The knowledge in the first function states that the minimal required irradiance for exterior walls is 30 minutes. The second function asserts that we123 is an exterior wall. Implicitly, the knowledge tells us that the fire-resistance requirements for exterior walls are less stringent than for interior walls. The attribute thermal insulation is not relevant for exterior walls and, under certain circumstances, this might yield a higher fire resistance for exterior walls. Now, the previous conclusion that there is one fire-resistant wall should be replaced by the *inference* that two walls are fire-resistant. The example can be made more realistic by incorporating more attributes of walls and by introducing concepts defining conditional relationships between fire-resistance requirements and different types of spaces adjoining the walls represented in the database. Once again, other conclusions will be reached!

This example illustrates that a view of data bases from a knowledge-level perspective permits the assessment of the semantic consequences of adding knowledge. These assessments enable us to make decisions about the symbol-level trade-off tractability/expressiveness. Actually, the example denotes a more universal problem caused by a symbol-level approach: the difficulties occurring while querying a database (see for instance Remmen 1985; De Jonge, Bruijning, Schoemaker and Otten 1988; Van de Riet 1990). In the next section, a specific approach to knowledge-level research will be discussed.

4. Functional Object-Types as Knowledge-Level Models

A knowledge-level model should contain adequate knowledge of its environment. Since concepts play an important role as classificatory and storage mechanisms for the organization of knowledge, they can be regarded as basic epistemological components of a knowledge-level model. A concept has an intension and an extension. The intension of a concept is a set of sufficient and necessary conditions which should be satisfied by an object in order to belong to the class covered by the concept. The extension of a concept is the set objects complying with the intension. The intension relegates to an *object-type*.

There are several basic views on how to reconstruct the object-type of a concept, one of which is the functional view (Hendriks 1986; Lucardie 1989; Reitsma 1990; Van der Smagt and Lucardie 1991). According to the functional view, an object-type of a concept is established by a goal-oriented reconstruction process wherein a *disjunction of conjunct sets* is modelled. Such a disjunction is depicted in the treelike schema of Figure 3. This disjunction consists of three conjunct sets, all leading to goal 1:

{(*i1*;*a*),(*i2*;*k*)},
{(*i1*;*b*),(*i2*;*r*)} and
{(*i1*;*b*),(*i2*;*s*),(*i3*;*y*)}

An element of a conjunct set is an *INUS-condition*: an Insufficient but Necessary part of the conjunct set which is Unnecessary but Sufficient for the result. Within a conjunct set, an *INUS-*

condition is indispensable for achieving a goal, but the conjunct set itself, to which the *INUS*-condition belongs, is replaceable by other conjunct sets.[6]

As an object must satisfy one of the conjunct sets of an object-type in order to belong to the extension of a concept. The example, though greatly simplified, shows some interesting features of the functional approach. For instance, it is possible for objects, which at first sight are different, to be identical in the context of a goal. Or, in other words, objects having different attributes, but matching with a conjunct set of an object-type, are equivalent. An example of two 'different' but functionally equivalent objects is formed by an object 1 characterized by the attributes $\{(i1;a),(i2;k)\}$ and an object 2 having the attributes $\{(i1;b),(i2;s),(i3;y)\}$. In the context of goal 1 both objects are similar. Here, the notion of *functional equivalence* is essential: objects are identical, fall in the same concept or are similar if they possess - even quite different - attributes to perform the same function. Three mechanisms are responsible for the fact that a goal or a function is attainable by quite different strategies.

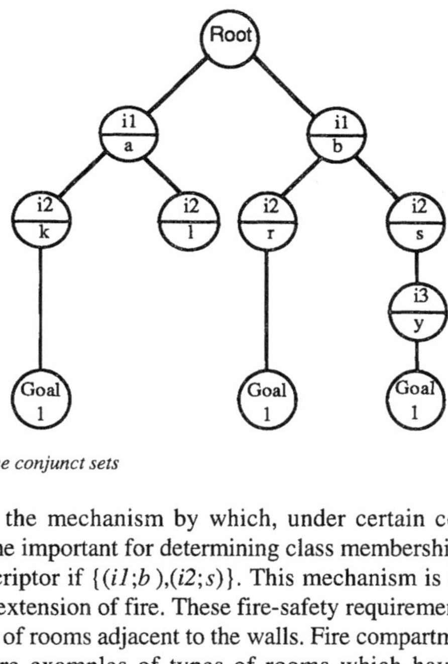

Figure 3. A disjunction of three conjunct sets

The first strategy is the mechanism by which, under certain conditions, other attributes (descriptors) may become important for determining class membership. In the third conjunct set, $\{(i3;y)\}$ becomes a descriptor if $\{(i1;b),(i2;s)\}$. This mechanism is effective in the fire-safety regulations limiting the extension of fire. These fire-safety requirements for walls are influenced by the conceptualization of rooms adjacent to the walls. Fire compartments, bathrooms, lavatories and traffic rooms all are examples of types of rooms which have different fire-resistance requirements. Therefore, it is important to have adequate definitions of these rooms at one's disposal. If we look at the conceptualization of a fire compartment, we can see that quite 'different' rooms can be classified as a fire compartment: both a heating room and a technical room can function as fire compartments. But the technical room should have a user surface

6 The definition of the *INUS*-conditionality is not free from discussion. Mackie defines an *INUS*-condition as follows: "*A* is an *INUS*-condition of a result *P* if and only if, for some *X* and some *Y* ((*A* and *X*) or *Y*) is a necessary and sufficient condition of *P*, but *A* is not a sufficient condition of *P* and *X* is not a sufficient condition of *P*." (Mackie 1965, p.246). The definition of Mackie is amended by Denise (1984) as follows: "*A* is an *INUS*-condition of a result *P* if and only if, for some *X* and some *Y*, ((*A* and *X*) or *Y*) is a necessary and sufficient condition of *P*, and *A* is necessary condition of (*P* and *X*), but *A* is not sufficient condition of *P* and *X* is not a sufficient condition of *P*." (Denise 1984, p.50).

exceeding 50 m^2. The user-surface attribute is virtually a new descriptor, which is only useful for the conceptualization of fire compartments in case we are dealing with a technical room.

The second strategy is that categorizations of attributes of objects influence each other. This phenomenon is called *conceptual interaction*. In Figure 3, conceptual interaction manifests itself in the mutual influence of the categorizations of the first and the second attributes. If $\{(i1;a)\}$, the classification of the second attribute is (k,l). If, on the other hand, $\{(i1;b)\}$ the classification of second attribute is (r, s). Referring again to the conceptualization of fire compartments, conceptual interaction is present between the type of room and the user surface. If we are dealing with an enclosed room, the adequate categorization of user surface is =< 500 m^2 or > 500m^2, because the user surface of an enclosed room should not exceed 500 m^2. However, if we are dealing with a technical room, the categorization of user surface is =< 50 m^2 or > 50 m^2. To be classified as a fire compartment, the user surface of a technical room should exceed 50 m^2.

The third strategy refers to the situation that objects may have different attribute values, but that this variation is limited to, or falls within, a goal-constructed category. Two objects characterized by the same $\{(i1;a)\}$, but with different values of $\{(i2;\_)\}$ - object 1 and object 2 respectively have the values $k1$ and $k2$ with $k1$ and $k2$ both falling in the category k of $\{(i2;\_)\}$ - are functionally equivalent. A technical room with a user surface of 55 m^2 and a technical room with a surface of 60 m^2 are functionally equivalent in the context of defining a fire compartment.

Functional equivalence shows that no a priori defined concepts are allowed to describe a knowledge-level model; only goal-constructed concepts are. From a functional viewpoint, it is a prerequisite that a knowledge-level model should avoid describing its application domain by a priori fixed categorizations. Rather, a knowledge-level model should avoid these reifications by accounting for goal-based dynamic, flexible categorizations of the environment. Under the influence of (changing) goals, continuously different descriptors will be needed to assess class membership. Not the object properties in their own right are relevant, but the functionally required properties. Functional equivalence stresses the structural heterogeneity of objects; in many situations, the presumption that objects are describable by fairly stable characteristics will be a misconception.

We have seen that one of the advantages of working at the knowledge level is the possibility to analyze representation formalisms at the symbol level. When we examine the basic properties of record-based information models from a functional viewpoint, we gain some interesting insights. If the essential configuration of the conjunct sets leading to the same goal is characterized by identical attributes, each attribute having the same kinds of values, records are excellent representation and processing tools. If, on the contrary, these conjunct sets are characterized by heterogeneity, which is caused by having to classify objects in a goal-oriented fashion accounting for different descriptors and new conceptual interactions, records are not appropriate (Kent 1979; Murdoch and Johnson 1990).

The more a domain deviates from homogeneity, the less appropriate the record configuration is. Referring again to our example, it is easy to see that many attributes play a role in limiting fire extension: external walls, internal walls, thermal insulation, irradiance, fire compartments, technical rooms, enclosed rooms, traffic rooms and so on. Furthermore, the definitions of different types of rooms yield also other attributes. For instance, the definition of a technical room states that it is an enclosed room for the installation of equipment necessary for the functioning of a building. This definition yields at least two extra attributes.

There are certain techniques for accommodating the variability in the description of an object-type in a record-based database design. We mention the design of multi-field tables, multi-meaning fields and multi-table databases. Unfortunately, each of these designs has its drawback. A database design in which all attributes are represented in multiple fields of one table is likely to have null-values in many fields. A simple illustration is the fact that if we are dealing with an external wall, the attribute thermal insulation is not relevant. This delivers a null-value in the corresponding record and field. A more serious problem is that no system facility will enforce a correct input, because this limited relevance is not defined to the system. So the multi-field format will introduce a data-integrity hazard.

A database design in which the same field is allowed to have different meanings is inefficient because every field value has the same meaning for the system. In this design, the system is unable to navigate through tables by matching key values. Only in the buried logic of an application program can the significance of field values be made clear. Also, a database design in which different multi-field tables represent a conjunct set of an object-type is not a workable solution. A user has to know the number and names of tables representing an object-type and be able to query them. Validation is still a problem. Again, data-integrity hazards endanger the suitability of this representation formalism. All these record-based solutions to cope with heterogeneous but functionally equivalent objects remain makeshift. For a more elaborate discussion on the limited capabilities of record-based information models, see Kent 1979.

5. Conclusions

This paper has explored the possibilities of integrating knowledge and database technology to manage the complexity of regulations for design and construction. In studying the integration of knowledge technology and database technology, the assumption is adopted that a clear conception of knowledge should logically precede that of representation. Knowledge-level research completely excludes user-interface and implementation aspects, but it includes semantic issues which can be effectively used for modelling, validation and design purposes. We have asserted that the theory of functional classifications is consistent with Newell's idea that "knowledge is to be characterised entirely functionally, in terms of what it does, not structurally in terms of physical objects with particular properties and relations" (Newell 1981, p.10).

At the knowledge level, the functional approach accounts for a rational goal-directed strategy by stressing the flexibility and dynamics responsible for the complexity in building regulations about which we wish to communicate knowledge and data. It enables the reconstruction of (deep) models of regulations, of complex objects and of the different users of the regulations.

At the symbol level, the functional view supports the examination and comparison of representation formalisms. For instance, the theory of functional concepts can explain why record-based information models are limited. In situations where quite different attributes or conceptualizations of objects can fulfil the same function, and thus are functionally equivalent, the assumption of vertical and horizontal homogeneity of record-based information models is not matched (Kent 1979).

At the systems level, the introduction of a knowledge-level perspective, in particular the theory of functional concepts, permits an effective development of a new class of DSS. A DSS constructed according to a functional approach could be of use for the design and modification of building regulations and for compliance-checking by applying these regulations to complex objects. For each of these modalities, a DSS could be used to check standards on completeness (regulations should be applicable in every situation within their scope), correctness (the meaning of the regulations is intended by the code designers), and consistency (regulations do not contradict each other). In particular, the phenomenon of different actors, each with their own goals and needs having to work with building regulations, requires extended capabilities for representing and inferencing complex knowledge and data.

Although research on functional object-types has evolved from a critical evaluation of the 'causal modelling approach' in quantitative scientific research, the principles of the functional approach can serve as a paradigm to reconstruct knowledge-level models and to design and implement DSS. Knowledge-level research conducted according to a functional viewpoint of the goals and the technical regulations to achieve them should reduce the complexity of the regulations and restore the logical consistency between the goals and the technical regulations.

References

Berg-Cross, G. and Price, M.E. (1989) "Acquiring and Managing Knowledge Using a Conceptual Structures Approach: Introduction and Framework", *IEEE Transactions on Systems, Man and Cybernetics* **19** , 513-527.

Brachman, R.J. (1983) "What IS-A Is and Isn't: An Analysis of Taxonomic Links in Semantic Networks", *IEEE Computer* **16**, 30-36.

Brachman, R.J. (1985) "“I Lied about the Trees” Or, Defaults and Definitions in Knowledge Representation", *AI Magazine* **6** , 80-93.

Brachman, R.J. and Levesque, H.J. (1986) "What Makes a Knowledge Base Knowledgeable? A View of Databases from the Knowledge Level", in L. Kerschberg (ed.), *Expert Database Systems*, Benjamin/Cummings, Menlo Park, pp. 69-78.

Brock, E.O. de (1989) *De Grondslagen van Semantische Databases*, Academic Service, Schoonhoven.

Brodie, M.L. and Jarke, M. (1986) "On Integrating Logic Programming and Databases", in L. Kerschberg (ed.), *Expert Database Systems*, Benjamin/Cummings, Menlo Park, pp. 191-207.

Chow, E. (1987) "Representing Databases in Frames", *Proceedings AAAI-87*, Sixth National Conference on Artificial Intelligence 2, pp. 405-409.

Denise, T.C. (1984) "On the Nature of INUS Conditionality", *Analysis* **44**, 49-52.

Di Battista, G., Kangasallo, H., and Tammasia, R. (1989) "Definition Libraries for Conceptual Modelling", *Data & Knowledge Engineering* **4**, 245-260.

Etherington, D. and Reiter, R. (1983) "On Inheritance Hierarchies with Exceptions", *Proceedings AAAI-83*, pp. 104-108.

Fahlman, S.E., Touretzky, D.S., and Roggen, W. van (1981) "Cancellation in a Parallel Semantic Network", IJCAI-81, pp. 257-263.

Hendriks, P.H.J. (1986) De Relationele Definitie van Begrippen: Een Relationeel Realistische Visie op het Operationaliseren en Representeren van Begrippen, Ph.D. Dissertation, Faculty of Policy Sciences, University of Nijmegen.

Herwijnen, J., Houten, E.G. van, Houtsma, M.A.W., and Romkema, H.M. (1990) "Implementatie van een Regel-gebaseerd Kennissysteem in een Relationele Database-Omgeving", *Information* **32**, 14-21.

Jonge, J. de, Bruijning, J., Schoemaker, W.H., and Otten, G.A.M. (1988) "Het Formuleren van Opdrachten in Relationele Vraagtalen", *Information* **30**, 393-460.

Kent, W. (1979) "Limitations of Record-Based Information Models", *ACM Transactions on Database Systems* **4** , 107-131.

Koster, A. and Stott Parker, D. (1990) "Regulation Management and Logic Programming", *Software-Practice and Experience* **20** , 79-107.

Levesque, H.J. (1984) "Foundations of a Functional Approach to Knowledge Representation", *Artificial Intelligence* **23** , 155-212.

Lucardie, G.L. (1989) Beslissingstabellen en Kennisgebaseerde Systemen: Een Conceptuele Benadering, TNO, Delft.

Mackie, J.L. (1965) "Causes and Conditions", *American Philosophical Quarterly* **2**, 245-264.

McCullough, J. (1991) Knowledge Based Systems in Architecture, Swedish Council for Building Research, Stockholm.

Murdoch, S. and Johnson, L. (1990) *Intelligent Data Handling*, T.J. Press, Padstow, Cornwall.

Newell, A. (1981) "The Knowledge Level", *AI Magazine* **2**, 1-20.

Reitsma, R.F. (1990) Functional Classification of Space: Aspects of Site Suitability Assessment in a Decision Support Environment, Ph.D. Dissertation, International Institute for Applied Systems Analysis, Laxenburg, Austria/Faculty of Policy Sciences, University of Nijmegen.

Remmen, F. (1985) "Hoe Vriendelijk zijn Vraagtalen in het Gebruik", *Information* **27**, 666-673.

Riet, R.P. van de (1990) "Kennisbanken", *Information* **32**, 105-232.

Risch, T., Reboh, R., Hart, P., and Duda, R. (1988) "A Functional Approach to Integrating Database and Expert Systems", *Communications of the ACM* **31** , 1424-1437.

Smagt, A.G.M. van der and Lucardie, G.L. (1991) "Decision-Making under not-Well-Defined Conditions: From Data Processing to Logical Modelling", *Tijdschrift voor Economische en Sociale Geografie* **82**, 295-304.

Smith, J.M. (1986) "Expert Database Systems: A Database Perspective", in L. Kerschberg (ed.), *Expert Database Systems*, Benjamin/Cummings, Menlo Park, pp. 3-15.

Steels, L. (1990) "Components of Expertise", *AI Magazine* **11**, 29-49.

Stonebraker, M. (1984) "Adding Semantic Knowledge to a Relational Database System", in M.L. Brodie, J. Mylopoulos, and J.W. Schmidt (eds.), *On Conceptual Modelling*, Springer Verlag, Berlin, pp. 333-356.

Twine, S. (1989) "Mapping between a NIAM Conceptual Schema and KEE Frames", *Data & Knowledge Engineering* **4**, 125-155.

KNOWLEDGE AND DATA MODELLING IN CAD/CAM APPLICATIONS

J.T. DE GELDER and G.L. LUCARDIE
TNO Building and Construction Research
P.O. Box 49
2600 AA Delft
The Netherlands

ABSTRACT. Modelling knowledge and data in CAD/CAM applications is complex because different goals and contexts have to be taken into account. This complexity makes particular demands upon representation formalisms. Today many modelling tools are based on record structures. By analyzing the requirements for a product model of a portal structure in steel, this paper shows that in many situations record structures are not well suited as a representation formalism for storing knowledge and data in CAD/CAM applications. This is illustrated by performing a knowledge-level analysis of the knowledge and data generated in the design and manufacturing process of a portal structure in steel.

1. Introduction

A large number of DataBase Management Systems (DBMS) applied in many commercial applications are implementations of record-based models (Date 1985). Record structures are mainly successful because of their efficiency and ease of manipulation. Record structures are excellent data-processing tools in application areas where objects have identical attributes, attributes always have the same type of value, and relationships between objects are relatively static. These application areas are called homogeneous. However, record structures are not very well suited for so-called heterogeneous application areas. In such cases, objects sometimes have to be described by many different and frequently changing attributes. Moreover, attributes can have different types of values, changing interactions between attributes can exist, and relationships between objects can be dynamic.

This paper performs a knowledge-level analysis of the knowledge and data generated in the design and manufacturing process of a portal structure in steel. This analysis demonstrates that CAD/CAM knowledge and data is heterogeneous, implying that CAD/CAM applications belong to the area of heterogeneous applications. The knowledge-level analysis has been conducted using the theory of functional classifications (Van der Smagt 1985; Hendriks 1986; Reitsma 1990; Van der Smagt and Lucardie 1991). This paper is organized as follows. First, the knowledge-level analysis itself is described in Section 2. Section 3 then describes an evaluation of record structures as a representation formalism for storing CAD/CAM knowledge and data. Next, in Section 4, some conclusions and alternative research directions are presented.

2. A Knowledge-Level Analysis

In his presidential address to the American Association of Artificial Intelligence (AAAI), Newell introduced a new modelling level: the knowledge level (Newell 1981). At this level, knowledge about an application area can be defined (Newell 1981; Brachman and Levesque 1986; Berg-Cross and Price 1989; Twine 1989). A knowledge-level analysis of an application area permits the development of implementation-independent conceptual (product) models, which

H. Timmermans (ed.), Design and Decision Support Systems in Architecture, 111–121.

better reflect the reality of an application area (Levesque 1984; Brachman and Levesque 1986; Steels 1990). These models can then be used to evaluate representation formalisms like record structures to see whether they can be used to store the knowledge and data of the conceptual model (Brachman and Levesque 1986).

In order to perform a knowledge-level analysis, a theory has to be adopted that prescribes how the analysis should be performed. In this paper, the theory of functional classifications is used (Van der Smagt 1985; Hendriks 1986; Reitsma 1990; Van der Smagt and Lucardie 1991). The theory of functional classifications considers knowledge as a set of concepts. Each concept has an intension and an extension. The intension of a concept is a set of attributes and constraints, also called an object-type. The extension of a concept is a set of objects satisfying the intension. Several theories deal with the modelling of object-types. The theory of functional classifications, however, is unique in the way the set of attributes and constraints is reconstructed. In reconstructing object-types, two aspects are essential: the functional aspect and the relational aspect.

2.1. THE FUNCTIONAL ASPECT

A key element in the theory of functional classifications is that the reconstruction of object-types is always governed by a specific function or goal. This means that the reconstruction of an object-type always underlies a certain function. For instance, the function of a connection in a portal structure in steel is: joining steel profiles. With respect to this function, it is conceivable that mechanical requirements determine the reconstruction of the intension of the concept 'connection'. All objects meeting these requirements are part of the extension of the concept 'connection'. It is possible that objects satisfying the intension are quite different. Notwithstanding the differences, these objects are functionally equivalent. Functional equivalence of objects is a main element in the theory of functional classifications. Objects which at first sight seem to differ may be equivalent if we consider the function they have to fulfil. The occurrence of different objects which appear to be functionally equivalent will be demonstrated by a simplified example of the design process of a portal structure in steel.

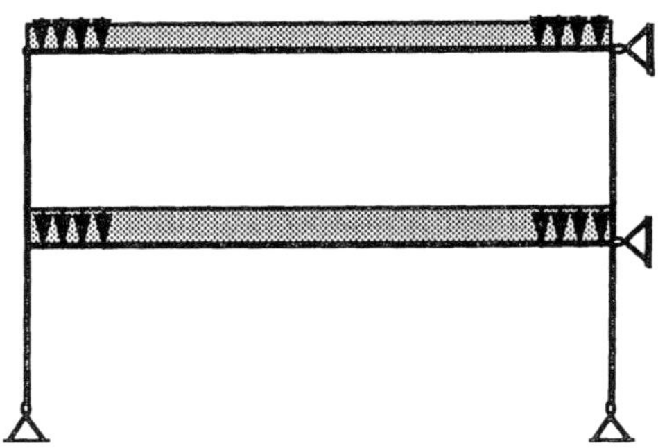

Figure 1. Portal structure with loads

In the design process of a portal structure in steel, three main activities can be recognized (Steenhuis and Dol 1991): design of the portal structure and the loads, design of the profiles, and design of the connections. The starting point is the portal structure displayed in Figure 1. The portal structure consists of two columns and two beams. Figure 1 also shows the loads on the two beams. The portal structure and loads are the basis for designing the profiles. The required profile types for the columns and the beams can be calculated. Some results of this calculation are, for instance, that profile type HE220A is satisfactory for both columns, profile type IPE220 is satisfactory for the roof beam, and profile type IPE330 for the floor beam. Figure 2 displays the portal structure again, but now after the design of the profiles. Since no connections have been designed yet, the profiles overlap at the place of the nodes. After designing the profiles, the required moment capacity in the nodes can be calculated. The moment capacity is an indication of the strength of a connection. Based on the results of this calculation, connections are designed at each node to meet the required moment capacity. We will concentrate on node 2 and design an end plate connection for it. In column-to-beam end-plate connections, the actual joint between the column and the beam is provided by an end plate that is welded to the beam and bolted to the column.

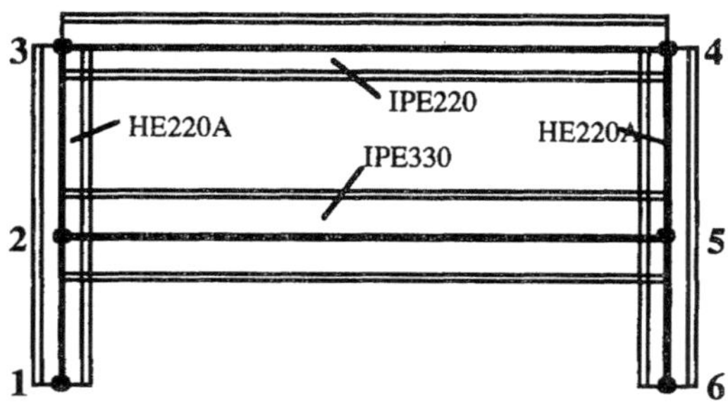

Figure 2. Portal structure with profiles

Above, the function of a connection in a portal structure was defined as: joining steel profiles. For the particular connection in node 2, a more specific function can be defined namely: connecting a beam (profile type IPE330) to a column (profile type HE220A) by means of an end plate. The required moment capacity in node 2 governs the reconstruction of the intension of the concept 'column-to-beam end-plate connection'. Based on the profile design, we found, for instance, that the moment capacity should be at least 60 kNm. The extension of the concept 'column-to-beam end-plate connection' is the set of all column-to-beam end-plate connections between an HE220A and an IPE330, having a moment capacity of at least 60 kNm. In practice, it is possible to design several different column-to-beam end-plate connections having a moment capacity of at least 60 kNm.

Figure 3a displays the first connection design including an end plate with a thickness of 5 mm. The moment capacity appears to be insufficient (25kNm), and the determining collapse mechanism is yielding of the end plate. This means that if the moment capacity exceeds a value of 25kNm, the first element in the connection that will collapse is the end plate. In a situation where yielding of the end plate is the determining collapse mechanism, there are two alternatives to increase the moment capacity of the connection: (i) including a thicker end plate to increase the strength of the end plate, which is the weakest element of the connection, and (ii) including a haunch to spread the moment force over a longer distance and so reduce the moment force on the end plate.

If the end plate, with a thickness of 5 mm, is replaced by an end plate with a thickness of 10 mm (Figure 3b), the moment capacity is still insufficient (56kNm) and the determining collapse mechanism is changed into shear of the column web. This means that if the moment capacity exceeds a value of 56kNm, the first element in the connection that will collapse is the column web. In a situation where shear of the column web is the determining collapse mechanism, including an end plate thicker than 10 mm makes no sense. The moment capacity of the connection in Figure 3a can also be increased by including a haunch. It was found that by including a haunch with a height of 330 mm, the moment capacity can be increased to 60kNm. The determining collapse mechanism remains yielding of the end plate. Figure 3c displays the connection when a haunch with a height of 330 mm is applied.

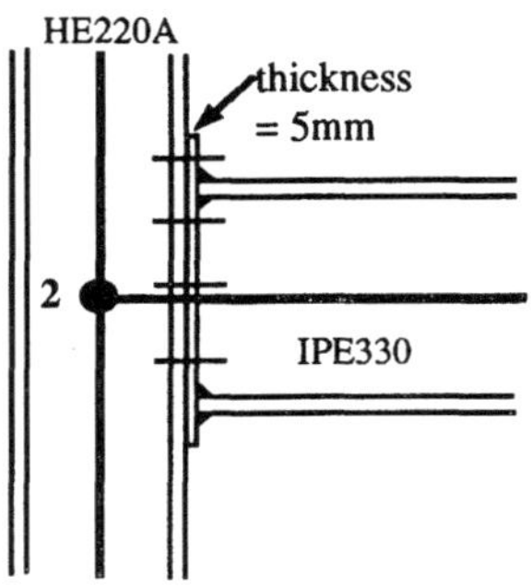

Figure 3a. Moment capacity = 25kNm

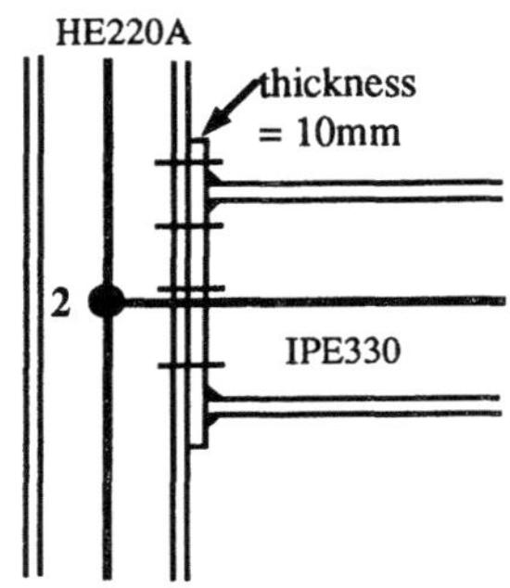

Figure 3b. Moment capacity = 56kNm

The solution in Figure 3c meets the requirements, since the moment capacity should be at least 60kNm. The number of bolt rows has to be raised from four to seven because of the haunch. The moment capacity of the connection in Figure 3b appeared to be not sufficient yet. Shear of the column web was found to be the determining collapse mechanism of this connection. Considering this collapse mechanism, there are two alternatives to increase the moment capacity of the connection: (i) including a web plate to strengthen the column web, which is the weakest element of the connection, and (ii) including a haunch to spread the moment force over a longer distance and so reduce the moment force on the column web. Figure 3d displays the connection after including a haunch with a height of 60 mm.

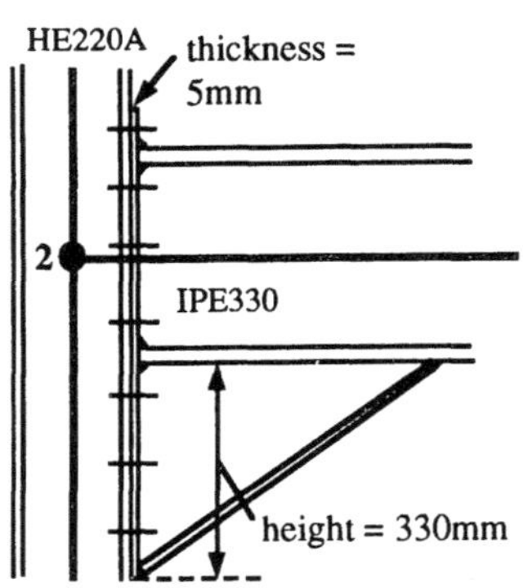

Figure 3c. Moment capacity = 60kNm

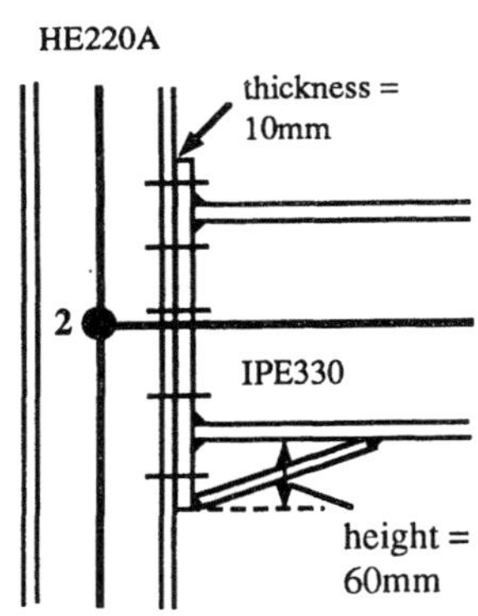

Figure 3d. Moment capacity = 66kNm

The moment capacity of this connection is 66kNm, and the determining collapse mechanism remains shear of the column web.The solution in Figure 3d meets the requirements, since the moment capacity should be at least 60kNm. The number of bolt rows has to be raised from four to five because of the haunch. If, instead of including a haunch, a column web with a thickness of 7 mm is included, the moment capacity turns out to be 82kNm. The determining collapse mechanism remains shear of the column web. Figure 3e displays the connection with a column web applied.

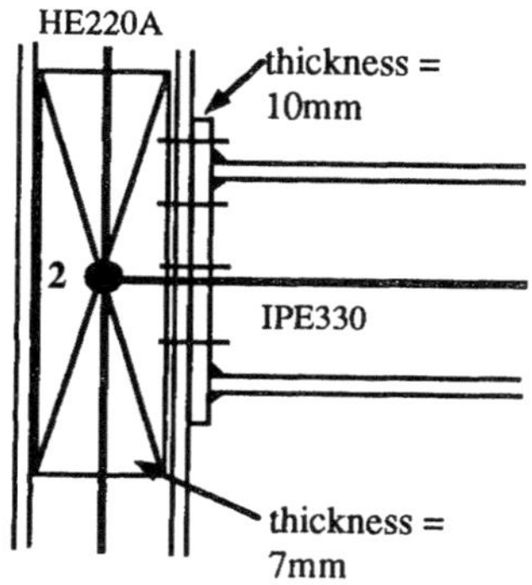

Figure 3e. Moment capacity = 82kNm

Figure 4 describes the connection design process by means of a disjunction of conjunct sets. Only the main attributes are included to this model. These conjunct sets all lead to the same goal: required moment capacity >= 60 kNm. Every element of a conjunct set is an *INUS-condition* (Denise 1984): an Insufficient but Necessary part of a set which is Unnecessary but Sufficient for the result. Within a conjunct set, an *INUS-condition* is indispensable for achieving a goal. But theconjunct set to which the *INUS-condition* belongs is replaceable by other conjunct sets. The conjunct sets in Figure 4 are:

1. {(end plate;[5,..]),(haunch;[330,..]), (bolt;[7,16,8.8,..])},
2. {(end plate;[10,..]),(web plate;[7,..]), (bolt;[4,16,8.8,..])}
3. {(end plate;[10,..]),(haunch;[60,..]), (bolt;[5,16,8.8,..])}

For example, *INUS-condition* (haunch;[330,..]) is an Insufficient part of conjunct set 1, because the *INUS-conditions* (end plate;[5,..]) and (bolt;[7,16,8.8,..]) are necessary too. However, *INUS-condition* (haunch:[330,..]) is Necessary for conjunct set 1, because a conjunct set only consisting of *INUS-conditions* (end plate:(5,..)) and (bolt:(7,16,8.8,..)) is insufficient for reaching the goal (see Figure 3a). Conjunct set 1 is Unnecessary for reaching the goal, because the goal can also be reached by conjunct set 2 or conjunct set 3. However, conjunct set 1 is Sufficient for reaching the goal. An object must satisfy one of the conjunct sets in order to be classified as a column-to-beam end-plate connection between an IPE330 beam and an HE220A column with moment capacity >= 60 kNm.

This example demonstrates that geometrically completely different objects are functional equivalents if the function is: connecting a beam member (profile type IPE330) to a column member (profile type HE220A) by means of an end plate such that the moment capacity >= 60 kNm. This example also demonstrates that the set of attributes and constraints forming the intension of the concept 'column-to-beam end-plate connection' is rather heterogeneous. The object-type 'column-to-beam end-plate connection' has many different and frequently changing attributes, which are actually object-types in themselves, since they also have their own attributes.

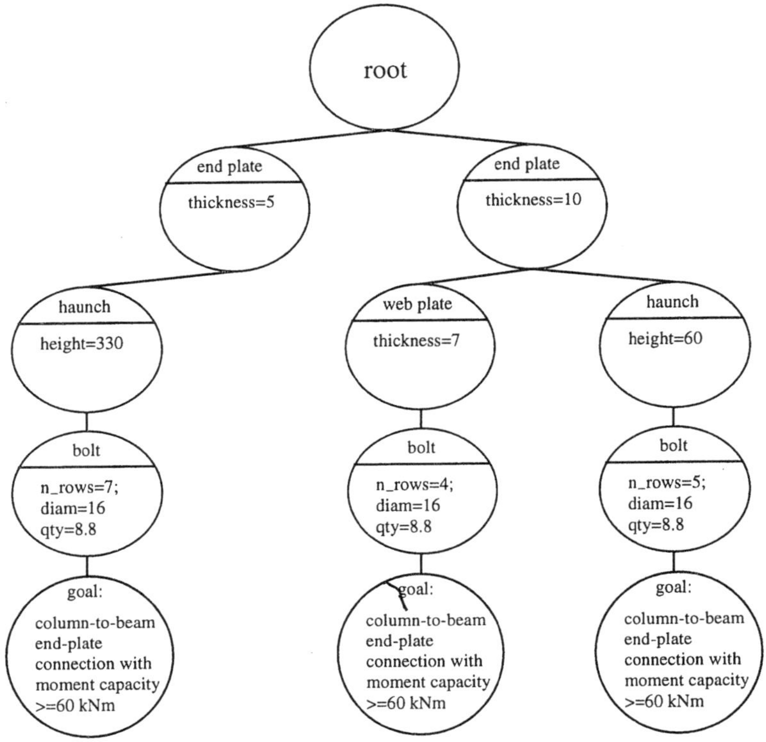

Figure 4. A disjunction of conjunct sets

2.2. THE RELATIONAL ASPECT

Another element in the theory of functional classifications is that classifications of object-types are only valid within one specific context. The intension of a concept is governed not only by a function or goal, but also by a context. This is demonstrated by comparing two phases in realizing a portal structure in steel: the design phase and the manufacturing phase. These phases can be considered as two different contexts. In the design phase, a set of mechanical requirements governs the reconstruction of the intension of the concept 'portal structure'. In order to meet these requirements, the designer designs profiles and connections between profiles (Figure 5).

In the manufacturing phase, a set of economical requirements governs the reconstruction of the intension of the object-type 'portal structure'. In order to meet these requirements the manufacturer produces assemblies in house, which are connected on site. Generally an assembly is a profile including all the parts welded to the profile. Because of a different context, a manufacturer describes the portal structure in a completely different way (Figure 6).

A different description means different attributes of object-types and different relationships between object-types. For example, in the design context, the end plate is part of the column-to-beam end-plate connection, in the manufacturing context, the end plate is part of the beam assembly. Likewise, in the design context, holes are not defined explicitly, because the existence of a hole is implicitly defined by the existence of a bolt. In the manufacturing context, a hole has to be defined explicitly and related to the column profile and the end plate, because holes have to be drilled into these parts in the manufacturing phase. Similarly, in the design context, the end

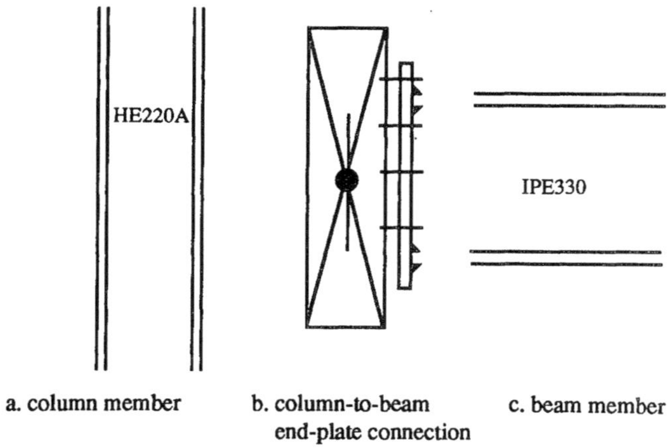

Figure 5. Design context

plates, haunches and web plates are located relative to the connection node. In the manufacturing context, these parts are located relative to either one end of the beam profile or one end of the column profile. Also, in the design context, a number of part dimensions are only implicitly defined, since they are mechanically unimportant or dependent on other dimensions. The length of the beam profile, for instance, is partly dependent on the height of the column and the thickness of the end plate. In the manufacturing context, all dimensions are explicitly defined, because parts have to be manufactured.

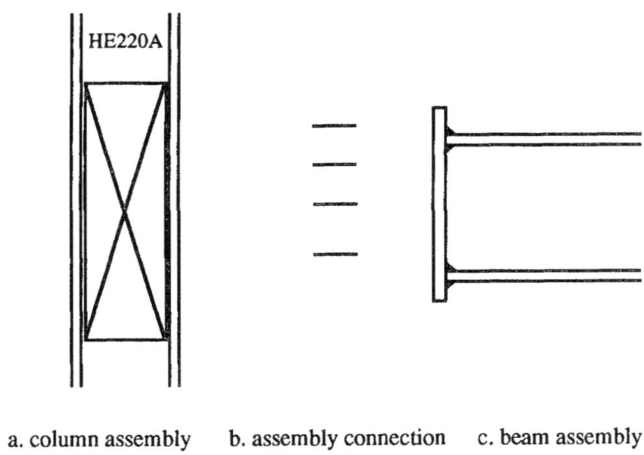

Figure 6. Manufacturing context

2.3. COMMON PRACTICE IN PRODUCT MODELLING

In product modelling practice, the functional and relational aspect in reconstructing the intension of a concept is often not recognized. The general problems that can be found in product modelling are no clear distinction of levels, no clear distinction of goals, and mixing of contexts. Product models are often developed using modelling techniques which make no clear distinction between knowledge level and symbol level. Often knowledge of reality is adapted to the limitations of representation formalism. The functional aspect stated that the reconstruction of object-types is always governed by a specific function or goal. In practice, product models are often developed without a specific modelling goal in mind. One just starts modelling a product without considering the goal of modelling. Since the reconstruction of object-types is always governed by a specific goal, modelling without a specific goal leads to general models, which are not useful for specific purposes. The relational aspect stated that the classification of object-types is only valid within one specific context. In the development of a product model for integrated CAD/CAM applications, often only one classification is reconstructed for all contexts. Either a classification for one of the contexts is selected for the product model or a mixture of all required classifications is developed for the product model. Again this leads to general models, which are not useful in a specific context.

3. Limitations of Record-Based Models

At the symbol level, object-types, which are reconstructed at the knowledge level, are reduced to representation formalisms. These consist of structures to store knowledge and data as well as structures to access mechanisms in order to operate on knowledge and data. For storing and accessing knowledge and data, record-based models are often applied, especially so-called relational models. In a relational model, object-types are stored as rather simple record structures. The intension is defined by a fixed number of fields, each field having a predefined value type. The extension consists of a number of records. A record is a fixed number of values, each corresponding with a field of the intension. Each value must be part of the range defined by the value type of the corresponding field. Value types are generally restricted to numeric and character string data types. Figure 7 shows an example of records in a relational model.

CONNECTION:

con.id	end plate.id	haunch. id	web plate.id	moment-capacity
01	11	21	NULL	60
02	12	NULL	31	82
03	12	22	NULL	66

END PLATE:

end plate.id	thick-ness
11	5
12	10

HAUNCH:

haunch. id	end plate.id
21	330
22	60

WEB PLATE:

web plate.id	thick-ness
31	7

Figure 7. Record structures in a relational model

The boxes in Figure 7 are called tables. Relationships between records can be defined by accessing the records and searching for identical attribute values. By accessing, for instance, the first record of an object-type 'connection', the related record of object-type 'end plate' can be found by searching the set of records for the attribute value '11' of the common attribute 'end plate.id'. How can the object-types, reconstructed at the knowledge level for the portal structure in steel, be stored by record structures at the symbol level? In Section 2.1 it was found that in CAD/CAM applications, functionally equivalent objects can be geometrically different. It was found, too, that these functional equivalent objects are complex objects consisting of frequently changing sub-objects. Modelling complex object-types in a relational model is difficult, because the access mechanism of a relational model is not able to deal with hierarchical ('part of') relationships. In a hierarchical relationship, one object-type is existence-dependent on another object-type. In Figure 7, end plate '11' is existence-dependent on connection '01'. If connection '01' is deleted from the model, the model should react automatically by deleting end plate '11' and haunch '21' as well. The relational model, however, would react to this request by deleting only connection '01' and leaving end plate '11' and haunch '21' untouched.

Functionally equivalent objects consist of frequently changing sub-objects. In Figure 4, the object 'column-to-beam end-plate connection' is at one time related to the sub-object 'web plate' and another time to the sub-object 'haunch'. In a relational model, relationships are defined by means of common attributes, so one time the object-type 'column-to-beam end-plate connection' has a common attribute with the object-type 'web plate' and another time with the object-type 'haunch'. It is possible to solve this problem of different attributes by including supertype/subtype relationships between object-types into the relational model. A supertype and its subtypes are functionally equivalent object-types. The supertype holds the common attributes of the functional equivalent object-types. A subtype holds the unique attributes, which differ from attributes of the other object-types. In the example, the supertype 'connection part' could be defined, having two subtypes 'haunch' and 'web plate'. The relationship between the supertype 'connection part' and its subtypes 'haunch' and 'web plate' defines that a connection part is either a haunch or a web plate. The problem of changing relationships between the object-type 'column-to-beam end-plate connection' and either the object-type 'web plate' or the object-type 'haunch' can be solved then by defining a fixed relationship between the object-type 'column-to-beam end-plate connection' and the object-type 'connection part'. This solution of defining a supertype/subtype relationship between on the one hand the object-type 'connection part' and on the other hand the object-types 'haunch' and 'web plate' is only appropriate when the objects 'haunch' and 'web plate' are functionally equivalent in any situation. In Section 2.1, however, it was demonstrated that in case of a determining collapse mechanism indicated as yielding of the end plate, haunch and web plate are not functionally equivalent. Thus, in that case they cannot be represented as two subtypes of the same supertype.

In Section 2.2, it was found that attributes of object-types and relationships between object-types in CAD/CAM applications can change dramatically after changing the context from design to manufacturing. In Kent (1979), several possibilities for modelling changing attributes are presented, all of which have their drawbacks. For all attributes of an object-type required in any context, a field can be defined. The drawback of this so called *multiple fields* solution is that for a particular context, only part of the attributes is relevant and the rest should be assigned null values. Another possibility is to create *multi-functional fields.,* indicating that an attribute of an object-type can have different meanings, depending on the context. The drawback of this solution is that such a practice is never defined to the system. With respect to any processing done by the system, that field appears to have the same significance in every record occurrence. The different meanings these fields have is only known in application programs. Finally, *multiple object-types*, suggesting that an object-type can be spread over several tables. Each table stores a set of attributes for one specific context. The drawback of this solution is that one object-type is split into different object-types. These are not real object-types concerning the reality of the application area, but only created because of the limitations of the representation formalism.

4. Conclusion

Record-based models are excellent modelling tools for application domains in which data and knowledge have a homogeneous nature. Both object-types and relationships should be homogeneous. This paper showed that knowledge and data in a CAD/CAM domain can hardly be represented by records, because of the heterogeneous nature of knowledge and data in both object-types and relationships. This has been demonstrated by performing a knowledge-level analysis of the knowledge and data generated in the design and manufacturing process of a portal structure in steel.

Since record-based models appear not be suitable for representing knowledge and data in a CAD/CAM domain, a lot of research has been conducted on alternative representation formalisms. These have more flexibility, as well as the dynamics to represent knowledge and data in heterogeneous application areas. This research has not only been conducted in the field of database technology (Ter Bekke 1983; Hull and King 1987), but also in the field of Artificial Intelligence (Loucopoulos and Karakostas 1989).

Recently, research has been started on the integration of knowledge bases and databases (Lucardie 1992). An important aspect of this research is that integration takes place at the knowledge level. Conceptual (product) models have to developed independent of any implementation strategy. After developing these knowledge-level models, attention will have to be paid to representation formalisms to represent these models at the symbol level. Hopefully the research on alternative representation formalisms will lead to representation formalisms which are able to represent all knowledge and data reconstructed at the knowledge level without reducing the semantics and dynamics of this knowledge and data. If so, the gap between knowledge-level and symbol-level models can be closed. Then information systems can be developed which better meet the requirements of the application domain.

References

Bekke, J.H. ter (1983) *Database Ontwerp*, H.E. Stenfert Kroese, Leiden-Antwerpen.

Berg-Cross, G. and Price, M.E. (1989) "Acquiring and Managing Knowledge Using a Conceptual Structures Approach: Introduction and Framework", *IEEE Transactions on Systems, Man and Cybernetics* **19**, 513-527.

Brachman, R.J. and Levesque, H.J. (1986) "What Makes a Knowledge Base Knowledgeable? A View of Databases from the Knowledge Level", in L. Kerschberg (ed.), *Expert Database Systems*, Benjamin/Cummings, Menlo Park, pp. 69-78.

Date, C. (1985) *Database: Een Inleiding*, Academic Service, Den Haag.

Denise, T.C. (1984) "On the Nature of INUS Conditionality", *Analysis* **44**, 49-52.

Gelder, J.T. de (1993) Design of a Central Database as Part of an Integrated System for Computer Aided Design of Steel Structures, TNO Building and Construction Research, report no. BI-88-076.

Hendriks, P.H.J. (1986) De Relationele Definitie van Begrippen: Een Relationeel Realistische Visie op het Operationaliseren en Representeren van Begrippen, Ph.D. Dissertation, Faculty of Policy Sciences, University of Nijmegen.

Hull, R. and King, R. (1987) "Semantic Database Modeling: Survey, Applications and Research Issues", *ACM Computing Surveys* **19**, 201-260.

Kent, W. (1979) "Limitations of Record-Based Information Models", *ACM Transactions on Database Systems* **4**, 107-131.

Levesque, H.J. (1984) "Foundations of a Functional Approach to Knowledge Representation", *Artificial Intelligence* **23**, 155-212.

Loucopoulos, P. and Karakostas, V. (1989) "Modeling and Validating Office Information Systems: An Object and Logic Oriented Approach", *Software Engineering Journal* **4**, 87-94.

Lucardie, G.L. (1992) "A Functional Approach to Realizing Decision Support Systems in Technical Regulation Management for Design and Construction", in H. Timmermans (ed.), *Design and Decision Support Systems in Architecture*, Kluwer, Dordrecht.

Newell, A. (1981) "The Knowledge Level", *AI Magazine* **2**, 1-20.

Reitsma, R.F. (1990) Functional Classification of Space: Aspects of Site Suitability Assessment in a Decision Support Environment, Ph.D. Dissertation, International Institute for Applied Systems Analysis, Laxenburg, Austria / Faculty of Policy Sciences, University of Nijmegen.

Smagt, A.G.M. van der (1985) Definiëren en Relateren in Sociaal-Wetenschappelijk Onderzoek, Ph.D. Dissertation, Faculty of Policy Sciences, University of Nijmegen.

Smagt, A.G.M. van der and Lucardie, G.L. (1991) "Decision-Making Under Not-Well-Defined Conditions: From Data Processing to Logical Modelling", *Tijdschrift voor Economische en Sociale Geografie* **82**, 295-304.

Steels, L. (1990) "Components of Expertise", *AI Magazine* **11**, 29-49.

Steenhuis, C.M. and Dol, C. (1991) "Geboute Kopplaatverbindingen: Gebruik van NEN 6772 tijdens het Ontwerpproces van een Staalconstructie", *Bouwen met Staal* **103**, 3-7.

Twine, S. (1989) "Mapping Between a NIAM Conceptual Schema and KEE Frames", *Data & Knowledge Engineering* **4**, 125-155.

RATIOS FOR COST CONTROL

E.R. POORTMAN
Department of Production and Construction
Faculty of Architecture, Building and Planning
Eindhoven University of Technology,
P.O. Box 513, 5600 MB Eindhoven
The Netherlands

ABSTRACT. The design of buildings takes place in phases representing a development from rough to precision planning. Estimates are made in order to test whether the result is still within the budget set by the client or developer. In this way, the decisions taken during the design phase can be quantified and expressed in monetary terms. To prevent blaming the wrong person when an overrun is discovered, the cost control process has to be improved. For that purpose, two new procedures have been developed: (i) a new 'translation' activity; and (ii) ratios by which quantities can be characterized. 'Translation' is the opposite of estimation. A monetary budget is converted - 'translated' - into quantities, reflecting the desired quality of the building materials. The financial constraints of the client are thus converted into quantities - the building components used by the designers. Characteristic quantity figures play an important role in this activity. In working out an estimate, the form factor (i.e., the ratio between two characteristic values of a building component) has to be determined. The unit cost is then tested against that ratio. The introduction of the 'translation' activity and the use of characteristic quantity figures and form factors enhance existing estimation methods. By implementing these procedures, cost control becomes considerably more reliable.

1. Introduction

In an overview of the development of approaches to building cost and price forecasting, Raftery (1987) identified first, second and third-generation models. Models of the second (use of regression analysis) and third (probabilistic estimates and knowledge-based computer systems) generation represent attempts by researchers to improve the quality of cost data. But the practitioners in the field in the UK still prefer first-generation models, such as the floor-area model and the elemental model. Ter Haar (1991) comes to the same conclusion for the Netherlands in his comparative analysis of methods used for costing in the early planning stages.

Skitmore (1992) compared more than 30 estimating techniques. He found that the reliability of estimates generated by the traditional model is a function of the reliability of each quantity value, the reliability of each rate value, the number of items, and the collinearity of the quantity and rate values. He concluded that the last factor is often overlooked in assessments of reliability; these tend to assume that the errors made in estimating quantity and cost value are independent. To enhance the reliability of cost values, ratios called characteristic cost figures have been developed. In this paper, I will present ratios that can be used to characterize quantities. This particular use of ratios was first introduced in the 1950s in the Storey Enclosure method (Bathurst and Butler 1980). To make the cubic method more realistic, this estimation method employs factors that account for storey heights, basements, and areas of enclosing walls.

In the United Kingdom, the Building Cost Information Service (BCIS) was set up in 1962. This organization created a data bank of building cost analysis with ratios that took the gross floor area into account. In the Netherlands, Misset (Poortman 1990) publishes cost analyses with ratios also related to the gross floor area and the cubic metre. Bathurst and Butler also give examples in the chapter describing cost analysis because "the relations between the costs and the method of

H. Timmermans (ed.), Design and Decision Support Systems in Architecture, 123–136.

comparing the values would hold true if the figures had been produced from the tender sum by the addition of the costs of those items associated with quantities of elements. ... The full effect of plan shape is not always easy to establish, but a simple method can be devised of relating the amount of external cladding elements to the superficial floor area of the building so as to reveal the effect of different plan shapes. The relationship can be expressed as a ratio of the area of the external cladding elements to the floor area of the building, and is commonly referred to as the external cladding to floor ratio" (Bathurst and Butler 1980).

All this was sufficient reason to examine the total process of designing buildings, including the ratios that may play an important role in cost control. In doing so, I have developed a new activity, which I call 'translation', where a budget couched in monetary terms is converted - 'translated' - into quantities based on the desired quality of the building materials. This operation is the opposite of making an estimate. In other words, it is the opposite of the check with hindsight, i.e., after the decisions have been taken, which can only reveal whether the decisions were right or wrong. In 'translation', characteristic quantity figures play an important role; in preparing an estimate, the form factor has to be used. Characteristic quantity figures and form factors will be discussed in the next sections.

2. Cost Control

The term 'budget' refers to the assets available for a particular item or activity. The term 'assets' has a different connotation for the diverse parties involved in the building process. To the client, the budget is a sum of money that he is prepared to spend on the building. To those implementing the order (contractors, installers), it is the means of production - labour, materials, and equipment. For both parties, the budget is expressed in terms of the assets allocated to the object. We could say that the client buys the building from parties who have used their means of production to construct it.

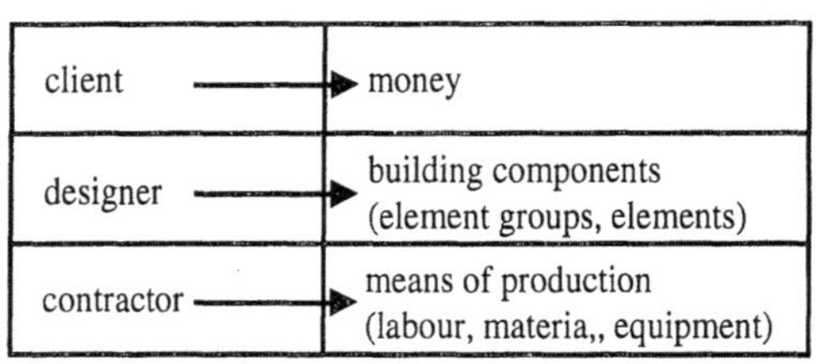

Figure 1. The parties involved and their budgets
Source: Poortman (1990)

But what are the assets of the designers (architects, builders, installation consultants)? What do they actually spend these assets on? The answer is straightforward: building components! During the design process, they take all kinds of building components from a storage cupboard, so to speak. These components are used to construct a technically, functionally and aesthetically acceptable product. Of course, this is a conceptual procedure; their product - the drawing - shows what building components have been used. These items play an important role in cost control. In principle, the designers should be able to give full attention to the selection of building components. However, as they are accountable to and have to communicate with the client, they must also take the budget into consideration. Cost-conscious design is mainly a question of making good use of the tool of cost control. In the beginning of the design process, the overall making good use of the tool of cost control. In the beginning of the design process, the overall costs can be influenced to a large extent. Towards the end, the influence of cost-conscious design is slight (UGCB 1985). Figure 2 illustrates this change in its impact on cost.

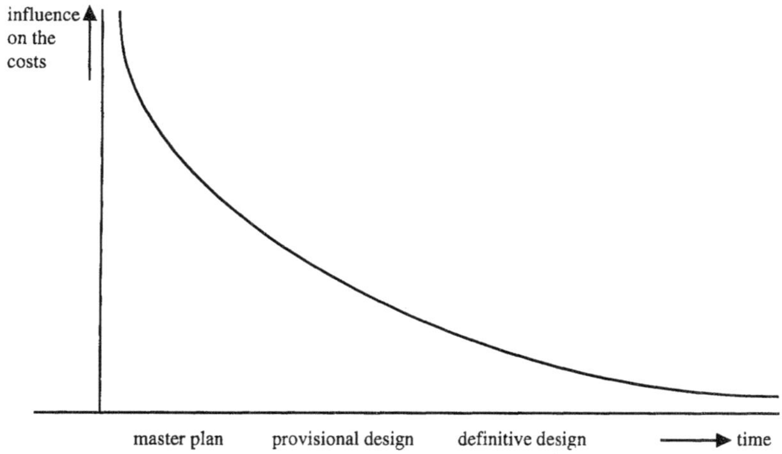

Figure 2. The effect on costs during the design process

Cost control can be reduced to two essentially different activities: estimation and 'translation'. The design process proceeds in three steps (BNA 1988): master plan, provisional design, and definitive design. In the course of these steps, these two activities are carried out repeatedly. In the meantime, the number of cost centres keeps increasing. The question recurring throughout the design process concerns how much of the assets will be required in relation to the budget, i.e., the client's money that is supposed to be spent on building components made from building materials. The quantities are specified in terms of building components. These items represent the cost per unit; in turn, these costs are made up of quantities and unit costs of the means of production.

The most familiar activity is the preparation of an estimate. Unfortunately, cost control is often restricted to this. The estimate is merely a financial representation of the design. After decisions have been taken, they are quantified and expressed in monetary terms. This is known as 'pricing'. The main purpose of making an estimate is to test whether the design is feasible within the (client's) budget.

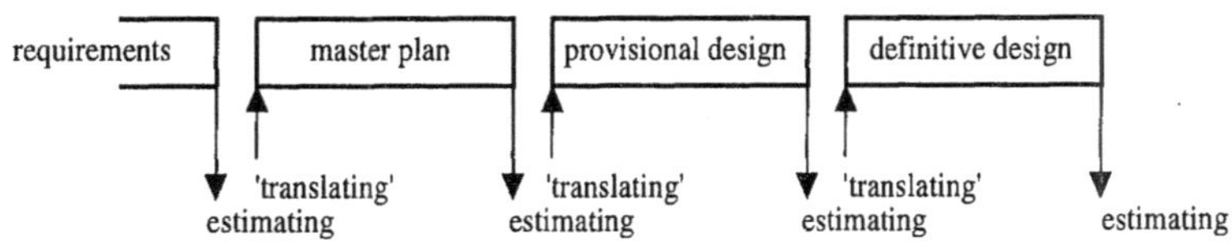

Figure 3. The place of estimation and 'translation' in the design process

'Translation' occurs prior to each step in the design process. It can be regarded as one of the planning activities. The requirements set for space components are translated into characteristics of building components. For instance, an air temperature of 20^0 C in a room determines the necessary R value of walls and roof, as well as the power kW of a heating system. In regard to cost, the budgets, expressed in Dutch guilders, are 'translated' into quantities of building components. The purpose of 'translation' is to relate the technical and financial requirements of the client to the units of the designer, i.e., the building components.

Estimation and 'translation' may be considered as processes with input data, output data, control data, and means.

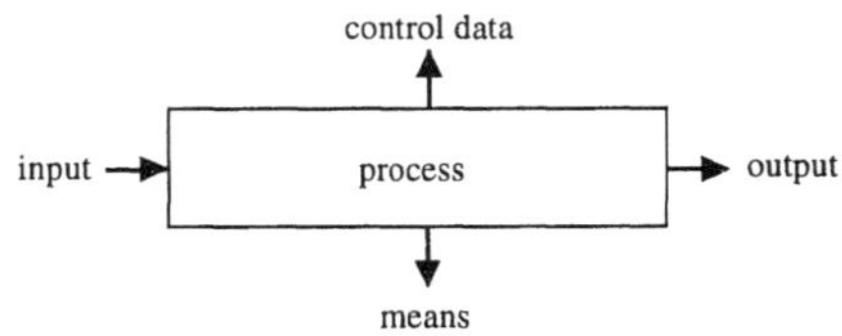

Figure 4. The process of estimation and 'translation'

The input and output data pertain to the project that is being designed. The input data can always be taken from the drawings or descriptions of the building in question. The output is the estimate or 'translation' for that building. In order to carry out the process, criteria, data, methods, knowledge, etc. are necessary. Together, these constitute the control or management data. They are independent of the project being undertaken; they must be applicable to any project. Finally, the means are the people and the machines that carry out the process; hence, both the costing expert, with his calculator, and the computer.

One and the same simple equation plays an important part in such processes. According to this equation, the quantity of the cost centres is multiplied by the cost per unit and the results are added together.

$$\sum_{\text{cost centres}} \text{quantity x costs per unit}$$

The remainder of this article elaborates on how this equation should be used in practice. The input, output and management data are treated separately for the estimation and 'translation' activities.

3. Estimation

The input data for this activity consist of quantities. These can be derived from written documents, such as a programme of requirements, or drawings, such as a floor plan. The data related to the design can be read or measured. Some further calculation is necessary, but generally it is not difficult to determine the quantities, i.e., the number of cost units. The result of the process, the output data, is best represented by the term 'building sum'. This is the amount which will probably be spent. The building sum may refer to a separate cost centre as well as to the total for all cost centres. The building sum can be calculated on the basis of the control or management data. That data consists of the cost per unit, on which basis the quality of a cost centre has been determined.

This procedure represents the process of estimating in its most elementary form. The estimation of the definitive design is generally carried out in this way. By that stage, the design has progressed so far that estimates can be made at the most detailed level; that is, the level of the building file (see Appendix 1). In determining the price, allowance is only made for factors affecting the cost per unit of the means of production. Factors such as series and routine effect, quantity discount, regional differences in time rates and cost of materials, and reference date can be determined on the basis of characteristic figures. These may include routine lines, discount percentages, correlation coefficients, and index figures, for instance.

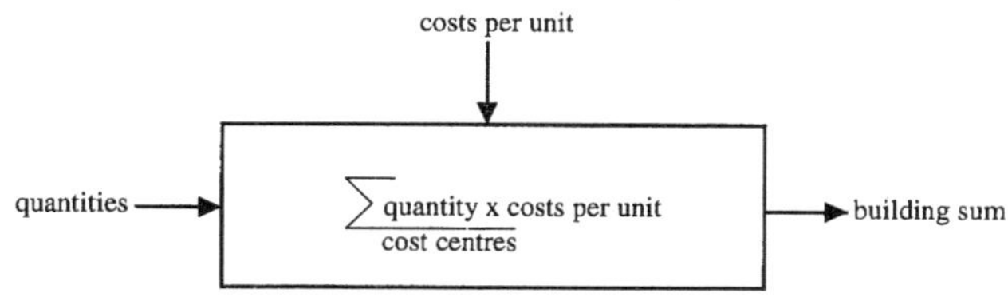

Figure 5. Estimation in its most elementary form

That does not apply to the estimates which have to be made earlier in the design process. There, the cost information is frequently obtained from so-called cost analyses (see Appendix 2). The form factor, which affects the cost per unit of building components, is of great importance. Thus, before the cost per unit can be entered into the equation, the form factor has to be checked. If discrepancies are found, it must be adjusted accordingly in the cost per unit. The following example will clarify the role of the form factor.

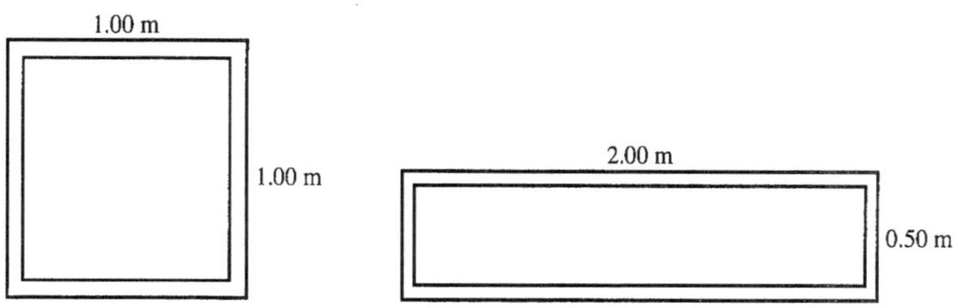

Figure 6. Two windows of the same surface area and different form factors

The cost per unit of a building component is expressed in the unit most convenient for that element. The cost of the square window in Figure 6, measuring 1 x 1 metre, was analyzed. It was found to be NLG 550 per square metre, specified as follows: 1 m^2 of glass at NLG 150 per m^2 = NLG 150; and 4 m of frame at NLG 100 per metre = NLG 400. Should this amount be used in an estimate for another project in which windows also occur with a surface area of 1 m^2 but are rectangular in shape (2 x 0.50 m)? The answer is no. The form factor, in this example the ratio *periphery* (m^1) / *surface area* (m^2) is 4.00 for the window analyzed and 5.00 for the new window to be estimated. This shows a difference of + 1.00. That outcome requires correction of the cost per unit, which was said to be + 1.00 x NLG 100 = + NLG 100. A sum of NLG 650 must therefore be included in the estimate as the cost per unit.

For an accurate estimate, it is necessary to take the form factors into account. This estimation process is shown schematically in Figure 7.

4. 'Translation'

The newly developed activity of 'translation' converts the budget - i.e., the amount of money made available by the client - into quantities of building components. The same equation can be used as for making an estimate, while the cost per unit can be used for control purposes (Fig. 8). 'Translation' in its most elementary form only applies if the budget can be spent freely. In reality,

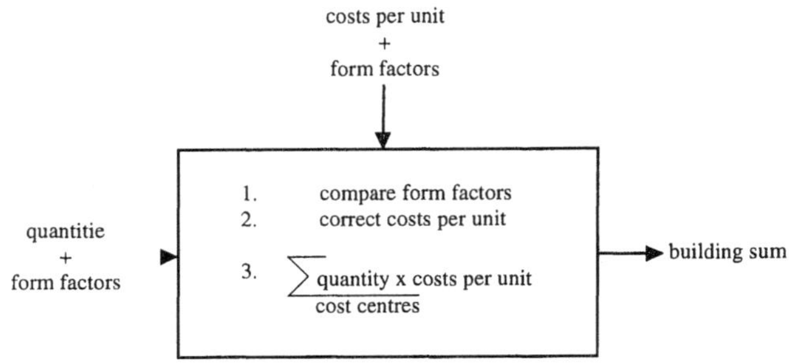

Figure 7. Estimation with form factors

however, that is rarely the case. From the outset, the quantities are specified which have to be achieved within that budget (e.g., the gross floor area in m^2 at a sum per m^2). The budget has always been composed of quantities and cost per unit of those quantities. 'Translation' must therefore convert those quantities into quantities of cost centres at a more detailed level. This can be done by using characteristic quantity figures as additional control data. The following example will make this clear.

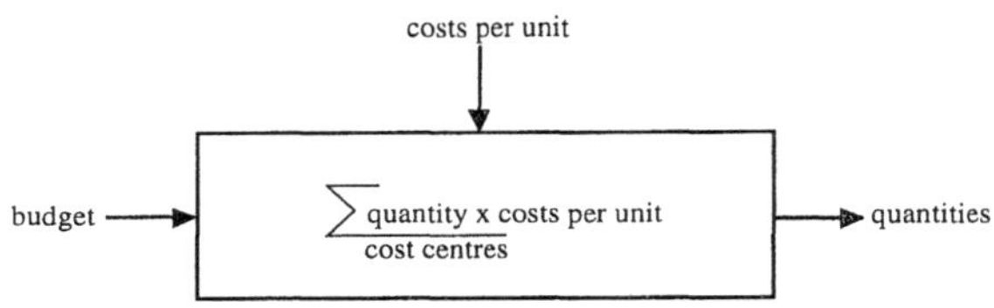

Figure 8. 'Translation' in its most elementary form

The estimate of a master plan states that the quantity of external wall structures - 800 m^2 - was measured in the plan; the cost per unit - NLG 250 - is taken from a cost analysis. The external wall structures of the project analyzed are drawn in Figure 10, together with the specification of the cost per unit. The cost per unit is thus based on an external wall structure with the following characteristic quantity figures:

surface area of external wall openings (m^2) /
surface area of external wall structures (m^2) = 0.286
surface area of external walls (m^2) /
surface area of external wall structures (m^2) = 0.714

code	designation	quantity plus unit	costs per unit	costs total	costs per m^2	%
A	location facilities	m^2				
B	ground floors	m^2				
C	upper floors	m^2				
D	roof 'floors'	m^2				
E	external wall structures	m^2	250	200,00		
F	internal wall structures	m^2				
G	transport facilities	each				
	building works					

Figure 9. Part of an element-group estimate

If the client accepts the element-group estimate, he also agrees with the amount allowed for the external wall structures, NLG 200. This is the budget for the external wall structures. Using the

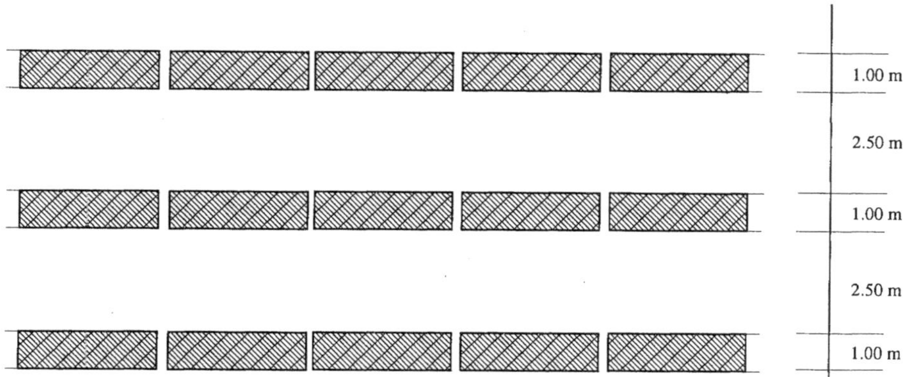

external walls	250 m² at NLG 150 =	NLG 37,500
external wall openings	100 m² at NLG 500 =	NLG 50,000
total	350 m²	NLG 87,500

costs per m² of frontage: NLG 250

Figure 10. Fragment of a frontage

characteristic quantity figures, it can then be calculated how much of that amount applies to external wall openings and how much to external walls: 0.286 x 800 = 229 m^2 external wall openings; 0.714 x 800 = 571 m^2 external walls. A 'translation' has thus been carried out in accordance with the model in Figure 11. The 800 m^2 of external wall structures has been 'translated' into 229 m^2 of external wall openings and 571 m^2 of external wall.

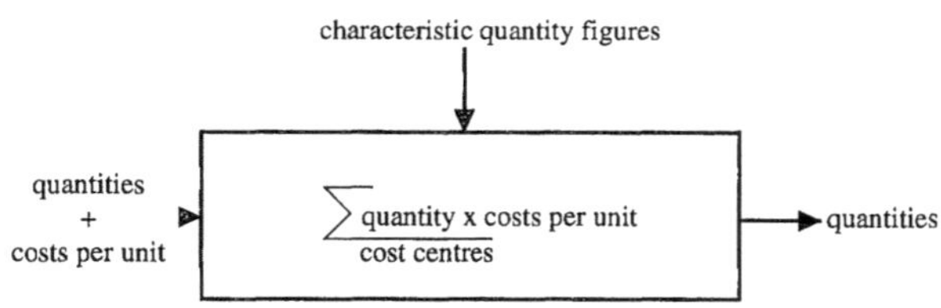

Figure 11. 'Translation' using characteristic quantity figures

The results of 'translation' can be regarded as one of the countless ways to stay within the budget. This is, however, not an arbitrary solution but one which is based on data derived from the analysis of a project. Therefore, this result is a goal to aim for.

5. Characteristic Quantity Figures and Form Factors

The distinction between estimation and 'translation' demonstrates that they are different activities. In the estimating process, the quantities of the cost centres are expressed in monetary terms. In other words, "what is the building sum for the design?" 'Translation' determines the quantities of the cost centres which have to be incorporated in the next step in the design process. The result of estimation - i.e., the building sum - is compared in this process with a previously set budget. After approval, it serves as a new, possibly adjusted, but in any case more detailed budget.

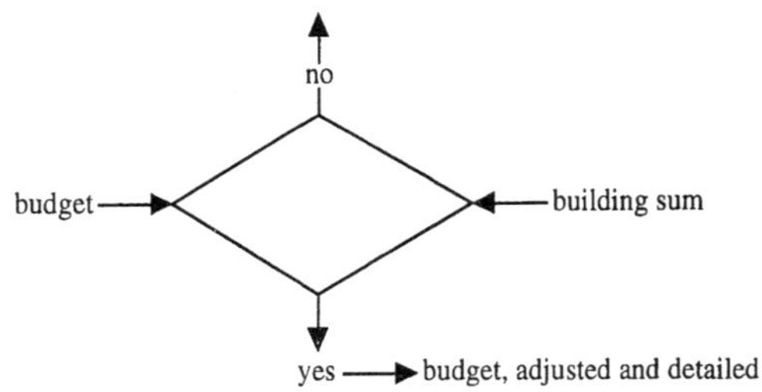

Figure 12. From building sum to budget

In practice, estimation and 'translation' cannot be performed separately. These activities have to be regarded as 'Siamese twins'. Form factors are used in making an estimate in order to check the cost per unit. The characteristic quantity figures are used in 'translation' for a further breakdown of the quantities. Sometimes, if the form factors cannot be verified, estimation yields not just a building sum but also quantities.

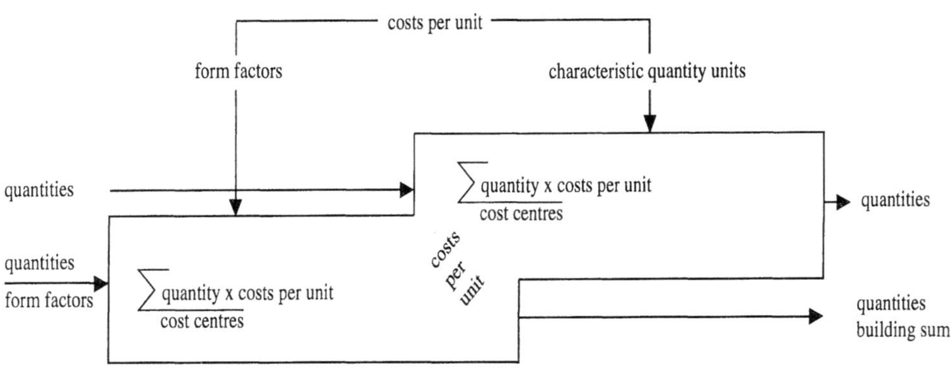

Figure 13. Estimation and translation combined

Jointly, estimation and 'translation' play a significant role in the m^2 estimate, the element-group estimate, and the element estimate. Different characteristic figures and form factors are important for each estimate. The ratios of the gross and net floor space in Figure 14 are significant for the m^2 estimate at the start of the master-plan phase. In preparing the element-group estimate, after the completion of the master plan, the characteristic figures and factors with the element groups in the denominator are important. The characteristic quantity figures with which the outside and inside wall openings can be quantified are also used (Figure 15).

Finally, the element estimate is associated with a large number of form factors. The elements and varying element groups which form part of the building work are expressed in m^2, pieces (number) or linear metres. For each building component with m^2 as its unit, the ratio *peripher* (m^1) */ surface area* (m^2) is used as the form factor. This factor is important for the following

cost centre	characteristic quantity figure	form factor
gross floor space	net floor space (m^2) / gross floor space (m^2)	location facilities (m^2) / gross floor space (m^2)
	structural surface (m^2) / gross floor space (m^2)	roof 'floors' (m^2) / gross floor space (m^2)
	ground floor (m^2) / gross floor space (m^2)	external- and internal wall structures (m^2) / gross floor space (m^2)
	upper floor space (m^2) / gross floor space (m^2)	transport facilities (no.) / gross floor space (m^2)
		heating- and cooling facilities (kW) / gross floor space (m^2)
		liquid and drainage facilities (no.) / gross floor space (m^2)
		lighting systems (kVA) / gross floor space (m^2)
net floor space	traffic area (m^2) / net floor space (m^2)	
	services area (m^2) / net floor space (m^2)	
	'lost' surface area (m^2) / net floor space (m^2)	
	useful surface area (m^2) / net floor space (m^2)	

Figure14. Characteristic quantity figures and form factors for the m^2 estimate

element or varying element groups: ground facilities (dirt), ground facilities (water), floor beds, external walls, internal walls, ground floors, upper floors, roofs, external wall openings, windows in external wall openings, doors in external wall openings, fixed external wall openings, internal wall openings, windows in internal wall openings, doors in internal wall openings, fixed internal wall openings, roof openings, floor finishes (ground floor), floor finishes (upper floors), ceiling finishes (under floors), ceiling finishes (under roofs), and roof finishes. For building components which are quantified on a piece basis, the ratio *length m^1 / quantity (no.)* is the form factor. This applies to pile foundations, main load-bearing structures (columns), stair structures, and transport. The building components with m^1 (linear metres) as their unit have no form factor.

The installation parts in the element estimate are expressed in units such as kW, m^3/h, kVA and pieces. The first three of these are associated with the elements for generating heat, cold, air, and possibly electricity. These are concentrated in one or two appliances. Form factors play scarcely any part in the cost per unit of those appliances. However, the building elements which are stated in units (off), i.e., the distribution parts of these appliances, always consist of two parts: pipes and outlet points.

For the cost per unit, the ratio *piping (m^1) / outlet points (off)* is extremely important. This ratio, which has already been described for the building elements expressed in units (off), applies to outlets (rain water), outlets (waste water), outlets (solid waste), water, gases, cold-air distribution, heat distribution, ventilation (distribution), and lighting. 'Translation' plays a considerably smaller role in the element estimate than in the element-group estimate. It does not determine quantities of building components at a lower level than estimated. The results can be described as quantities of building materials.

In principle, 'translation' plays an important role for all building components comprising more than one fixed combination of building materials. In particular, the elements subsumed under the

cost centre	characteristic quantity figure	form factor
location facilities	ground facilities; soil (m^2) / location facilities (m^2)	periphery at normal level (m^1) / location facilities (m^2)
	ground facilities; water (m^2) / location facilities (m^2)	pile foundations (no.) / location facilities (m^2)
ground floors	floor beds (m^2) / ground floors (m^2)	foundation beams (m^1) / ground floors (m^2)
	floors (m^2) / ground floors (m^2)	column bases (no.) / ground floors (m^2)
	floor finishes; ground floor (m^2) / ground floors (m^2)	
upper floors	floor finishes; upper floors (m^2) / upper floors (m^2)	beams under floors (m^1) / upper floors (m^2)
	ceiling finishes; under floors (m^2) / upper floors (m^2)	columns under floors (st) / upper floors (m^2)
roof 'floors'	roofs (m^2) / roof 'floors' (m^2)	beams under roofs (m^1) / roof 'floors' (m^2)
	roof openings (m^2) / roof 'floors' (m^2)	columns under roofs (st) / roof 'floors' (m^2)
	ceiling finishes; under roofs (m^2) / roof 'floors' (m^2)	
	roof finishes (m^2) / roof 'floors' (m^2)	
external wall structures	external walls (m^2) / external wall structures (m^2)	conn. external wall structures (m^1) / external wall structures (m^2)
	external wall openings (m^2) / external wall structures (m^2)	
internal wall structures	internal walls (m^2) / internal wall structures (m^2)	conn. internal wall structures (m^1) / internal wall structures (m^2)
	internal wall openings (m^2) / internal wall structures (m^2)	
transport facilities	stair structures (no.) / transport facilities (no.)	transport facilities (fl) / transport facilities (no.)
	transport (no.) / transport facilities (no.)	
heating and cooling facilities	heat systems (kW) / heating and cooling facilities (kW)	heat distribution (no.) / heat generation (kW)
	cooling systems (kW) / heating and cooling facilities (kW)	cold-air distribution (no.) / cold generation (kW)
		ventilation (m^3/h) / heating and cooling facilities (kW)
		ventilation; distribution (no.) / ventilation; generation (m^3/h)
liquid and drainage facilities	liquid facilities (no.) / liquid and drainage facilities (no.)	
	drainage facilities (no.) / liquid and drainage facilities (no.)	
lighting systems	lighting power (kVA) / lighting systems (kVA)	lighting (no.) / central electr. supplies (kVA)
	power current (kVA) / lighting systems (kVA)	
external wall openings	windows in external wall openings (m^2) / external wall openings (m^2)	
	doors in external wall openings (m^2) / external wall openings (m^2)	
	fixed external wall openings (m^2) / external wall openings (m^2)	
internal wall openings	windows in internal wall openings (m^2) / internal wall openings (m^2)	
	doors in internal wall openings (m^2) / internal wall openings (m^2)	
	fixed internal wall openings (m^2) / internal wall openings (m^2)	

Figure 15. Characteristic quantity figures and form factors for element-group estimation

heading of finishing are 'translated' into the share that each of thebuilding materials(e.g.,linoleum tiles and project carpet) has in the total (the floor finishing). The varying element groups for the outside and inside wall openings may consist of different building products (e.g., side-hung windows and casement-stay windows or fire-resistant and normal doors). In that case, they require a 'translation'.

With building components, it is still possible to give a concrete description of the characteristic quantity figures. This is no longer possible when we pass from building elements to building components. The number of building elements is limited, and they are more or less accurately classified. The number of building materials, on the other hand, is practically unlimited. Even the low number of building components which are still 'translated' in this phase will lead to an unacceptably long list of characteristic quantity figures. Standard characteristic values cannot be developed for the element estimate.

Estimation and 'translation' together build a bridge between the subsequent steps in the design process. The budget is regularly tested and given greater detail. At the same time, information is obtained in the form of a possible solution for the following step. In this process, characteristic quantity figures and form factors are important management data. As the estimates / 'translations' become more detailed, the number of characteristic values and factors increases. Up to and including the element estimate, the standard ratios do not depend on the individual project. The values associated with these ratios can be determined from analyses of projects that are considered 'right'. Non-standardized characteristic quantity figures can still play a role in the 'translation' element. These figures can only be obtained from one single comparable project and are therefore somewhat less reliable. At that point, however, we are at the end of the process. There - as indicated in Figure 2 - the influence on the total cost is slight. The use of less precise values will then certainly not lead to failure of the cost-control exercise.

References

Bathurst, P.E. and Butler, D.A. (1980) *Building Cost Control Techniques and Economics*, Heinemann, London.

BNA (1988) SR 1988, Amsterdam.

Haar, E. ter (1991) *Gebouwkostenkennis - Een Analyse van Instrumenten voor de Vroege Planfasen,* Delftse Universitaire Pers, Delft.

Naar een Ordening op de Bouwmarkt (1982) Government Publishing Office, The Hague.

NEN 2630 (1979) Oppervlakten en Inhouden van Gebouwen, Delft.

NEN 2631 (1979) Investeringskosten van Gebouwen, Delft.

NEN 2632 (1980) Exploitatiekosten van Gebouwen, Delft.

Poortman, E.R. (1990) *Budgetbewaking in de Bouw*, Samsom, Alphen a/d Rijn.

Raftery, J. (1987) "The State of Cost/Price Modelling in the UK Construction Industry: A Multicriteria Approach", in P.S. Brandon (ed.), *Building Cost Modelling and Computers: The Transactions of the Building Cost Research Conference*, E.&F.N. Spon, London, pp. 49-71.

Schaefer, W.F. and Erkelens, P.A. (1988) Structurering en Verwerking van Tijdgegevens voor de Uitvoering van Bouwwerken, Bouwstenen 9, Faculty of Building and Architecture, Eindhoven University of Technology, Eindhoven.

Skitmore, M. (1992) "Developments in Contract Price Forecasting and Bidding Techniques", in P.S. Brandon (ed.), *Quantity Surveying Techniques: New Directions*, Blackwell Scientific Publications, Oxford, pp. 75-120.

UGCB et al. (1985) Elementenmethode 1985, UGCB, Ede.

Uniform Prijsregelend Reglement (1988) Association of Cooperating Price-Control Organisations in the Building Industry.

Appendix 1

Specification of building costs (NLG), ref. date April 1, 1989

(32) Inside-wall openings

Code	Designation	Unit	Quan-tity	Man-hours per unit	Man-hours total	Costs per unit	Costs total
(32)011	**Inside-wall openings, inside doors, veneered doors, oak, stub, up to 25 off**						
	LABOUR						
	Hanging inside door, stub	piece	1	0.850	0.850	32.21	32.21
	Fitting closers+metalwork	piece	1	0.250	0.250	9.48	9.48
	Subtotal	piece	1		1.100		41.69
	MATERIAL						
	Hinge, st.st., 76x76mm	piece	1			5.04	5.04
	Latch, brass, front plate	piece	3			13.03	13.03
	......	piece	1			4.47	4.47
	....	piece	1			8.49	8.49
	......	piece	1			133.65	133.65
	....	piece	1			5.65	5.65
	Subtotal	piece	1				180.41
	SUBCONTRACTORS						
	Top coat sealed door	piece	1			31.30	31.30
	Subtotal	piece	1				31.30
	TOTAL						253.40
	COSTS PER UNIT	piece				253.40	

COSTS (NLG)				reference date May 1985			
code	designation	quantity per unit		costs per unit	costs total	costs per m²	%
(11)	ground facilities	867	m^2	17.01	14,749	5.43	0.6
(13)	floor beds	675	m^2	53.93	36,405	13.39	1.4
(16)	foundation	290	m^1	176.98	51,323	18.88	2.0
(17)	pile foundations	150	st	157.18	23,577	8.67	0.9
(1-)	**GROUND SUBSTRUCTURE**				**126,054**	**46.38**	**5.0**
(21)	external walls	1351	m^2	188.73	254,979	93.81	10.1
(22)	internal walls	2243	m^2	161.29	361,767	133.10	14.3
(23)	floors; ground floor	165	m^2	52.61	8,680	3.19	0.3
(23)	floors; upper floor	1788	m^2	117.61	210,282	77.37	8.3
(24)	stairs, ramps	3	st	14025.67	42,077	15.48	1.7
(27)	roofs	844	m^2	121.81	102,808	37.82	4.1
(28)	main-load bearing structure	388	st	170.88	66,300	24.39	2.6
(2-)	**STRUCTURE**				**1,046,893**	**385.17**	**41.5**
(31)	external wall openings	502	m^2	603.52	302,965	111.47	12.0
(32)	internal wall openings	246	m^2	444.94	109,456	40.27	4.3
(33)	floor openings		m^2				
(34)	balustrade and banister	76	m^1	167.76	12,750	4.69	0.5
(37)	roof openings		m^2				
(38)							
(3-)	**COMPLETION**				**425,171**	**156.43**	**16.9**
(41)	external wall finishes		m^2				
(42)	internal wall finishes		m^2				
(43)	floor finishes	2426	m^2	47.71	115,741	42.58	4.6
(44)	stair finishes		m^2				
(45)	ceiling finishes	2421	m^2	45.74	110,746	40.75	4.4
(47)	roof finishes	844	m^2	59.01	49,801	18.32	2.0
(48)							
(4-)	**FINISHES**				**276,288**	**101.65**	**11.0**
(51)	service centre	210	kW	109.81	23,060	8.48	0.9
(52)	drainage	44	off	554.14	24,382	8.97	1.0
(53)	liquids supply services	30	off	207.77	6,233	2.29	0.2
(54)	gas supply services	30	m^1	117.17	3,515	1.29	0.1
(55)	space cooling services		kW				
(56)	space heating services	138	off	490.28	67,658	24.89	2.7
(57)	air cond. services	16820	m^3/h	13.95	234,565	86.30	9.3
(58)	control	1		23709.00	23,709	8.72	0.9
(5-)	**SERVICES (MAINLY PIPED AND DUCTED)**				**383,122**	**140.96**	**15.2**
(61)	electrical centre	38	kVA	2263.68	86,020	31.65	3.4
(62)	power distribution services	15	kVA	264.07	3,961	1.46	0.2
(63)	lighting services	392	off	218.12	85,503	31.46	3.4
(64)	communication services	1		32600.00	32,600	11.99	1.3
(65)	security services	1		15442.00	15,442	5.68	0.6
(66)	transport services	1	off	41867.00	41,867	15.40	1.7
(6-)	**SERVICES (MAINLY ELECTRICAL)**				**265,393**	**97.64**	**10.5**
(71)	display, circulation fittings		off				
(72)	rest, work, play fittings		off				
(73)	culinary fittings		off				
(74)	sanitary, hygiene fittings		off				
(75)	cleaning, maintenance fittings		off				
(76)	storage, screening fittings		off				
(7-)	**FITTINGS**						
(-0)	**DIRECT PROJECT FACILITIES**				**2,522,921**	**928.23**	**100.0**
(0-)	**INDIRECT PROJECT FACILITIES**	**25**	**%**		**630,730**	**232.06**	**25.0**
(--)	**COSTS OF THE BUILDING**	**2718**	**m^2**		**3,153,651**	**1160.28**	**125.0**

Appendix 2

COST ANALYSIS	**Project:** Office building in Eindhoven	**CI/SfB-building code** 32
	Useful surface area: 1625 m^2	**Office building**
	This analysis was compiled by : ir. E.R Poortman	**Reference date:** May 1985

BUILDING PARTNERS

Client/management:
Maasveste BV, Eindhoven

Architects:
L.I.J. de Bever and L.J.H.F. de Bever Architecten, Eindhoven

Advisors:
- installations: Raadgevend Technisch Buro Van Heugten BV, Nijmegen
- structures: Ingenieursburo Schiebroek-Struik BV, Eindhoven

Contractor:
Nelissen van Egteren Bouwgroep BV, Venray

Installer:
Eikemans-Nobra BV, Eindhoven

DESCRIPTION OF PROJECT

The office building was planned for a site with a total area of approx. 2000 square metres situated on the ring road in Eindhoven, near the Evoluon. The location was the reason for a "bent shape" with an entrance, vertical transport and general facilities in the bend and two wings with user rooms, each ending in built-in emergency staircases. The wings are 11.40 m deep and consist of a 1.80 m wide passage situated in the middle, with 4.80 m deep rooms on either side. On the second floor one wing is fitted for instruction purposes, with a passage on the frontage and rooms 9.60 m deep. In the longitudinal direction the building is divided into 3.60-m bays, a dimension with is also repeated in the disc-shaped column on the frontage. Within the building, for the sake of ready divisibility, only (round) columns at center distances of 7.20 m have been built. The building consists of three floors and a roof superstructure.

CONTRACT PARTICULARS

Duration of operations:	180 working days
Form of tendering:	single price offer
Risk arrangement:	not applicable
Start of operations:	20th May 1985
Completion date:	April 1986

DRAWINGS

....

....

QUANTITIES

....

DESCRIPTION OF ELEMENTS

....

GRAPHICAL INTERFACING TO A CONCEPTUAL MODEL FOR ESTIMATING THE COST OF RESIDENTIAL CONSTRUCTION

J. GILLEARD
Hong Kong Polytechnic
Kowloon
HongKong

and

O. OLATIDOYE
Georgia Tech
Atlanta, GA 30332
USA

ABSTRACT. This paper presents a method for determining elemental square foot costs and cost significance for residential construction. Using AutoCAD's 'icon menu' and 'dialogue box' facilities, a non-expert may graphically select (i) residential configuration; (ii) construction quality level; (iii) geographical location; (iv) square foot area; and finally, (v) add-ons, e.g. porches and decks, basement, heating and cooling equipment, garages and carports etc. in order to determine on-site builder's costs. Subsequent AutoLisp routines facilitate data transfer to a Lotus 1-2-3 spreadsheet where an elemental cost breakdown for the project may be determined. Finally, using Lotus 1-2-3 macros, computed data is transferred back to AutoCAD, where all cost significant items are graphically highlighted.

1. Introduction

Custom builders are often required to produce 'conceptual' cost estimates of houses for prospective clients. These are usually based on historical figures and presented to the client as a total project cost, e.g. $250,000 at $100 per square foot. Factors such as job complexity, quality, location, add-ons etc., are normally consolidated into the overall job price. Normally, the contractor does not provide a detailed elemental breakdown of these figures to the client. Hence, when changes to the original specifications and blueprints are requested, clients are usually unable to determine the impact of those changes on cost.

However, home builders typically determine the basic building costs by reference to tabulated in-house company records or by recourse to published data. For example, let us assume that the gross floor area of a proposed six-cornered single-family residence is 1567 square feet and that the construction quality level has been specified as 'good'. Then, by reference to, say, the Berger Cost Manual, Table 1, a cost per square foot of $49.46 may be determined, giving a conceptual basic cost to the client of (1567 * $49.46), or $77,504.

In addition, supplementary costs for add-ons, e.g. provision for basements, heating and cooling equipment, garages etc. may also be consolidated into the overall sum cost by incorporating extra line-item costs. Thus the home builder is able to determine a cost estimate for the builder's cost associated with a proposed house. These costs typically include all design fees and construction supervisory expenses, although the land value, builder's profit and additional cost associated with unusual site conditions are typically excluded.

2. Cost Estimating by 'Rules-of-Thumb'

Once the residential contractor has determined a global construction cost for the project, a % elemental breakdown of costs may be produced. These %-cost 'rules-of-thumb' are typically tabulated according to historical company accountancy records (Table 2). When presented in a spreadsheet format, these figures may subsequently be used interrogatively in a 'what-if' mode, i.e. the (%) rule-of-thumb breakdown may be modified in order (i) to assess variations in

H. Timmermans (ed.), Design and Decision Support Systems in Architecture, 137–146.

subcontractor bid prices (+ or -) over anticipated costs, or (ii) to modify specific line items in terms of anticipated cost adjustments. In this way the home builder is able to validate subcontractor prices as well as establish company-specific broad-based elemental cost breakdown records.

TABLE 1. Single-family residences: six corners

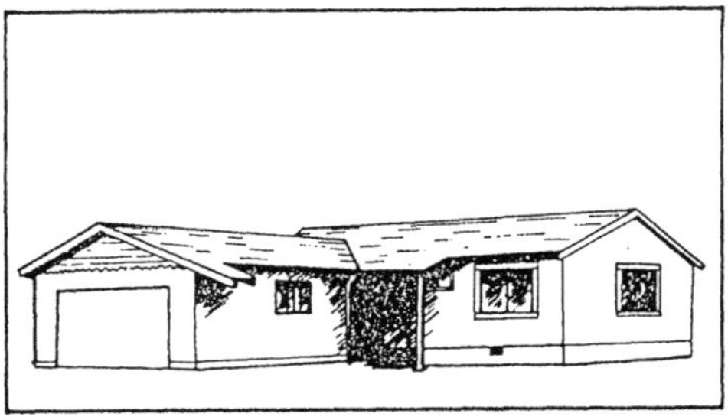

Single Family Residence, Class 4

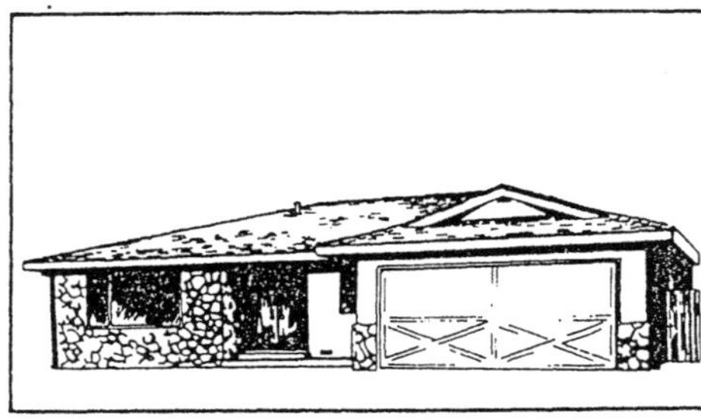

Single Family Residence, Class 2

	Square Foot Area										
Quality Class	700	800	900	1,000	1,100	1,200	1,300	1,400	1,500	1,600	1,700
1, Best	74.03	70.48	67.70	65.44	63.58	62.01	60.67	59.52	58.51	57.62	56.83
1 & 2	68.05	64.79	62.22	60.15	58.44	57.00	55.77	54.71	53.78	52.96	52.23
2, Good	62.58	59.58	57.22	55.32	53.74	52.42	51.29	50.31	49.46	48.70	48.03
2 & 3	57.28	54.54	52.38	50.63	49.19	47.98	46.95	46.05	45.27	44.58	43.97
3, Average	52.26	49.75	47.78	46.19	44.88	43.77	42.83	42.01	41.30	40.67	40.11
3 & 4	47.11	44.86	43.08	41.65	40.46	39.47	38.61	37.88	37.24	36.67	36.17
4, Low	42.47	40.44	38.84	37.54	36.47	35.57	34.81	34.14	33.57	33.05	32.60

With his 'expert knowledge' of construction terminology and wealth of previous experience, the elemental cost breakdown spreadsheet may be relatively simple. However, if the system were to be used by a client, or 'non-expert', modifications to the current system would be necessary. The remainder of the paper describes how these amendments may be made to facilitate greater ease of operation for a 'non-expert' user.

TABLE 2. Elemental cost breakdown 'rules-of-thumb'

Rules-of-Thumb (RoT) Cost Breakdown			
		Unit Cost 100%	Atlanta Cost 79%
ITEM	RoT %	ITEM COST ($)	ATLANTA COST ($)
Excavation	0.70%	$1,610	$1,272
Flatwork (drive & walk)	1.40%	$3,219	$2,543
Foundation, slab, piers	2.10%	$4,829	$3,815
Brick hearth & veneer	0.40%	$920	$727
Rough hardware	0.20%	$460	$363
Finish hardware	0.10%	$230	$182
Rough lumber	4.25%	$9,773	$7,720
Finish lumber	0.25%	$575	$454

Rules-of-Thumb (RoT) Cost Breakdown			
		Unit Cost 100%	Atlanta Cost 79%
ITEM	RoT %	ITEM COST ($)	ATLANTA COST ($)
Rough carpentry labour	4.75%	$10,922	$8,629
Finish carpentry labour	1.00%	$2,299	$1,817
Counter-tops	1.00%	$2,299	$1,817
Cabinets	2.00%	$4,599	$3,633
Insulation (R19 ceiling)	1.30%	$2,989	$2,362
Roofing	3.25%	$7,473	$5,904
Painting	2.50%	$5,749	$4,541
Shower & tub enclosure	0.30%	$690	$545
Prefabricated fireplace	0.15%	$345	$272
Bath accessories	0.20%	$460	$363
Built-in appliances	0.50%	$1,150	$908
Heating and ducting	2.75%	$6,324	$4,996
Plumb. & sewer connect	4.25%	$9,773	$7,720
Doors	1.25%	$2,874	$2,271
Garage door	0.25%	$575	$454
Alum. windows, door	0.75%	$1,725	$1,362
Exterior stucco	3.75%	$8,623	$6,812
Gypsum wallboard	2.50%	$5,749	$4,541
Resilient flooring	1.15%	$2,644	$2,089
Carpeting	1.50%	$3,449	$2,725
Wiring (Romex)	1.75%	$4,024	$3,179
Lighting fixtures	0.75%	$1,725	$1,362
Insurance, payroll tax	1.50%	$3,449	$2,725
Plans and specs	0.25%	$575	$454
Permits & utilities	1.00%	$2,299	$1,817
Final cleanup	0.25%	$575	$454
Total builder percentage	50.00%	$114,973	$90,829
Materials	34.0%	$78,182	$61,764
On-site labour	16.0%	$36,791	$29,065
Land and site improvements	25.0%	$57,487	$45,414
Sales and marketing	5.0%	$11,497	$9,083
Financing	8.0%	$18,396	$14,533
Overhead	7.0%	$16,096	$12,716
Profit	5.0%	$11,497	$9,083
Developer's sale price=	100.0%	$229,946	$181,657

3. AutoCAD to Lotus 1-2-3 Interface

Using AutoCAD as the primary interface to present data choices and to report cost significance provides several key benefits to non-expert users. For instance, (i) input data may be represented graphically; (ii) data transfer may be facilitated by simple AutoLisp routines; (iii) early cost estimates may become a powerful tool in determining likely cost projections for the project; and, finally, (iv) amendments to the building design may be automated.

Some of the steps taken in arriving at the model are enumerated below:

STEP 1. Create AutoCAD drawing files, including associated attribute data, for each residence configuration and add-on.

STEP 2. Use the AutoCAD BLOCK command to SAVE each named drawing file.

STEP 3. Create an ASCII template file that will control the (i) drawing data extraction, (ii) output data format, and (iii) field spacing.

STEP 4. Extract component data from the AutoCAD drawing file.

STEP 5. Integrate the Lotus 1-2-3 spreadsheet program with AutoCAD.

STEP 6. Set up spreadsheet facility to accept data from AutoCAD, import square foot cost data, and compute elemental construction costs in Lotus 1-2-3.

STEP 7. Establish Lotus 1-2-3 macros to automate data manipulation.

STEP 8. Send the spreadsheet data file back to AutoCAD using an AutoLISP routine to facilitate identification of cost-significant items.

This may be illustrated by reference to Figure 1.

4. Creating Icon Menus

4.1. AutoCAD slide libraries

Icon menus help in the selection of the residence configuration, add-ons and other cost-related items. Icon menus use slide libraries that are created by using AutoCAD's SLIDELIB.EXE program. Before invoking SLIDELIB.EXE, the user first makes all the required slides. This is done by invoking AutoCAD's MSLIDE command followed by a file name. Each slide represents a pre-drawn block, e.g. a construction element, equipment, symbol etc. The suffix DWG is automatically added to each file name. Continuing, an ASCII text file is created. The text file lists each slide name on a separate line. Care must be exercised not to include extra <returns> or <spaces> in the text file. Finally, SLIDELIB.EXE program is executed. The resulting slide file will contain a snapshot of the drawing displayed on the AutoCAD drawing screen. Slide files created are automatically given the file extension SLD, e.g. SLIDELIB ST < STYPE creates a slide library called ST from the ASCII list file STYPE.

4.2. Icon libraries

An icon menu is written in a manner similar to other pull-down menus except for its labels. That is, the user needs to load the device with a $I=pagename and then display it with a $I=*. However, icons use the label [ ] to identify the name of the slide to show in the box. Each label corresponds to one label on the screen. These boxes are automatically arranged in groups of 4, 9, and 16. If the label is in the form of libraryname(slidename), then the named slide from the library will be displayed as an icon.

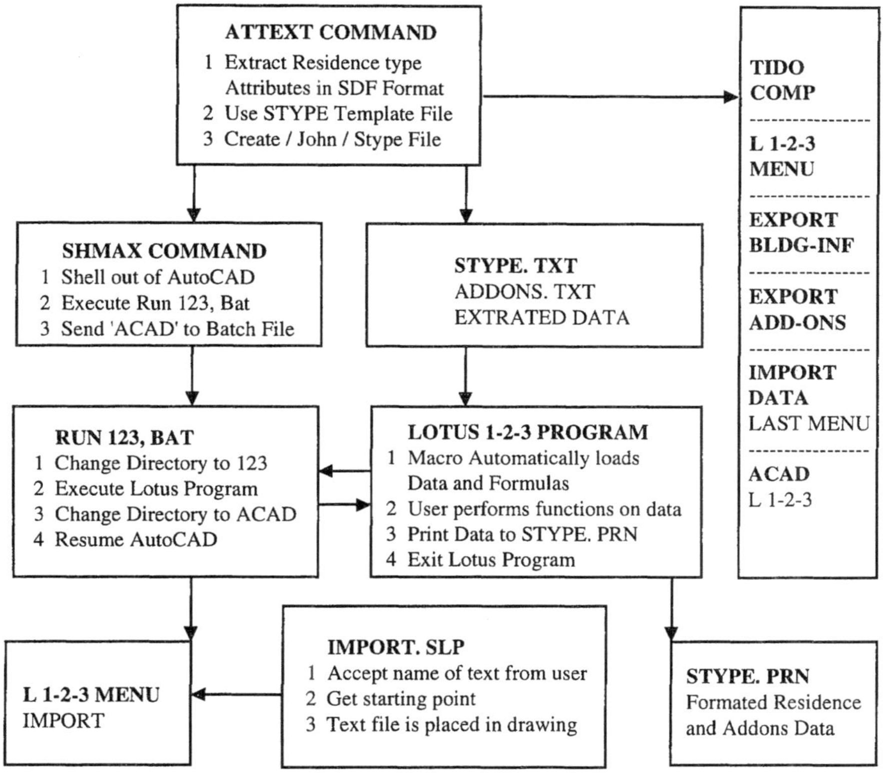

Figure 1. General flow diagram of problem solution

For example, the syntax for displaying ST may be added to ACAD.MNU, e.g.

```
***ICON
**ST
[JOHN WINDOWS]
[st(4-corn)]^c^cinsert 4-corn \;;;zoom e;zoom .9x;^c$s=wspecs
[st(6-corn)]^c^cinsert 6-corn \;;;zoom e;zoom .9x;^c$s=wspecs
[st(8-corn)]^c^cinsert 8-corn \;;;zoom e;zoom .9x;^c$s=wspecs
[st(10-corn)]^c^cinsert 10-corn \;;;zoom e;zoom .9x;^c$s=wspecs
[ close]^c^c$s=
```

For example, Figure 2, displays the residence configuration and add-on icon menus.

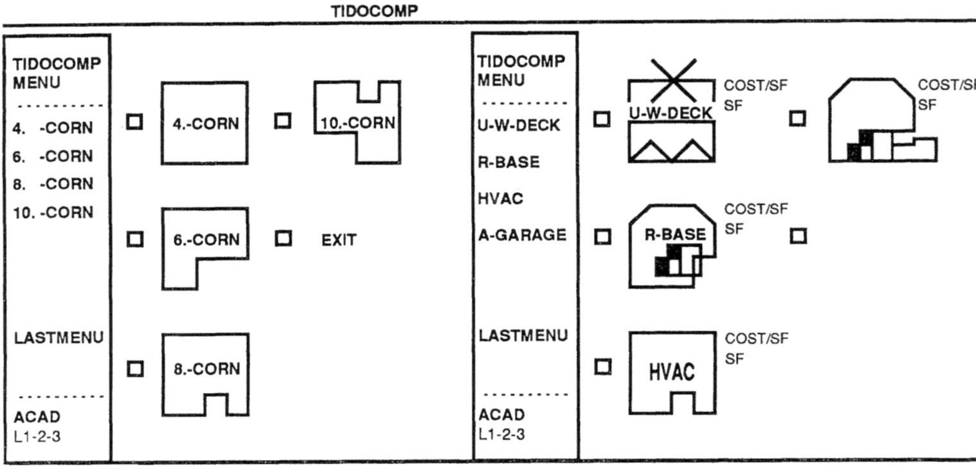

Figure 2. Residential configuration and add-on icon menus

5. AutoCAD Attribute Functions

5.1. DRAWING AND TEXT CHARACTERISTICS

Attributes are special drawing entities that contain text. They help in describing building parts through text information. Attributes may subsequently be extracted from a drawing file and transferred to a separate program, e.g. Lotus 1-2-3. Attribute data may be assigned to any drawing entity provided that it is part of a block. Thus, drawing files may contain useful attribute (e.g. cost and specification) data, information that complements the graphical data. In addition, a complex drawing file may be composed of a number of blocks; each block may also consist of separate blocks.

Attribute data attached to each drawing entity is designed to hold text and numeric data linked to each of the building configurations and add-on blocks constructed. Therefore, each time a component is inserted into a drawing, the user is prompted to add textual information to the component, e.g. construction quality class; anticipated square footage for the proposed house; city cost index. In addition, add-on information is also requested from the user with respect to construction modifications, e.g. provision of basement, porch, garage etc. After each drawing file, or icon, is selected, a dialogue screen will appear. Information requested at each dialogue box prompt may be defaulted, edited, or inserted.

5.2. ATTRIBUTE VARIABLES

To create the list items for the building components, attribute variables were created.
Each of these variables must have a line in the Template File that extracts the information, including the name and its numerical precision, if necessary. The Template Files used to create the building component list are shown below, i.e. files named STYPE.TXT, LOCATION_TXT and ADD_ONS.TXT.

TEMPLATE FILES

I. STYPE.TXT

Q_CLASS   N001000

SPACE 1 C002000
SQFT_AREA N040000

II. ADD_ONS.TXT
COST/SF N005002
SPACE 1 C020000
SF N004000

III. LOCATION.TXT
GEORGIA C009000
FACTORS N003002

The C and N following the field's name extract (N)umerical or (C)haracter information. The next three digits define the number of character spaces for the extracted information. For example, the first three digits (005) after COST/SF means that five Numeric spaces were provided for the cost of the building components. If the component's cost is longer than five numeric values, it is automatically truncated at the sixth character. The last three digits specify the decimal precision for numeric information. All of the numeric information necessary for the building component's file is given in whole numbers. Three zeros are used in this Template file.

The Utility sub-command ATTEXT in AutoCAD extracts attribute information from a drawing. AutoCAD provides three different formats that can be used by database programs, spreadsheets, and text editors. These formats are CDF, SDF, and DXF. If the CDF is picked, AutoCAD produces a standard Comma Delimited File format. That file contains, at most, one record for each block reference in the drawing file. Each field of each record is separated by a comma with character fields enclosed in quotes. However, for this exercise the SDF was picked. A Space Delimited File is created which is the most standard file and is basically a columnar list of information. Lotus 1-2-3 reads this by using the IMPORT command. One record is written here for each block reference. Fields are affixed so that the program knows where one stops and the other one starts. If DXF is selected, AutoCAD creates a subset of the DXF file with only Block References, Attribute, and End of Sequence entities. Generally, DXF file extraction is much more complicated than CDF or SDF, and extracted data may be difficult to interpret.

6. AutoCAD to Lotus 1-2-3 Interface

6.1. AutoLisp routines

Two AutoLisp routines, IMPORT.LSP and EXPORT.LSP, have been written to facilitate data import and export from the AutoCAD drawing file directly to Lotus 1-2-3 and vice versa. Each routine acts as a bridge. During the initial data transfer, basic parametric cost information is shelled into the Lotus 1-2-3 spreadsheet. In order to be able to use the shell capability of AutoCAD, a command was defined in the ACAD.PGP file. That is, the ACAD.PGP file is modified to include the SHMAX shell command, a command that enables access to the Lotus 1-2-3 program from within AutoCAD. The following line may thus be added to the ACAD.PGP file:

```
SHMAX,,480000,*DOS Command: ,4
```

The link between AutoCAD and Lotus 1-2-3 has also been automated with the addition of several Lotus 1-2-3 menu macros. Therefore data import and export to Lotus may be made without direct user input.

7. Sample Problem

7.1. INSIDE AUTOCAD

STEP 1. Pick from the screen menu RES-CONF.
STEP 2. List of drawing icons is displayed.
STEP 3. Pick a house floor plan, e.g. 4-cornered, 6-cornered etc.
STEP 4. Initially, accept all the defaults.
STEP 5. Dialogue box displays data. Data may be edited at this stage.
STEP 6. Select data transfer icon. Data is automatically transferred to Lotus 1-2-3.

7.2. INSIDE LOTUS 1-2-3

STEP 1. Enter / (slash) to obtain main Lotus 1-2-3 menu.
STEP 2. Select FR (File Retrieve), named file BUDGET.WK1.
STEP 3. Press <ALT>I. Lotus 1-2-3 macro that updates the spreadsheet finds the last line of text brought in from the import command; positions the cursor at the range location STYPE and imports the file; a DATA PARSE format string is created; input and output ranges are identified and the data is parsed.
STEP 4. Add-on data allocated.
STEP 5. Cost-significant items determined.
STEP 6. Print spreadsheet.
STEP 7. SQ (Save and Quit)

7.3. RETURN TO AUTOCAD

STEP 1. Select [IMPORT DATA] from custom side menu.
STEP 2. From prompt, select name of Lotus 1-2-3 file to import.
STEP 3. Review data, and repeat iteration as necessary.

Table 3 shows the elemental cost breakdown returned to AutoCAD with each element highlighted either as a significant or non-significant cost item.

TABLE 3. Significant and non-significant cost elements

ITEM		
Excavation	$1,272	Non-significant
Flatwork (drive & walk)	$2,543	Non-significant
Foundation, slab, piers	$3,815	Significant
Brick hearth & veneer	$727	Non-significant
Rough hardware	$363	Non-significant
Finish hardware	$182	Non-significant
Rough lumber	$7,720	Significant
Finish lumber	$454	Non-significant
Rough carpentry labour	$8,629	Significant
Finish carpentry labour	$1,817	Non-significant
Counter-tops	$1,817	Non-significant
Cabinets	$3,633	Significant
Insulation (R19 ceiling)	$2,362	Non-significant

ITEM		
Roofing	$5,904	Significant
Painting	$4,541	Significant
Shower & tub enclosure	$545	Non-significant
Prefabricated fireplace	$272	Non-significant
Bath accessories	$363	Non-significant
Built-in appliances	$908	Non-significant
Heating and ducting	$4,996	Significant
Plumb. & sewer connect	$7,720	Significant
Doors	$2,271	Non-significant
Garage door	$454	Non-significant
Alum. windows, door	$1,362	Non-significant
Exterior stucco	$6,812	Significant
Gypsum wallboard	$4,541	Significant
Resilient flooring	$2,089	Non-significant
Carpeting	$2,725	Non-significant
Wiring (Romex)	$3,179	Significant
Lighting fixtures	$1,362	Non-significant
Insurance, payroll tax	$2,725	Significant
Plans and specs	$454	Non-significant
Permits & utilities	$1,817	Non-significant
Final cleanup	$454	Non-significant
Total builder percentage	$90,829	

8. Discussion

8.1. EXPERT KNOWLEDGE

No two computer-aided drafting (CADr) programs are identical, and their capabilities and ease of operation vary greatly. However, ease of use can be improved with most CADr packages by customizing the user interface. This can be done by modifying the manner in which commands are accessed to the program. This has been clearly demonstrated with the conceptual estimating routines described above. By modifying the AutoCAD default screen and pull-down menus, as well as introducing custom icon menus and dialogue boxes, non-expert CAD users are able to work interactively with a relatively complex cost model.

The user first selects from the floor plan icon menu the building layout that characterizes the proposed house, e.g. four-cornered, six-cornered etc. A dialogue box requests the user to specify the overall gross floor area and construction quality for the dwelling. Subsequently, the user is required to specify all add-ons. Finally, the user simply selects the 'export' data icon. The data is then transferred to Lotus 1-2-3. Lotus 1-2-3 macros are invoked, initially setting a @Lookup operation to determine the square foot cost given the house complexity (in terms of configuration) and specified construction quality. Continuing, the elemental cost breakdown routine is activated, thus enabling a conceptual cost breakdown for the project to be determined. Lastly, cost significance is established. The rules for cost significance are relatively simple, i.e. the mean cost of all construction elements is calculated. If the sum cost of each construction element is determined to be greater than the mean value, then that element is deemed to be cost-significant. This information is then transferred back to AutoCAD, where a listing of significant and non-significant cost items are depicted on the screen.

The cost model described in the paper is not complex. However, it does demonstrate the advantages of linking compatible software. Further work in the field is under investigation in order to (i) allow for different construction forms, e.g. commercial, industrial etc.; (ii) accommodate actual CAD construction plans (other than the current simple icon configurations); and (iii) link with more complex cost estimation models.

References

Gibson, S. (ed.), (*1990) Berger Cost Manual*, Craftsman Books, Hongkong.

A DESIGN DECISION SUPPORT SYSTEM FOR BUILDING FLEXIBILITY AND COSTS

M. PRINS, M.F.TH. BAX, J.C. CARP and H. TEMPELMANS PLAT
Eindhoven University of Technology
P.O. Box 513
5600 MB Eindhoven
The Netherlands

ABSTRACT. Because of possible changes in demand, buildings must have some flexibility. In this paper a building model, a financial-economic model and a process model will be presented, which together constitute a design decision support system. This system may be used to decide on flexibility and costs of building variants in all phases of the design process.

1. Introduction

Quality control is becoming increasingly important in the face of declining quantitative demands and rising qualitative demands for differentiation in buildings. This shift is occurring just when the available financial means are shrinking. Investments have to be put to better use. The quality of a building has to be monitored over a long period of time. And the building must be able to adapt to changing needs. To decide between new construction and reconstruction, a designer has to be able to assess the lifetime of a building. The designer will then be able to analyze the effect of long-term design decisions. A flexible building system is only useful if it is recognized that buildings have a time dimension.

The objective of this research project is to develop a decision support system which can help the designer weigh the financial-economic and physical/spatial factors in a decision-making process. This system can be used as a design tool to facilitate decisions regarding the programming, design, construction and management of flexible buildings. In particular, we are concerned with support structures having infill elements with differing lifetimes, based on dwelling performance and costs. To realize this objective, we have to quantitatively define flexible buildings for decision-making purposes.

2. Flexibility

Flexibility mainly concerns the time dimension of buildings. In fact, the time dimension (which does justice to the building dynamics) distinguishes this study from many others in the field of life-cycle costing. The time dimension involves not only the durability of a building but also the changes that take place during its lifetime. Buildings have both a duration and a tempo of change. The building is described independently of the levels of perception. Thus, at least in formal terms, the situation is identical at every level. This improves the general applicability of our system. The system is defined in terms of form, function and process; moreover, in accordance with the research objective, it is also defined in terms of time and cost. We define the subject of our research as follows: a building - alternatively, parts of a building (building elements), or groups of parts of a building, where there is a formal, functional, or temporal relationship between these parts - manifesting itself in reality (empirically), in an abstract representation (a

H. Timmermans (ed.), Design and Decision Support Systems in Architecture, 147–163.

model or an image) or in a design (an instruction), which can be expressed in terms of time and cost.

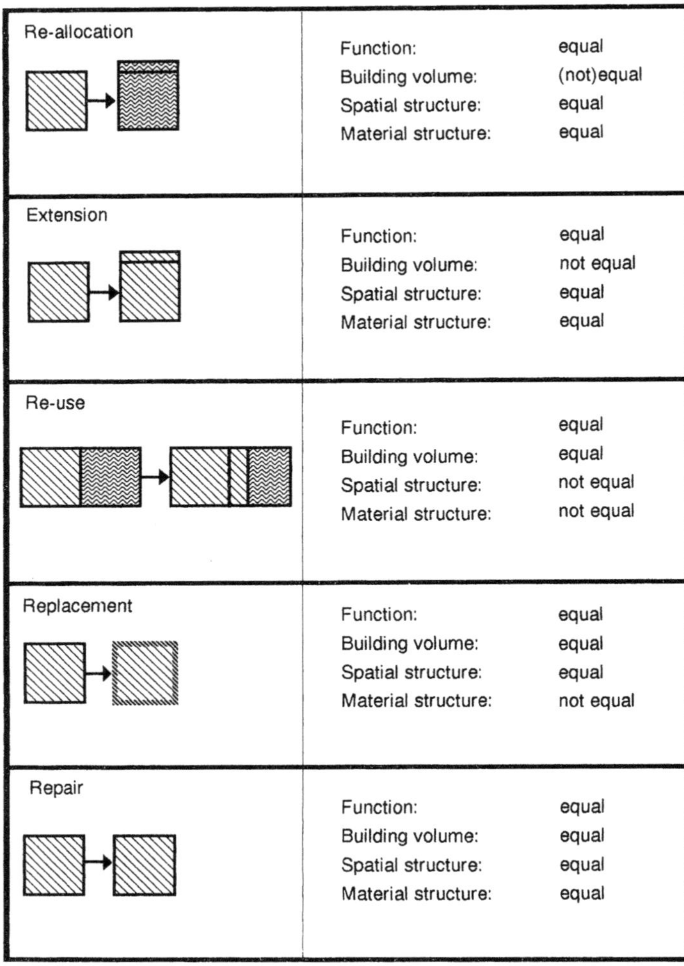

Figure 1. Five flexibility types

The concept of flexibility is limited to the ability of buildings to adapt themselves to events (either incidental or intermittent) whose statistical probability can be (more or less) modelled. The (boundary) constraints given in the definition refer to the limitations needed to make the term operational so that measurements and calculations can be carried out. Flexibility can be defined as the characteristic of a building which makes it able, on the basis of its physical composition (its spatial/material characteristics), to adapt or modify itself to changes (its use-oriented/functional and spatial/material characteristics) - which can be characterized as incidental or intermittent events - within technical, legal, temporal, and financial constraints (Brinkman 1989; Prins 1989). This study is concerned with the flexibility to be achieved by changing the material elements of a building. Thus, multifunctionality as a specific form of flexibility is not discussed here. The diversity in the supply of building types, and the consequent capacity to respond to changing demand on the building market, is also omitted from this discussion (Tempelmans Plat 1991). Flexibility can be classified into five types, according to the demand for change, by assigning the values 'changes' or 'does not change' to the following parameters: function, building volume, spatial structure and material structure. These five types are the result of operations which can be quantified in terms of material, labour and research tools. These operations form the foundation of our method. We note that this classification is meant to be complete.

We define flexibility types according to the kind of flexibility which is determined by the kind of change (its use-oriented/functional and spatial/material characteristics), as expressed in the following types: re-allocation, extension, re-use, replacement and repair. The content of the five flexibility types distinguished here is summarized in Figure 1. In this figure, each flexibility type is characterized by an icon and a description of the amount of change. The right column identifies which of the given parameters change. The two rectangles in the left column represent a building before and after a change. The surface of the rectangles represents the building volume. The division of the rectangles represents the spatial structure. The lines which form the rectangles represent the material structure. And the filled-in surface of the rectangles represents the function.

Flexibility operations are those leading to the various flexibility types. For the purposes of determining the resulting expenditures, these can be classified as follows: retaining the structure as is, repositioning, upgrading, equivalent replacement, non-equivalent replacement, or various combinations of the above. In Figure 2 the content of these operations is summarized. This figure depicts a building on a time axis on which three moments are distinguished. The filling of the surface represents a service level. A change in form represents replacement; a change in place denotes a non-equivalent replacement.

In our research, the most general statements about change pertain to the flexibility scenario; the most detailed statements refer to flexibility operations. Both of these involve potential events which occur while the building is in use and which initiate the demand for change. Our research covers only those changes that can be predicted and therefore modelled. To define a scenario, we first have to understand what a 'period' is. A period is a time interval between two moments in a scenario and a building's time scale. It is the time span in which the events occurring while a building is in use may be expected to initiate a demand for change. That, in turn, will create a demand for decisions about change to be made in the building.

A flexibility scenario can now be defined as a part of the program of requirements for a building. This part refers to the desired flexibility and predicts future incidental or intermittent events. In this part, the total lifetime of a building is classified into periods on a time scale. Those periods are related to the events initiated by change. The flexibility scenario is part of the planning approach aimed at combining the social, economic and physical plans. The physical component is initially small but gradually gains in content in subsequent stages of the scenarios. During the design phase, the flexibility scenario is developed into the flexibility demand. Both the flexibility scenario and the flexibility demand are placed on a scale, moving from a general program of requirements to concrete performance specifications.

The flexibility supply of a building is the opposite of its flexibility demand. The supply cannot be calculated without first ascertaining the demand. After the demand is known, it can be assessed whether, or to what extent, there is a response to the demand. This response is determined by the intrinsic characteristics of the building: the elements of which it is composed, their lifetimes, and the relationship between the two. The flexibility demand of a building is the articulated need for flexibility. It is expressed as the physical implications which a flexibility scenario has on a specific building. It is also expressed as a series of design alternatives classified according time series broken down into periods. These expressions can be translated directly into flexibility operations (Prins and Bax 1992). The flexibility supply of a building consists of the characteristics of a building in physical (material) and legal terms in as much as they answer the flexibility demand.

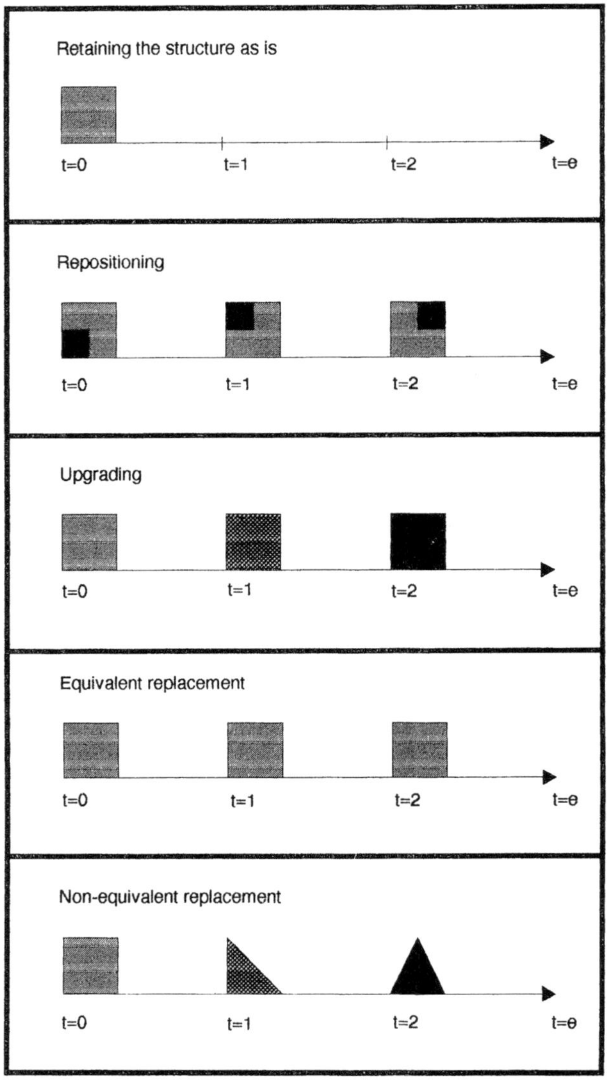

Figure 2. Five types of flexibility operations

Buildings and building elements decay in the course of time under the influence of use-oriented and contextual factors. Thus, they no longer meet the original performance specifications. The technical lifetime is the time interval in which a building is expected to meet stated performance specifications. These are based on its intrinsic, initially physical and material characteristics, in which specific maintenance factors are taken into account. The technical lifetime of materials and products can be determined objectively by empirical means. The technical lifetime does not depend on the scenario. The (economic) lifetime is the time interval in which a building is expected to function in an economic sense (maximally equal to the technical lifetime), measured in periods according to the flexibility scenario. The economic lifetime is determined by deciding whether the element is still able to perform its function in an economic sense.

It is possible for an element to be replaced before its technical lifetime is over, for reasons related to its use. In this paper, we consider economic lifetime and use-oriented lifetime to be synonymous. Whenever we refer to 'lifetime' (without using an adjective), we mean economic lifetime. The economic lifetime does depend on the scenario. The technical and economic lifetimes determine when it is time to decide whether or not to adapt a building. A decision moment is any moment marking the beginning or end of a period in which adaptions have to be made to a building or its elements. An element may have to be replaced when it reaches the end of its technical or economic lifetime. A building may have to be adapted to a changing demand for dwelling performance.

3. Building-Oriented Model

In this section, we describe buildings in terms of systems and models. We emphasize their ability to change, i.e. their flexibility. In even the simplest definition of systems, we apply a hierarchical classification of sub-systems, which are also denoted as elements. In general systems theory, an object (and therefore also a building) is seen in terms of the interplay between three partial systems: a hierarchical system, an aspect-oriented system and a phase-oriented system (Leeuwen 1976). In other words, an element in a building can belong to all three systems at the same time. Each of the three systems describes, from its unique viewpoint, the building in its entirety.

We define a building-oriented system as the way of viewing a building as a system apart from its context (system boundary) as consisting of individual elements, each having their own functionality, which are characterized by their internal and external relationships. Each element is simultaneously part of three partial systems: a hierarchical system, an aspect-oriented system and a phase-oriented system. In Figure 3 these dimensions are shown in a three-dimensional space in which a building can be defined. In line with our research objective, we develop the building-oriented system into a flexible building-oriented system in three steps. Each step illustrates a special relationship between the three partial systems.

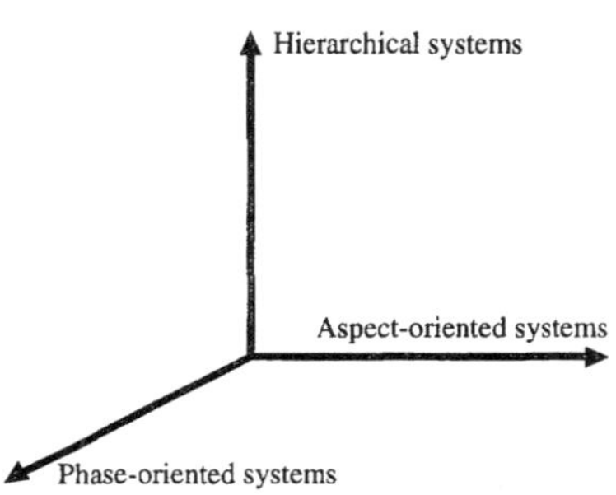

Figure 3. Three dimensions in which a building can be defined in (partial) systems

Step 1: A special relationship between the aspect-oriented system and the hierarchical system
In the aspect-oriented system, the building is classified according to a number of objective-oriented ways of looking at the building. In this research into flexibility, aspect-oriented systems are determined by the categories of changes that may occur in the building. These potential changes are reduced to the sources of these changes. Because of the (axiomatic) definition of flexibility, these sources are defined in terms of their material/technical, spatial/functional, and

social/legal characteristics. Consequently, we can distinguish three types of aspect-oriented systems: technically oriented (aspect-oriented) systems, use-oriented (aspect-oriented) systems and socially oriented (aspect-oriented) systems. Theoretically, the boundaries of the elements on which the three aspect-oriented systems are based will not coincide. This can be seen in the order of magnitude of the elements per aspect-oriented system and consequently in the number of elements needed per system to describe the entire building. Because of the clustering of elements in the various aspect-oriented systems, we can theoretically divide these into the following three cases: (i) interdependence (the number of elements in an aspect-oriented system is not equal to the number in another aspect-oriented system, and elements in aspect-oriented systems cannot be expressed as clusters of elements in other aspect-oriented systems), (ii) unity (the number of elements in an aspect-oriented system is not equal to the number in another aspect-oriented system, but elements in aspect-oriented systems can be expressed as clusters of elements in other aspect-oriented systems), and (iii) similarity (the number of elements in an aspect-oriented system is equal to the number in another aspect-oriented system, and elements in aspect-oriented systems can be expressed as elements in other aspect-oriented systems).

These three cases are shown in Figure 4. The two rectangles represent an aspect-system of a building. The division of the two rectangles represents the division in elements within a specific aspect-system.

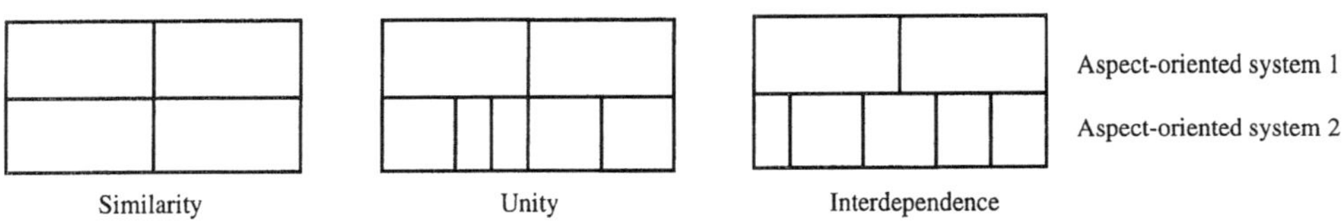

Figure 4. Three cases for the clustering of elements between aspect-oriented systems

In actual building practice, the typical situation is that of interdependence; unity and similarity only occur when flexibility has been the implicit or explicit objective in the design and building phase. When there is unity or similarity between all aspect-oriented systems, a situation arises in which a socially oriented system consists of a whole number of elements in the use-oriented system, which in turn consists of a whole number of elements in the technically oriented system. In that case, we can state that a socially oriented system arises out of an aggregation of elements in the use-oriented system. And we can also state that the use-oriented system arises out of the aggregation of elements in the technically oriented system.

Although we are dealing with aspect-oriented systems here, this description clearly resembles a level-based arrangement that is characteristic of hierarchical systems. We will make use of this characteristic to express the above-mentioned special relationship between aspect-oriented and hierarchical systems. When there is unity or similarity between the three aspect-oriented systems, the aspect-oriented classifications may correspond to the levels in hierarchical systems. We refer to these levels as aggregation levels. Figure 5 depicts a two-dimensional space in which aspect-oriented systems respectively aggregation levels are set along the axes. A point in this graph represents a system which belongs to an aggregation level as well as to an aspect-system. In the special case of unity or similarity, all these points can be found on the diagonal.

Step 2: A special relationship between two types of hierarchical systems

We make use of domain theory to interpret the three-dimensional system-theory model with its three partial systems in terms of decision-making. This means that the three dimensions maintain their own character even though they are interrelated. In this case, the building is also viewed in terms of the interplay between three types of systems, namely formal, functional and procedural systems (Trum and Bax 1990; Bax 1979). Here, too, a description is given of the building in its

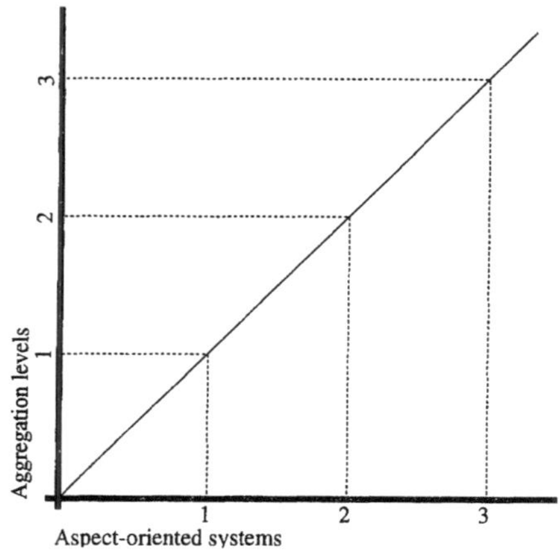

Figure 5. Aspect-oriented systems and aggregation levels

entirety at each level. The formal system is arranged hierarchically into levels of magnitude. The functional system involves the performance specifications of the building in terms of use, technique and construction (domains). The procedural system involves the status of the system, which follows from the position in the design and/or decision-making process. The position in that process is determined in part by those persons authorized to make decisions (social/political factors). Figure 6 displays this so-called GOM model. (The name 'GOM' is derived from the Dutch name of the 'Design Methods Group' at Eindhoven University of Technology, by whom this model was developed.)

The formal system consists of generalization or resolution levels, characterized in part by the magnitude of the measurement unit (the module) used to measure the elements at that level. Note that we are still dealing with systems in a static state. Inside the building, we classify the levels according to the fact that an element at a higher level can be developed into one or more alternatives at a lower level in the same hierarchy.

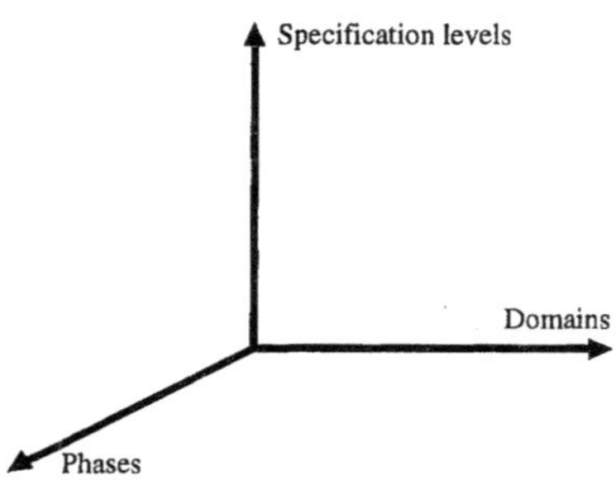

Figure 6. The GOM model

To do this, we must further specify both the spatial and the material components of an element which is, as far as its level is concerned, inherently complex. To put some kind of order into the decision-making process, we must fully understand the concept of a level. This is necessary, since a decision at a higher level must leave the designer free to make decisions at lower levels. Moreover, decisions at higher levels do not automatically lead to 'decisions' at lower levels. Thus, there are two different ways to hierarchically arrange a building into sub-systems. Although the two are defined differently, there is definitely a correspondence ('match') between them. Aggregation levels and generalization levels can be drawn on the two axes of an orthogonal matrix. This matrix is depicted in Figure 7. One point in this matrix represents an element that is part of both a generalization and an aggregation level. In this matrix, elements at any aggregation level can, in principle, be represented at any generalization level. In practice, it is only useful to represent the (composite) elements from higher aggregation levels at more than one generalization level.

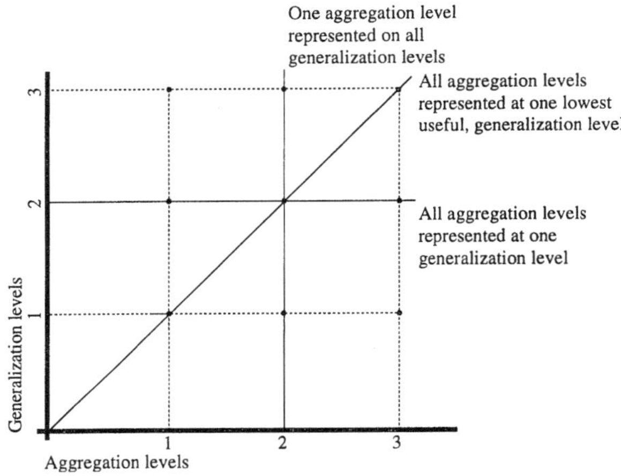

Figure 7. Aggregation levels and generalization levels

For (single) elements from lower aggregation levels, it is only useful to have one generalization level. If the space in the matrix is divided into two parts by a diagonal from the origin, only the elements of one sector are important. Those lying along the diagonal are particularly important. We then seek, given a certain aggregation level, the lowest generalization level for these elements. The characteristics of the elements in the aggregation level, and therefore also the expenditures and costs, can therefore be assigned to the elements in the generalization levels in the form of indices. In this way, we have taken the objective-oriented specification of the original model a step further, leaving the designer some freedom to make decisions.

Step 3: A special relationship between the aspect-oriented system and the hierarchical system

The generalization levels are supposed to be used for structuring and decision-making. In general, these levels go from the general to the detailed, or from a high to a low level of generalization. Decision-making is seen as a formalized process in which the persons who are authorized to do so make decisions about the elements in a building. Those decisions are arranged into levels and domains in successive phases. The systems-theory models described above reveal an intrinsic relationship between phase-oriented systems (arranged in phases) and hierarchical systems (arranged in levels). The level-based hierarchical arrangement can be used to design a decision-making process. The phase-based arrangement of the decision-making process, including the persons authorized to make decisions, can be used to design the building. This relationship between phases and specification levels is illustrated in Figure 8. Flexibility involves change and the sources of change. In the design, building, and management phases, the persons authorized to make decisions are sources of change by virtue of their decisions. Therefore, it is necessary to define a building system such that a relationship exists between the elements determining the generalization and aggregation levels and those determining the phases occurring during the process. Consequently, we transform the original building-oriented system into a flexibility-oriented system. A building-oriented system is a way of viewing a building as a system apart from its context (system boundary). It consists of individual elements, each having its own functionality. The elements are characterized by their internal and external relationships. Each element is simultaneously part of the three partial systems: a hierarchical system, an aspect-oriented system and a phase-oriented system.

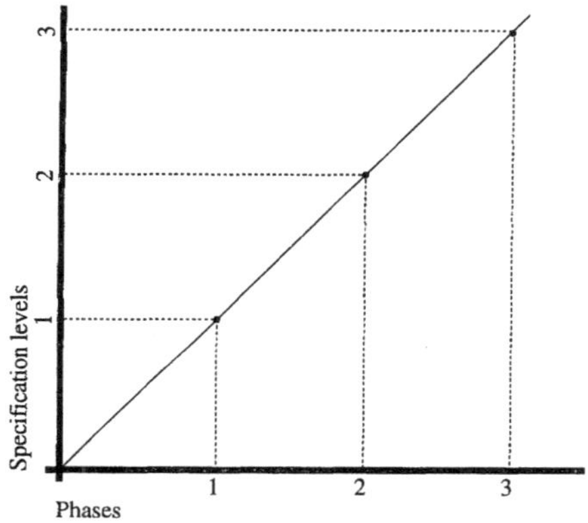

Figure 8. Phases and levels

The aspect-oriented system can be classified in terms of the technically oriented system, the use-oriented system, and the socially oriented system into three aggregation levels. These correspond to the levels in the hierarchical system: technically oriented level, use-oriented level, and socially oriented level. The hierarchical system can be classified into aggregation levels. These correspond to the generalization and order-of-magnitude levels, according to the following scale: technically oriented level 1:1 - 1:20; use-oriented level 1:10 - 1:100, and socially oriented level 1:100 - 1:500. The hierarchical system can be classified into generalization levels. These correspond to the phases in the phase-based system (which depend on the persons authorized to make decisions).

The term 'system' is used to denote the examination of reality. The term 'model' refers to the pictorial representation of reality. A building which is depicted according to the rules of the building-oriented system is a building-oriented model. Because we examine building-oriented (sub-)systems in terms of flexibility, it is useful to take part of this system and develop it into a new type of system. In this case, we focus on the hierarchical system: the level-based formal part of the building-oriented system, whose elements all change at the same moment of decision. A (building-oriented) flexibility system can now be defined as the level-based system, seen as part of a building-oriented (sub-)system (2), whose elements (sub-systems) all change synchronously. They react to the same source of change, as predicted in a flexibility scenario.

With the aim of making the flexibility systems operational so that design decisions can be made, this system is differentiated into aspect-oriented systems and defined accordingly. We can distinguish various types of systems and therefore various models, depending on the categories into which the sources of change are classified. These systems and models fit into an overall hierarchically arranged system and model. In other words, a particular system or model has to be used as a building block to form a system or model at a higher aggregation level. The technically oriented system is at the lowest level, with the use-oriented and socially oriented levels coming next. The socially oriented level corresponds to the generalization level of the support structures with infill elements, the highest level defining the field of study. In analogy to the three types of aspect-oriented systems, we can define three types of flexibility systems. A *technically oriented flexibility system* is a system, the elements of which change synchronously as a result of their corresponding technical lifetime and their material/technical relationship. In contrast, if the elements change synchronously at the end of a period as a result of their (subjective) spatial/use-oriented relationship and their corresponding use-oriented/economic lifetime we have an *use-oriented flexibility system*. Finally, the elements of a *socially oriented flexibility system*, falling under the jurisdiction of the same authorized persons, change synchronously at the end of a period (Habraken 1985).

To make design decisions about buildings involving flexibility/cost relationships, it is necessary to use a time scale to determine which elements can (synchronously) change at a certain moment. This involves a set of elements in the above-defined systems. However, this set is mainly characterized by its structure rather than its content. Here, too, we are looking at the building in its entirety and distinguishing a set of elements (sub-systems) primarily determined by their lifetime. A lifetime selection set is a set of all elements in a flexibility system which change synchronously at the end of a chosen period. The total costs of a building are determined by the amount and moment at which the expenditures are made. Thus, the way in which a building is arranged into a lifetime selection set is a determining factor. In the flexibility system, the lifetime selection set can be viewed as one of the aggregation levels. Given the definition of the lifetime selection set, these levels have a temporal character. We note that this particular level cannot be linked to the customary decision-making and/or order-of-magnitude level. Moreover, it is impossible to specify any causal source of change which they have in common. Up to now, the systems and models presented here have been abstract. We can model a specific building according to the rules of these systems in reaction to a likewise specific flexibility demand. In that case, these systems and models are made concrete and can be subjected to quantification. Because of the inherent causal hierarchy of change and the synchronism of the change, we refer to the four flexibility models as 'strings of changes' or merely 'strings'.

A flexibility model (string of change; string) can be defined as a specified flexibility system or specified lifetime selection set of a specified building at the technically oriented, use-oriented, or socially oriented level, initiated by the technical lifetime or the flexibility demand at these levels. The string of change is a key concept in this research. In analogy to the flexibility systems, the strings of change can be labelled as follows (Prins and Bax 1991): (i) *technically oriented string* (this string of change takes the form of a technically oriented flexibility model), (ii) *use-oriented string* (this string of change takes the form of a use-oriented flexibility model), and (iii) *socially oriented string* (this string of change takes the form of a socially oriented flexibility model). The three distinguished strings of change are described below.

A technically oriented string is a collection of elements which are changing simultaneously because of the change of one of them. The collection is determined by the material/technical connections between the elements. All elements of the technically oriented string have a technically inseparable mutual connection. This means that if one of the elements is removed, all the others will be damaged. So the technically oriented string is by definition the smallest changing element of a building. To specify a technically oriented string, one first has to detect the so-called 'prime-mover', the element which initiates the change. The technical connections of this prime-mover with other elements defines the size of the technically oriented string. A use-oriented string is a collection of elements which are changing simultaneously because of the change of one of them. The collection is determined by the process of use and the way the building is composed of technically oriented strings. The use-oriented strings can only be specified if a flexibility demand is expressed. Because the technically oriented string is the smallest changing element of a building, the elements of the use-oriented string are technically oriented strings. The relationship between elements of the use-oriented string has primarily a functional character. A socially oriented string is a collection of elements of which changes are initiated by the decision competence of a certain social group. Besides, the collection of elements is determined by the coherence in life span, use and technical interdependence. All elements of a socially oriented string are subservient to the same mandate. Because by definition all the changing elements in a building are technically oriented or use-oriented strings, these strings form the elements of the socially oriented string. The common distinction of a building in a support structure and in an infill structure is an example of the division of a building in two socially oriented strings. We can now define the terms 'building' and 'flexibility supply'. A building consists of the entire string of change. The flexibility supply is the description of a building in terms of strings of change.

4. The Calculation Model

In this section we add a financial-economic dimension to the building which was modelled in the previous chapter. There, we first described the building abstractly and then more concretely, in physical terms and in terms of time (the duration and tempo of change). The financial-economic dimension is based on the expenditures and costs involved in constructing and maintaining buildings (Prins and Tempelmans Plat 1990). We define expenditures as concrete cash flows needed to construct and maintain a building in a desired state in all design, implementation and management phases. Expenditure is calculated in terms of material, labour and research tools. These can be broken down into: investment expenditures, maintenance expenditures and reconstruction expenditures. Costs can be defined as all conceivable cash flows (derived from the expenditures) in a flexibility scenario. Cost is calculated in terms of strings of change. These include changes due to economics which take place during the anticipated lifetime of the building.

To control expenditures, it is necessary to make decisions based on the costs involved during the entire lifetime of the building and therefore over all the periods on the scenario's time scale. Thus, we have to be able to calculate the costs per period or over several periods (Tempelmans Plat 1982). Effective period costs can be defined as costs which are calculated in terms of a string of change over one or more periods in the flexibility scenario of the flexibility supply of a building. An important research tool is the financial-economic comparison of qualitatively equal design alternatives (i.e. designs with identical performance levels). Over the entire period of development, they constitute the totalled (and subsequently averaged) period costs. These total (average) period costs are determining factors when a choice has to be made from several design alternatives. Total (average) period costs can be defined as period costs totalled and averaged for all relevant strings over all periods in the flexibility scenario of a series. These costs are determining factors when a designer has to select a design alternative for a building. The total (average) period costs are a good basis on which to determine the design. However, they do not indicate the relationship between dwelling performance and costs. Since the period costs do show this relationship, we add the term 'actual' to emphasize this fact.

In fact the costs are determined by a calculation method in the form of a computer program that is both input and output-dependent. The input consists of data about a building, modelled as a flexibility supply consisting of specified strings; the output consists of actual or total (average) period costs. This calculation method can be applied in a variety of ways in design strategies geared to the design problem when there is a specific flexibility demand. In this research we use the term calculation method for an algorithm (computer program) which can calculate several values. Specifically, these are the period costs involved in a given flexibility supply, the flexibility operations derived from this, and the expenditures based on these operations. The calculations are performed for all strings of change where several explicitly stated calculation assumptions are made. In the research project, this calculation method is developed as a user-friendly, fully interactive and menu-guided computer program called 'Calculation Model Period Costs' (CMPC).

5. Process Model and Design Decision Support System

In this section, the modelling process is concluded by combining the flexibility model, the calculation model, and the process model into a decision support model. This process model is a systematic description of the basic activities a designer has to execute when applying the flexibility model and calculation method to the design process. Subject to these activities are the states of the flexibility model and calculation model. The process model is a model for implementation of the flexibility model and the calculation method in a decision support system. The elements of this process model are activities, ordered according to levels obtained from the flexibility model and calculation method. The following activities can be distinguished: generate physical plans, aggregate strings of change, generalize expenditure data, specify lifetime selection set data, and analyze costs. In Figure 9, the process model is summarized.

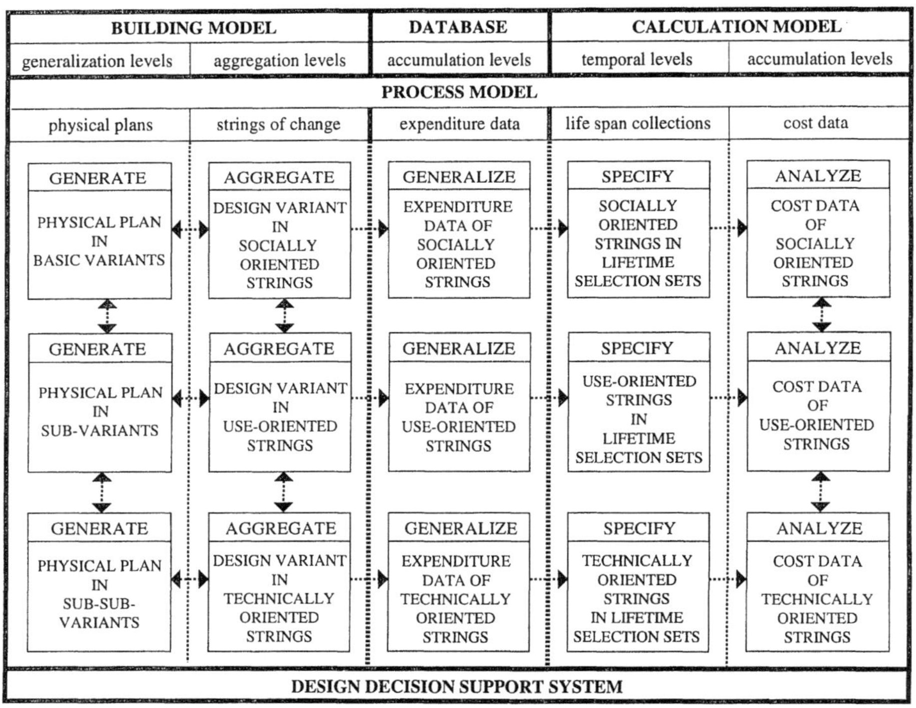

Figure 9. Process model and design decision support system; a systematic description

The first column represents activities which can be characterized as generating physical plans. These activities are distinguished in three levels, characterized by their scale and their information density. Another characteristic is that an element which is subject to the designer's activities on a certain level is manifest as the situation for the design activities on the lower level. In the first column, the building is modelled as a set of generalization levels, of which it is assumed they can be matched with phases in a design process. This level concept of the first column is illustrated by an example in Figure 10. At the highest generalization level, a zoning concept is made in which sectors (functional units) can be placed. On the level of the basic variant, these sectors are the situation which can be detailed in design variants. In the example, the sector for the sanitary functions is detailed into a bathroom. On the next level (sub-variant level), this bathroom is detailed into sanitary equipment. On the lowest level (sub-sub-variant level), the technical details are given. For each level, the situation and the elements are specified in the figure. The second column represents activities which can be characterized as aggregating strings of change. These strings have already been discussed in detail in Section 3. The third column represents activities which can be characterized as generating expenditure data in a database. This database can be seen as part of the calculation model. Every time the calculation model is used, new expenditure data can be calculated, generalized and saved. The ordering in levels of the database corresponds with the aggregation levels of the building model.

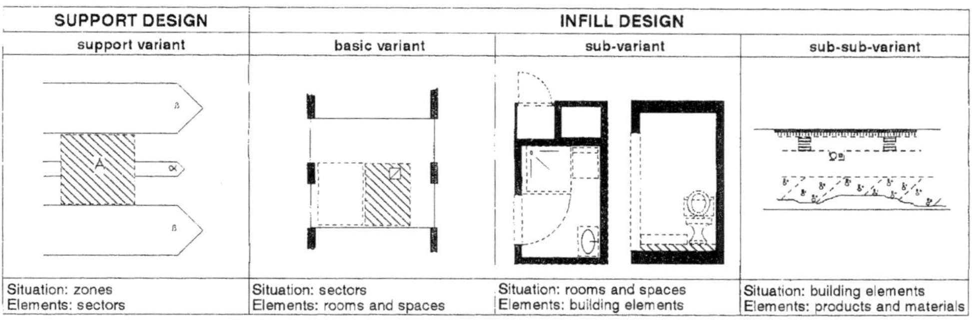

Figure 10. The three generalization levels in the design of an infill system

At the first use of the model, only data on the aggregation level of the technically oriented string can be obtained. Thus, only bottom-up processes are possible. After repeated use in several design processes, a database can be created which is also filled with expenditure data of strings of change on the higher levels. This enables the user to execute top-down processes as well. As long as the database is not filled with data on all levels, we prefer to use the term 'design decision support model' instead of design decision support system'. In the fourth column, the life span properties of the strings of change, defined in the third column, have to be specified according to the flexibility demand. The subject of the activities is the strings of change. By specifying their life span properties, these strings are transformed in lifetime selection sets. The levels in this column correspond with the database levels and the levels of the building model. They are of a temporal character. In the fifth column, the costs of the lifetime selection sets defined in the fourth column are calculated and analyzed. Here the levels are accumulation levels, corresponding with the levels of all previous columns. On the basis of the total average period costs of the highest accumulation level, decisions can be made about two design variants.

Together, the flexibility model, the calculation method, the filled database, and the process model constitute the decision support system, developed in line with our research objective. Thus the term decision support system can now be defined as a system that consists of three components: a flexibility model for modelling the buildings; a calculation method for calculating the period costs of the design alternatives, according to a flexibility demand (including the database); and a process model for the implementation of the flexibility model and the calculation method. This model supports the following four types of processes: (i) decision-making by enabling one to choose among design alternatives in a bottom-up process based on comparison of the calculated period costs, (ii) realization of the optimal design, based on repeated comparison, (iii) calculation of the indices in the form of period costs of the standardized strings of change, in which the aggregation levels in the flexibility system correspond to specific generalization levels, and (iv) phase-based development of buildings in a top-down process with the help of indices.

6. Flexibility and Costs in the Design Process

Depending on the data available in the database, three design strategies can be distinguished. When the database is not filled, decisions can only be based on (traditional) calculations of expenditure data of technically oriented strings and on the calculations of the total average period costs of these strings of change. Figure 11 illustrates this strategy within the process model (strategy A). After sufficient runs when expenditure data on the other generalization levels are also available, strategies of type B and C (see Figure 11) are also possible. In those cases, costs are calculated on generalized expenditure data of use-oriented strings and socially oriented strings. Then the designer can make decisions about flexibility and costs in the early phases of the design process as well, when only general information about the design variants is available. Before one can start designing a flexible building with the aid of the developed decision support system, a relevant flexibility scenario has to be formulated. After this is written, one can choose for a bottom-up or a top-down strategy (Tempelmans Plat and Prins 1991). A mix of these two strategies is also possible. We can now define a general strategy for a design process of a building.

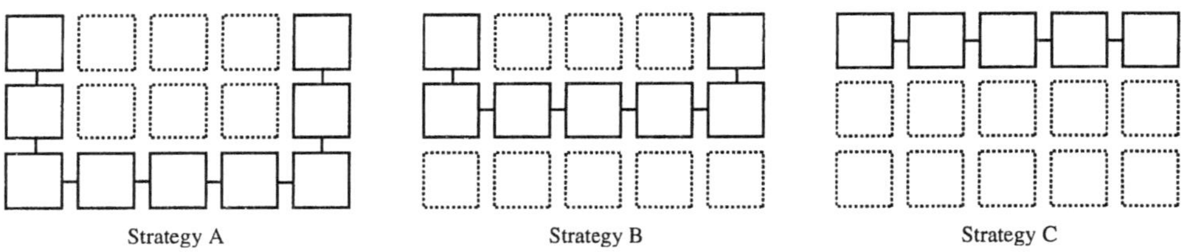

Figure 11. The three strategies within the process model

In this strategy, an attempt is made to optimize building flexibility and to minimize life-cycle costs with the aid of the developed design decision support system. The steps of this general strategy are given below. At the end of the description of each step, the activities in the process model are indicated by the letters A-E (A refers to the activity 'generate', B refers to the activity 'aggregate' etc.).

0) Write a flexibility scenario as part of the brief, in which assumptions are made about all relevant social, political and cultural events which may influence the use of the building. Also write a financial-economic scenario in which the expected developments of the relevant parameters are specified. (A)
1) Design a physical plan which is the basis for the further design of the building. Depending on the design problem, this physical plan can be represented on several generalization levels. (A)
2) Write the flexibility demand for the chosen level. This means that the designed physical plan has to be worked out in a series of plans in time, according to the flexibility scenario. (A)
3) Specify the flexibility operations for each plan in each period in terms of equivalent replacement, non-equivalent replacement etc. (A)
4) Model the physical plan as a set of strings of change corresponding to the relevant generalization level. Which strings of change are specified depends on the technical life span and the chosen connections. (B)
5) Estimate the expenditures if no relevant data is available in the database and generalize these data; otherwise, select the relevant data from the database. (C)

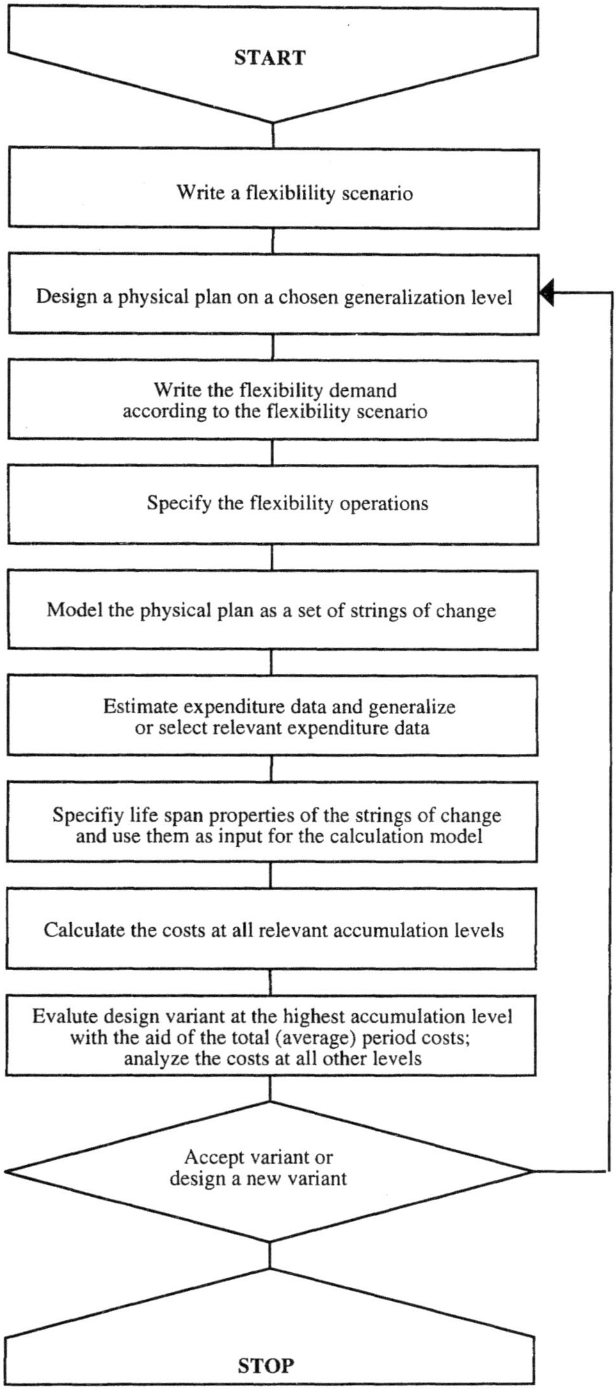

Figure 12. General design strategy, as a ten-step flow chart, when using the design decision support system

6) Specify the life span properties of the strings of change. Use the now available lifetime selection sets, together with the relevant financial economic parameters, as input for the calculation model (CMPC). (D)
7) Calculate the costs with the aid of the CMPC. Generate CMPC output on all relevant levels and decide whether the variant is acceptable or not, depending on the value of the total verage period costs on the highest accumulation level. (E)
8) If the variant fits the norm, then decide whether to continue the design process or stop. If the variant does not fit the norm, then analyze how the costs accumulate and try to find clues for lowering the costs in a new variant. (E)
9) Proceed to the next level. Generate a new variant and repeat steps 1 to 9. (E)
10) Select the collection of design variants of strings which together have the lowest total average period costs. (E)

Figure 12 shows this general design strategy as a flow chart. This procedure can be carried out in the design process for new buildings as well as for existing buildings. It is valid for the building as a whole as well as for its constituent elements.

7. Epilogue

We can formulate a few conclusions about the developed model. The most important of these are the following. It is only useful to analyze the flexibility of buildings when the analysis is based on a flexibility scenario. The flexibility scenario should be part of a building's program of requirements. The flexibility of a building is determined by the number of strings of change it contains, its lifetime, and the technical connections between the strings of change. It is impossible to include price increases in a design or investment decision when the design alternatives contain varying lifetimes. It is preferable to base the calculations on the effective interest and to use a relative (construction) price increase in special cases. Decisions should be made on the basis of the total (average) period costs. With the decision support system, it is possible to make decisions about the flexibility and costs of the buildings in all phases of the design process, by using 'feedback' and 'feedforward' techniques.

The most important suggestions for further research are summarized below. We recommend greatly expanding the number of case studies and applying the experience gained in the course of the case studies to the system. To be able to generate indices for the various program runs, it will be necessary to elaborate the theory linking the aggregation levels to the generalization levels in the flexibility system. This is also necessary to execute top-down design processes. An important research tool in the field of flexibility would be the development of an intelligent system of drawing and the determination of technically oriented strings based on building elements and their connections. The development of a standard method of describing scenarios and flexibility demands based on these scenarios would significantly accelerate the acceptance of flexibility as an indispensable part of a program of requirements.

Acknowledgement:
This research is supported by the Netherlands Technology Foundation (STW).

References

Bax, M.F.Th. (1979) "Domain Theory", *Open House International* **4**, 30-42.

Brinkman, S. (1989) Flexibiliteit: Beheersprocessen en Proces Beheersing, Ph.D. Dissertation, Eindhoven University of Technology, Eindhoven.

Habraken, N.J. (1985) *Supports - An Alternative to Mass Housing*, Architectural Press, London.

Leeuwen, A.C.J. (1976) Hiërarchische Systemen, Eindhoven University of Technology, Eindhoven.

Prins, M. (1989) "Decision Making and Flexibility: An Integral Design Decision Making Model", in N.Wilkinson (ed.), *Quality in the Built Environment*, Open House International Association, New Castle Upon Tyne, pp. 141-149.

Prins, M. and Bax, M.F.Th. (1991) "An Integral Relational Design Decision Support Model to Optimize the Flexibility of Buildings in Relation to Life Cycle Costs", in M. Zaja (ed.), *Organization and Management in Construction*, University of Dubrovnik, Dubrovnik, pp. 477-485.

Prins, M. and Bax, M.F.Th. (1992) "The Management of Building Flexibility in the Design Process", in M.P. Nicholson (ed.), *Architectural Management,* E.&F.N. Spon Ltd., London, pp. 65-76.

Prins, M. and Tempelmans Plat, H. (1990) "An Integral Design Decision Making Model Based on the Lives of Components", in V. Ireland, et al. (eds.), *Building Economics and Construction Management Vol. 2*, CIB/Sydney University of Technology, Sydney, pp. 234-242.

Tempelmans Plat, H. (1982) "Micro-Economics Analysis of the Process of Design, Construction and Operation of Houses", *IABSE Journal* **14**, pp. 1-14.

Tempelmans Plat, H. (1991) "Cost-Optimal Flexibility of Housing Supply", in A. Bezelga and P. Brandon (eds.), *Management, Quality and Economics in Building*, E.&F.N. Spon Ltd., London, pp. 1286-1295.

Tempelmans Plat, H. and Prins, M. (1989) "Reverse Planning of the Built Environment", *Open House International* **16**, pp. 19-23.

Trum, H.M.G.J. and Bax, M.F.Th. (1990) "The Applicability of Domain Theory in Technological Design", in V.J. McCabe (ed.), *Design in Engineering Educating* , *Proceedings SEFI Conference*, Dublin, pp. 397-402.

DEVELOPMENT OF OBJECTIVE METHODS FOR MEASURING FLEXIBILITY OF SCHOOL BUILDINGS

G.S. SHABHA
School of Estate Management, Faculty of Built Environment
University of Central England in Birmingham
Perry Barr, Birmingham B42 2 SU
United Kingdom

ABSTRACT. This research investigates many problems related to the design for change and flexibility of school buildings. The problem of change proves to be highly complex due to the unpredictability and the difficulty of measuring change. Many deficiencies have been identified in the previous methods concerning lack of indicators for measuring flexibility, reliability and insufficiency of data for indicating change in buildings over use. In the light of the above problems, two main objectives have been established: (i) to propose operational measures of the extent of incorporation of design variables in school designs, and (ii) to propose operational measures of the extent of flexibility of school buildings in use. However, due to the limitations this paper is bound to, the investigation will focus on the second objective. It is anticipated that such operational measures might provide a framework for both architects and researchers, during the early design stage, to ensure that their conjectures about the potential of flexibility might be enhanced during use; hence, improving their prediction of buildings performance over use. This might assist in developing a more coherent objective body of knowledge, which could be fruitfully manipulated during the early design stage to enhance the effectiveness of flexibility in use.

1. Problems of Measuring Flexibility

It has generally been recognised that the need for change and flexibility arises from the increasing rate of change in activity organisation, particularly its spatial and services aspects, which characterises particular building types, such as hospitals, laboratories, universities, housing and school buildings. In these buildings, the interface between space and technical services has evoked much controversy about how change can be accommodated cost-effectively during use, and to what extent the provision for change may improve the operational efficiency of buildings in use. The search for an optimum approach to flexibility has become a significant aspect in building design and the focus of many studies in the past. Several design ideas for optimising flexibility can be distinguished, which however differ in their scope and the aspects of change that have been investigated. The basic assumption underpinning these studies is that the extent of incorporation of particular design variables in building designs might enhance and optimise their flexibility in use. Only a few proposals concern us here: (i) the adaptability of design layout to accommodate change in the organisation and activities housed in them (Fawcett 1976a, 1976b; Steadman 1983) with particular emphasis on whether a prediction could be made about changes in activities to match their own spaces in the future (Fawcett 1978; Maver 1979; Vickery 1979), and (ii) the flexibility of buildings in relation to different design variables, such as uniformity of the spatial organisation and structure, and patterns of circulation and access systems (Weeks 1964, 1969; Habraken et al. 1976; Oxmon 1978; Oddie 1974; Moharram 1980; Al-Nijaidi 1985).

The above studies could be criticised as being parametrical rather than comprehensive, since they concentrated on particular facets of flexibility as related to certain variables of the design layout, which in turn detracts from applicability to other building types. These studies have in common the lack of operational definitions and measures of design variables and flexibility, the

H. Timmermans (ed.), Design and Decision Support Systems in Architecture, 165–177.

inadequacy of the previous methods for measuring flexibility, and the insufficiency of empirical data for measuring flexibility. Only a few attempts have been made in the above studies to assess the extent of flexibility advocated in buildings. These could be categorised as largely subjective; such studies tend to be descriptive, based on a morphological analysis of flexible design (Oxmon 1978; Moharram 1980). Alternatively, they could be categorised as objective. The latter fall into four main types: purely theoretical, without any empirical evidence (Aylward 1968; Aylward et al. 1974); theoretical, based on hypothetical models with some empirical evidence (Cowan and Nicholson 1965; Llewelyn Davies et al. 1973); theoretical framework, supported by empirical evidence (Al-Nijaidi 1985); and theoretical, based on mathematical models and predictive formula for measurement (Fawcett 1976a, 1976b, 1978; Vickery 1979).

The methods adopted in the subjective studies were based on arbitrary weighting scores for indicating flexibility (Moharram 1980); hence their findings can be considered to have a lower degree of reliability. Most of the quantitative methods suffer from a number of deficiencies, particularly a lack of objective criteria and lack of indicators for assessing flexibility; hence there is a genuine lack of data for indicating change in buildings. Though some indicators have been developed in the latter empirical studies, these are shown to be selective concerning certain types of flexibility related to certain parts, or aspects, of the building and organisation, which makes them unreliable for measuring flexibility.

For example, the method proposed by Fawcett can be criticised as highly probabilistic in tackling the issue of matching activity to space in the future from a purely abstract point of view (Fawcett 1978), as it measures future flexibility rather than past real flexibility. Moreover, the probability of being able to allocate teaching activities to their units of accommodation is highly dubious since the activities are themselves changeable. On the other hand, it hardly considered or presented any operational technique for measuring change in activities. It fell short of explaining the complexity of the nature of activities, particularly about how change has been carried out in other characteristics of the activities. It did not classify aspects of change, nor did it identify types of change or quantify their magnitude, which brings it beyond the scope of this study.

By the same token, the same problems can be identified in the Markov simulation model adopted by Vickery (1979). In this model, flexibility of design was measured according to a lower average of communication cost penalty over time, concentrating on organisational flexibility, and many operational problems can be detected in this respect. Therefore, a comprehensive method for assessing and measuring flexibility is needed to enable data about change to be obtained and applied. In the following, an attempt will be made to develop some operational definition and indicators for measuring flexibility of school buildings.

2. The Meaning of Flexibility

In the literature, several definitions of flexibility have been proposed. These differ according to the type of building and the emphasis that has been given in each study. This is particularly clear in cases where certain attributes or parts of these buildings were investigated. The controversy in describing flexibility in buildings could be envisaged through a diversity of terminologies, such as flexibility, adaptability, variability (Dluhosch 1973), expansion, versatility, convertibility and so on (Aylward and Lapthorne 1974; Pena 1977). Most of these definitions have described flexibility as the potential to accommodate change, indicating why flexibility is needed and how it was advocated in design.

Fawcett (1978) considered flexibility to take into account the uncertainty and changeability of the relationship between activities and space. He defined flexibility as "the probability that a building will not become obsolete, and it is a subjective probability based on state of knowledge about the activities that might occur in the building". Oddie (1974 p.115) distinguished between three aspects of flexibility in school buildings: the 'modular coarseness' or 'refinement' which responds to the diversity of accommodating activities; 'adaptability', which refers to capacity for physical alteration to meet changing circumstances; and 'versatility' or 'polyvalence'. He defined versatility as "the quality of building as planned and fitted out which allows for variation in patterns of activities, without the need for physical adaptation". He added that the greater the versatility, the longer the adaptation can be postponed (Oddie 1974 pp. 114-115).

The above distinction contradicts the definitions suggested by Dluhosch (1973 p.13), who distinguished between flexibility and variability. He defined flexibility as the ability to achieve a change of condition, which he claimed to be synonymous with adaptation and modification, whereas variability was defined as the ability to change the condition without recourse to adaptation.

Thus, it seems that flexibility is a multi-faceted concept and that there is a diversity of definitions that describe these facets, which are mostly concerned with the accommodation of change or adaptation or both. In some definitions flexibility was distinguished from adaptability, and both were considered to be extremes on one continuum, where the accommodation of change can be made with, or without, adaptation. In other definitions (Fawcett 1978) flexibility was mentioned interchangeably with adaptability and in relation to change only.

Despite the diversity of definitions, there seems to be a shared agreement that flexibility could be described according to attributes of both change and adaptation, in any part of a building or a building as a whole. Although such a statement was not explicitly formulated in the literature, it could be stated that the flexibility of buildings could be described along one continuum, depending on the type and the amount of change, in relation to the amount of adaptation that may occur over time. This will be the starting point for measuring flexibility in school buildings. Two operational definitions can be explored and refined in order to arrive at a system of measurement. First, "Flexibility is the quality of building, as it exists at any point in time, which allows for change in patterns of activities that can be accommodated, without recourse to physical change i.e. adaptation" (Oddie 1975 p.33). Secondly, "Flexibility denotes the potential of school building which permits change and variations in activities, time-tabling and class-size group without need for adaptation" (Pearson 1975 p.25).

It appears that flexibility refers to the accommodation for change of activities pursued at a particular point in time. Pearson (1975 p.26) argued that flexibility could be facilitated by a continuum of space and a well-designed system of portable and adaptable furniture, equipment and storage units, allowing teachers and pupils to rearrange their accommodation whenever needed. On the other hand, adaptability was defined as "the capacity of physical alteration by relocation, replacement and removal of components in respect to either the constructional elements or services of the buildings or by addition of further components towards increased adaptability" (Oddie 1975 p.35).

Most of the previous ideas dealt with relocation of partitions as a prime objective of adaptation. Very little attention was paid to other parts of the building. The demand for physical alteration arises from a need to alter the pattern of internal spaces, which partitions demarcate and separate. Nonetheless, it is evident that, in real cases, adaptation was made by pulling down partitions, to make a larger space, or inserted for subdivision. Two main assumptions can be pointed out: (i) both notions are inversely related and opposite in their objective, and (ii) the demand for flexibility of teaching space necessitates certain flexible provisions - the higher the degree of change, the lower the demand for redesigning or adapting the space to the new requirements.

During the course of this study, the term flexibility will be considered comprehensively and it will encompass attributes of both change and adaptation. Meanwhile, the overlap between the classes of flexibility which catered for change, with or without adaptation, can be clarified by further empirical evidence.

3. Development of Objective Methods for Measuring Flexibility

To measure flexibility, some criteria need to be developed and established. Such criteria must be reliable, valid and sensitive enough to allow an objective comparison of the flexibility between different school buildings. Two main measurement criteria can be distinguished in the literature: the objective and the comparative. The former should be based on a yardstick of shared understanding. The latter has to permit teaching spaces belonging to different samples to be assessed comparatively. In this case, the comparison can be carried out according to what was originally proposed so that any deviation or mismatch in use can be detected. This is exemplified when change in the activity patterns and in area/size of teaching spaces are examined. Both criteria will be deployed in this study to develop some indicators related to both change and adaptation.

3.1. INDICATORS OF ORGANISATIONAL CHANGE

Several indicators can be proposed in this study for the purpose of investigation. Though these indicators are interrelated and most likely associated with particular design aspects, they will be dealt with and examined separately in this study. However, it is of primary importance to cull out only few indicators: change in activity types, the use and function of the space, the size of learning group and the area of teaching spaces. These represent various interwoven aspects of change of activity requirements, in terms of their allocation in the space and their interrelationship as a whole. An attempt will be made to examine their reliability and the criterion for measuring each of them will be outlined separately.

3.2. CHANGE IN ACTIVITY CHARACTERISTICS

A major emphasis has been given in past studies of activities to investigate the change in activity patterns in buildings. It could be defined as a description of the typical makeup or flow of activities for a population group, a class or an organisation over time (Faberstein 1974 p.19). Most of the past studies of activity are selective in the properties of activities which are incorporated. They lack comprehensiveness since they emphasise parts of the institution rather than the whole. However, measuring change in activities was found to be constrained by many problems in the procedures of collecting data about activity requirements (Al-Nijaidi 1985 p.120). The availability of data will enable an objective assessment to be made during any limited period of time.

Only two approaches concern us here. The first emphasises activity as related to group size, percentage of meeting time and room sizes (DES 1972; Rawlinson 1978). The second focuses on the activity/space interface (Markus et al. 1972; Maver 1979). It appears that the first is highly relevant to real flexibility of school buildings as it provides an insight into how the organisation of activities housed in the building can fruitfully be manipulated over time to achieve improved use of space. A time-tabling and space allocation system has been developed and three important interdependent functions of activity were distinguished: (i) the nature of the activity, (ii) the type of facilities and environmental requirements, and (iii) the number of participants in the activities (i.e. size of groups related to areas of rooms).

The second approach was developed by the Building Performance Research Unit (BPRU) and centred around the temporal variations in the functional requirements of comprehensive schools (Markus et al. 1972). The degree of the mismatch between the need and the provision of both activity and space was measured. This was thought to start during the use of the building and to increase as time passes (Maver 1979 p.12). It was attributed to the change in the timetable of activities which may normally have a duration of one week or longer, and the change in the teaching methods and curriculum (Maver 1979; Davies and Th'ng 1972, 1975). A few beneficial points for measuring change in activities can be distinguished from the above: (i) describing how the initial schedule of accommodation in school organisation has been transformed and changed into another during use, (ii) revealing the behaviour setting of different group size and the way of performing activities, and (iii) indicating how different subjects within the curriculum could affect the size of learning groups and their distribution as time passes, particularly when different groups can perform the same types of activities (Maver 1979 pp. 13-14).

A few operational deficiencies could be envisaged: most of them were bound by the lack of information about activities and their time-tabling in the past as well as the existing patterns;this might affect the reliability of measurements. Even if some observational techniques are adopted for data collection, there are still some difficulties in achieving a desirable level of accuracy through continuous observations. Such difficulties could emerge, for instance, when detailed observation of teaching spaces is required, which makes it very laborious and time-consuming.

An exploratory approach through questionnaire, observation and discussion with users can be proposed. This would be more flexible and allow points of significance to emerge, but this may require more time as the sample becomes larger. Despite the usefulness of the questionnaire in some situations, detailed observation of an exploratory nature would be a much more adaptable method. This may enable a cross-checking of the layouts of the teaching spaces to be investigated. It would also allow observation of the users at work to take place, discussing with them the nature of their activities and change in their requirements. Such informal discussions with users may highlight several points about user problems and areas of dissatisfaction.

4. Proposal for Measuring Change in Activity

This study argues that change in activity can be measured in relation to teaching space, in terms of the degree of mismatch between the requirements of activities and the space, and how frequently that may occur over time. The availability of records of activities is very important. But when these records are not obtainable for whatever reason and for any of the school sample under investigation, this might result in unprecedented consequences. It could affect the representativeness of the sample and hence lower the number of comparable samples.

From the pilot study of selected schools, some clues about approaching the survey by reducing the number of schools were revealed, so that a higher degree of accuracy can be ensured. This implies assessing change of a selected teaching department rather than school buildings. It seems that the whole school flexibility in use would be very difficult to achieve. It would be very time-consuming and necessitate expensive resources, which are not available. There are considerable problems concerning the lack of criteria by which such flexibility can be assessed. These are ascribed to the variation of size of schools and diversity of their functions which have to be considered. On the other hand, though the criteria were identified and used in the investigation, flexibility of school buildings in use is a dynamic and changeable phenomenon, which would hardly allow such criteria to be adequate.

These arguments would be applicable to the teaching department as well, but to lesser extent than to the whole school. Although teaching departments are not totally representative of schools, they can be considered a satisfactory indicator of the overall flexibility of school buildings. Their significance could be envisaged throughout. First, they are the main organisational and teaching units of a school building. Secondly, they vary in their design characteristics, such as the shape of the layout, types of activities they house and their technical requirements; hence, teaching departments may enable us to compare their extent of flexibility in relation to these variations. Thirdly, they represent a basis for comparing different constructional methods and building systems.

Only two types of teaching departments will be chosen for the purpose of comparison in the empirical study. Both differ in terms of activities and their requirements. They fall into the generalised (e.g. Humanities/Languages departments) and the specialised (e.g. CDT/ Science departments). Three main characteristics will be assessed:

(i) *The type of activities performed.* It is evident that teaching activities can be categorised according to four main types: theoretical, light-practical, heavy-practical, and movement and physical education (P.E.) activities. Although this is very useful in indicating change in the level of services of teaching spaces, it would be of lower validity in transforming any detailed information about the nature of activities. Thus, an alternative categorisation can be proposed depending on broad types of activities, or clusters of activities which are most likely recurring and commonly in use. These include, for instance, general teaching, individualised activities, discussion and seminar, experimenting, project work activities, and so on. Such categorisation includes an implicit assumption about the size of learning groups. Nonetheless, the choice of learning group size is an arbitrary measure, which can be varied to suit specific requirements, but it is a more reliable criterion than others for indicating types of activities. By measuring change in type and number of activities, the extent of change in the characteristics of activities could be obtained for each space, which could then be averaged for all teaching spaces in the teaching department.

(ii) *The size of learning groups.* It is very important to identify whether any change in the size of the learning groups has occurred. By measuring the mismatch in the size of learning groups - comparing the actual size with those originally proposed by the timetable - a new indicator can be obtained. Such a mismatch could be examined in relation to room areas, which may indicate any inadequacy or inappropriateness of the size of areas in regard to their users.

(iii) *The time spent on performing activities.* This indicator measures the amount of change in the average period of time spent on performing each activity, or sets of activities, and the average number of daily and weekly periods of time-tabled activities. Some complementary information regarding change in teaching methods, curriculum and teaching-aids is needed in relation to change in the level of services of teaching activities.

5. Indicators of Adaptation

Previous studies of flexibility pointed to some manageable procedures for measuring adaptation in buildings in terms of their operational cost of adaptation (Bottle and Piper 1970; Bottle 1971). By indicating all expenditures on conversion and adaptation, it would be possible to arrive at a reliable measure of adaptation. This too would depend on the availability of records about cost for each particular part of the building. Even when these records are available, they often include the total adaptation cost of a school building as a whole rather than the departments, and hence fall short in measuring adaptation. An attempt was made to identify these relevant indicators of adaptation by an examination of the available records and drawings concerning school buildings and accounts of past situations, so a comparison could be made on a chronological sequence, and detailed observations and description of school buildings, which aimed to quantify the type and the amount of adaptation undergone during a fixed period of time. This will be supported by a questionnaire to gather information about the status quo of these buildings in use. That information will enable many changes in building elements which have not been annotated on the plans or on other records to be checked.

Two main indicators can be proposed. The first is adaptation in the area and size of teaching spaces, which might indicate the pattern of change in the spatial organisation of teaching departments as a whole. The second is internal adaptation in teaching space, which might indicate the types of adaptation in the technical/environmental services and facilities within the teaching spaces. It is evident that adaptation in areas could be made by addition of adjacent teaching spaces, contraction, subdivision and addition of completely new teaching spaces. The overall extent of adaptation in any particular department could be measured in terms of the average percentage scores for each space. However, the data about adaptation could be manipulated to enable development of a hypothetical model of cost of adaptation.

6. Alternative Measure of Cost of Adaptation

A hypothetical model for assessing flexibility in terms of cost of adaptation can be developed in this study as an alternative measure of adaptation. This model is based on extrapolation of percentages of adaptation of areas and internal adaptation into more meaningful cost indicators. It incorporates the parameter of cost of adaptation, as a variable, in allocating scores for adaptation rather than percentages. By allocating scores for each teaching space, a more reliable measure of adaptation may be arrived at. This will be a very difficult task to achieve, due to many interwoven aspects that might influence the estimation of such scores relating to: a) aspects and types of adaptation (e.g. constructional, services and contents characteristics); b) the amount and frequency of adaptation; c) the area and size of the space that underwent such adaptation; and d) the level of services provision of that particular space. Two main indicators of cost of adaptation can be identified as follows: (i) cost of adaptation of area, resulting from inserting or removing walls and/or partitions, according to one or more of the possible ways of adaptation, and (ii) cost of internal adaptation in the teaching space, which incorporates those aspects of adaptation in the contents and facilities within the space. However, this study argues that these indicators could be aggregated into spatial and services characteristics.

6.1. ASSUMPTIONS ON MEASURING COST OF ADAPTATION

The following assumptions have been made for measuring cost of adaptation: first, a system of weighting of different indicators of adaptation was established by deploying previous records about the cost of construction of superstructure, including different building elements and excluding preliminaries. This system has been based on bills of quantities prepared and deployed in the construction of some comprehensive schools which were indicated by cost per sq m of area for each basic element. The analysis proceeds to obtain the average percentage contributed by each element to the overall cost; later on, the relative cost per element can be identified as shown in Table 1. Secondly, a weighting of indicators was accomplished by creating an association matrix to determine which of the elemental costs were affected by any single indicator of adaptation examined, as indicated in Table 2. Thirdly, indicators of adaptation were summed and the resulting scores were normalised to the lowest weighting scores. Hence, relative weighting scores for each single indicator of adaptation could be identified, as shown in Table 3, so that a comparison and assessment of cost, with respect to adaptation of area of teaching spaces and the internal facilities within them, could be made. On the other hand, further aggregation of scores in all or in part of these indicators can be made by substituting the occurrence of adaptation in each particular space with its corresponding weighting scores. Given the variations in areas and the difficulty of identifying the extent of adaptation for each indicator, it becomes necessary to cross-check any possible type(s) of adaptation through detailed observation. A modified design layout indicating all alterations made during use will then be drawn and prepared. In this way, a reliable measure of cost of adaptation could be arrived at by incorporating the area of space and

TABLE 1. Analysis of percentages of elemental costs at the tender stage of four comprehensive schools

Building elements	Maiden Erlegh School	Sedgefield Secondary School	Guillemont Junior School	Farnborough College of Technology	(*) Scores	(X)
1. External walls	6.18	14.73	13.85	9.59	11.58	5.2
2. Columns, primary	10.39	7.72	10.39	11.13	9.50	4.2
beams and slabs	-	-	-	-	-	
3. Upper floors and	3.47	-	-	-	-	
roofing components	4.70	15.18	14.05	25.39	12.46	5.6
4. Floor finishing	6.84	3.38	6.50	3.30	5.57	2.5
5. Wall finishing	3.62	-	4.10	1.75	3.62	1.6
6. Ceiling finishing	6.22	6.27	4.98	2.11	5.82	2.6
7. Partitions	8.33	8.49	6.87	11.98	7.90	3.5
8. Openings (external	5.84	8.74	2.60	1.05	5.73	2.6
doors and windows)	-	-	-	-	-	
9. Internal doors	2.03	2.07	2.58	1.53	2.23	1.0
10. Fittings and equip.	6.80	-	7.30	0.97	7.05	3.2
11. Sanitary services	4.14	3.72	2.10	0.46	3.32	1.5
12. Water and gas serv.	3.52	3.45	3.48	2.06	3.50	1.6
13. Electrical/Telecom.	8.88	5.63	5.15	8.78	6.35	2.9
14. HVAC	16.03	16.76	14.33	18.54	15.70	7.0
15. Heat source	-	0.72	1.61	-	-	0.6

(*) The approximate average percentages of each elemental cost in four schools.

(X) The final average percentage of cost per building element has been calculated according to the percentages indicated in the four schools, except when there are greater variations in the design and building materials used.

Sources:

Col. 1. DES (1975) "Maiden Erlegh Secondary School", *Building Bulletin* No. 48, HMSO, p. 68

Col. 2. DES (1978) "Sedgefield Secondary School, Durham, Phase II, Base for first year and music", Design Note No. 21, Architects and Building Branch, HMSO, p. 26.

Col. 3. DES (1976) "Guillemont Junior School, Farnborough, Hampshire", *Building Bulletin* No. 53, HMSO, p. 22.

Col. 4. Barbrook, R. (1986) "Farnborough College of Technology; Cost Comments", *The Architect's Journal*, 26 Nov. 1986, p. 51.

the aspect(s) of adaptation expressed by its weighting scores. A mathematical formula was developed by multiplying the weighting adaptation scores (Ws) by the areas of each space and obtaining the average scores for all spaces examined. Consequently, the Index of Adaptation Scores (IAS) proportionate with area could be identified for each department expressed by the formula:

$$IAS = (Ws \times A) / N \qquad (1)$$

where,

Ws = weighting adaptation scores

A = area of teaching space

N = number of teaching spaces

TABLE 2. Association matrix between indicators of adaption and relative costs of different building elements

Building elements	Indicators of adaption of teaching space														
	A1 Scores	A2	A3	I1	I2	I3	I4	I5	I6	I7	I8	I9	I10	I11	I12
1. External walls	5.2			*											
2. Columns, beams, slabs	4.2			*					*						
3. Upper floors/roofing	5.6			*					*						
4. Floor finishing	2.5	*	*	*											
5. Wall finishing	1.6	*	*	*											
6. Ceiling finishing	2.6			*									*		
7. Internal partitions	3.5	*	*	*	*										
8. Openings (external)	2.6			*						*					
9. Internal doors	1.0	*	*												
10. Fittings and equip.	3.2	*	*	*		*						*			
11. Sanitary services	1.5							*							
12. Water and gas serv.	1.6								*			*			
13. Electrical/Telecom.	2.9	*	*			*									
14. Heating/ventilation	7.0	*	*	*										*	
15. Special install.	0.6					*									*
Derived weighting		18.9	7.7	37.4	3.5	3.2	2.5	3.1	6.6	2.6	2.9	3.1	6.0	7.1	3.3

A1: Addition of adjacent space
A2: Subdivision of teaching space
A3: Addition of new teaching space
I1: Change in movable partitions
I2: Change in interior cabinets
I3: Change in movable equipment
I4: Adaptation in place-fixed equipment
I5: Adaptation in the height of space
I6: Change in no. of openings
I7: Adaptation in utility service (A)
I8: Adaptation in utility service (B) (e.g. water supply, sanitary instal.)
I9: Change in lighting/false-ceiling
I10: Adaptation of environmental serv.
I11: Addition of computers/micros
I12: Addition of audio-visual/video

TABLE 3. Relative weighting scores of cost of adaptation

Indicators of flexibility	Summation of weighting scores of adaptation indicators													
	A1	A2	A3	I1	I2	I3	I4	I5	I6	I7	I8	I9	I10	I11
Derived weighting	18.9	7.7	37.4	3.5	3.2	2.5	3.1	6.6	2.6	2.9	3.1	6.0	7.1	3.3
Normalised weighting	7.4	3.0	14.5	1.4	1.2	1.0	1.2	2.5	1.0	1.1	1.2	2.3	2.4	1.33

7. Proposed Indicators for Measuring Flexibility

7.1. THE OBJECTIVE INDICATORS OF FLEXIBILITY

Given the type of information demanded about flexibility and the level of specificity required, two data collection methods can be deployed. The first is by examining and observing change and adaptation in the school buildings. The second is by asking users, who may provide supporting empirical data, about different types of change and adaptation which would occur in their particular teaching spaces. By adopting a questionnaire as a measuring instrument, two problems concerning the role of users can be identified as follows: (i) the limitations of the time interval which may highlight the issue of users' awareness of changes that have occurred in their particular spaces prior to their occupation, and (ii) the difficulties that emerged during the analysis and manipulation of the data in the pilot study, due to the multiplicity of the indicators. This might suggest further aggregation of those relevant indicators, if a questionnaire is used, resulting in further reduction in the number of questions. Four indicators are proposed and outlined below.

7.2. OBJECTIVE INDICATORS OF FLEXIBILITY

These indicators measure flexibility of teaching departments and incorporate the change in the function and use of the teaching departments; the adaptation in the area and size of teaching spaces, which incorporates an aggregation of the sub-indicators regarding adaptation by addition of adjacent teaching space, subdivision of the space, and addition of completely new space(s); the organisational change in teaching space, which deals with change in the requirements of activities housed in the teaching space, and, finally, the internal adaptation in teaching space, which investigates different possibilities for internal adaptation regarding the physical building aspects and the contents of the space. It incorporates two main aggregations of sub-indicators of the spatial and services characteristics.

8. Proposal for Data Collection Techniques

The search for proposals for data collection methods reveals that no single method of obtaining data to test hypotheses is perfect. Each method has its own inadequacies which could lead to the possibility of rival hypotheses explaining the findings (Van Delan 1979 p.128). Three methods have been adopted in this study: the questionnaire, detailed observation, interviews. Both questionnaire and detailed observation can be considered as the main measuring instruments for acquiring data, compared to the interview which could be considered supplementary. However, both could be criticised on the basis of reliability and subjectivity compared to interviews. Despite the above disadvantages, the interview seems to be more interactive and analytical by its nature, supporting the observation and the questionnaire and enabling a discussion about the issues to be researched in parallel. Also, it may allow points of significance to emerge which might otherwise require more time.

8.1. THE QUESTIONNAIRE

A questionnaire will be adopted to enable comparison between different teaching departments in terms of their past change and adaptation in relation to the present. Several advantages of the questionnaire could be foreseen, such as its efficiency in coping with large numbers of selected schools and users which have to be questioned. Meanwhile, to overcome the dispersion of schools around different regions, the questionnaires could be mailed directly to each particular group of respondents. There are also advantages relating to the particularity of the gathered data, as well as to its usefulness in the final refinement on a more numerical basis. These advantages may facilitate data analysis and manipulation in comparable way. Further advantages of the questionnaire lie in the comprehensiveness of the data that will be collected. Comprehensiveness will thus apply to the most relevant indicators, not only to those for which records are available. This might enable users to report detailed information about a change in activities, or about the extent of adaptation which is often reversible and not reported on the plans. A few disadvantages can be identified, such as the probable bias of the collected data which results from the users' misinterpretation, the construction and the understandability of the questions. These could be dealt with by careful restructuring and piloting of the questionnaire on a selective group of users before being sent to the whole school sample. Another disadvantage of the questionnaire was that it yielded soft rather than hard data. This, however could be compensated by other advantages such as comprehensiveness of the collected data, larger size of sample and a higher rate of response. It was decided that the questionnaire and the questions to be asked should be mainly of the multiple-choice and closed-ended type. Three main types of questions have been asked to obtain the necessary background, consisting of factual and perceptual data about flexibility.

8.2. PRETESTING THE QUESTIONNAIRE

A pilot study was undertaken and a preliminary version of two types of questionnaire for both groups of users was distributed to a number of respondents. The aim was to test the structure of the questionnaire and the understandability of the questions. Some very enthusiastic remarks and suggestions emerged in the pilot study concerning the contents, the clarity of the questions and the degree of their relevance to flexibility. These gave the necessary momentum for the next stage of elaborating the final version and refining the questions with a few minor amendments. Both questionnaires are self-administered; there was a concern initially, after reviewing several other instruments, that they might prove to be too lengthy. However, in the light of the amount of information available, it was decided to simplify the language and format. Checking the length of the questionnaire in the test has led to a reduction in the number of the questions to be addressed.

9. The Importance and the Applicability of the Study

Flexibility refers to the ability or the capacity of buildings to cater for change and adaptation in use. Flexibility is inherent in every building but to a different extent; the tendency to increase this capacity is a highly desirable characteristic. The flexibility of any building can only be measured in relation to the amount of both change and adaptation. An empirical study of a selected school sample has been undertaken to test the framework of measurement. A few methodological points of interest emerged from the test. First, more than one method of measurement is needed. The emphasis on new areas of investigation most likely necessitates developing some subjective methods compared to those already adopted in this study, highlighting the demand for the triangulation of methods. In doing so, the validity of the data collected by the questionnaire should under no circumstances be achieved at the expense of its reliability or representativeness.

Secondly, there is a large potential for manipulating school layouts to provide for flexibility, though on lines that are different from what was suggested before. This may lead to more exploratory research into other variables of the layout. This points to the importance of tackling uncertainty through some design decision support systems, rather than adopting a historical approach in predicting the development of organisations and designing for this development. That is to say that the problem of obsolescence due to unpredictable change could be encompassed in some way. This might also be very beneficial in predicting the performance of school buildings in use, and hence determining the limits of flexibility, or the degree of planning versus techniques of change, which could be manipulated during the initial design stage to enhance the potential of accommodating change during use. For architects, this is likely to limit the design considerations that architects have to encounter related to flexibility, concentrating on more demanding issues about the human and physical environment interface. For researchers in the field, this study demonstrates that flexibility could be manipulated more effectively during the early design stage.

This study draws their attention to the major role of design in enhancing flexibility in use. It shows that they could approach flexibility within a valid framework of measurement and deal with it empirically, rather than by relying on intuition and subjective judgement This study supports the growing belief in the necessity of adopting research in design, arguing that architects can no longer rely entirely upon intuition, personal experience and precedent. It acknowledges that practice must be based on empirical knowledge carefully acquired through research, as well as personal experience, which can improve both the design process and the final architectural product.

References

Al-Nijaidi, H.R. (1985) Flexibility of Design in Buildings, Ph.D. Thesis, Postgraduate Research School, Department of Architecture, Oxford Polytechnic.

Aylward, G. (1968/69) "Towards a Theory for Describing and Designing Adaptability in the Built Environment", *Transactions of Bartlett Society* **7**, 127-148.

Aylward, G.M. and Lapthorne, K. (1974) Designing for Stability in Designing for Change, Working Paper 3, Technical Research Division, Department of Architecture, University of Cambridge.

Bottle, P.F. (1971) The Economic of Adaptability: A Comparative Cost of the Initial and Life Costs of Partitions in Relation to Likely Frequency of Relocation , Laboratories Investigation Unit (LIU), Paper No. 4.

Bottle, P.F. and Piper, L.E. (1970) Deep or Shallow Building: A Comparison of Costs in Use, LIU, Paper No. 2.

Cowan, P. and Nicholson, J. (1965) "Growth and Change in Hospitals", *Transactions of Bartlett Society* **3**, 63-79.

Davies, M. and Th'ng, R. (1972) SPACES 3: A Programme for the Appraisal of School Building Designs, ABACUS Occasional Paper No. 24, University of Strathclyde, Glasgow.

Davies, M. and Th'ng, R. (1975) ECOLE 1 - The Basic Structure and Format of a Computer Programme for Accommodation Scheduling in Educational Environment, ABACUS Occasional Paper No. 40, University of Strathclyde, Glasgow.

Department of Education and Science/Architects and Building Branch (1972) Polytechnics: Planning for Development, Design Note 8.

Dluhosch, E. (1973) Flexibility / Variability in Prefabricated Housing, Ph.D. Dissertation, Dept. of Architecture and Urban Planning, University of California-Berkeley.

Dluhosch, E. (1974) "Flexibility / Variability in Programming", *Industrialisation Forum* **5**, 40-44.

Farbstein, J. (1974) "The Definition and Description of Activity", *Journal of Architectural Research* **3**, 18-25.

Fawcett, W.A. (1976a) "School Management and School Design", *Journal of Architectural Research* **5**, 10-12.

Fawcett, W.A. (1976b) "Measuring Adaptability in the Design of Buildings", *Martin Centre for Architectural and Urban Studies Transactions* **1**, 5-40.

Fawcett, W.A. (1978) A Mathematical Approach to Adaptability in Buildings, Ph.D. Thesis, Postgraduate Research School, Department of Architecture, University of Cambridge.

Habraken, N.J. et al. (1976) Variation: The Systematic Design for the Support, Laboratory of Architecture and Planning at MIT, MIT Press, Cambridge, Massachusetts.

Llewelyn-Davies et al. (1973) *Long-life, Loose-Fit: A Comparative Study of Change in Hospitals Building*, Llewelyn-Davies-Weeks-Forestier-Walker-Bu,London.

Markus, T.A. et al. (1972) BuildingAppraisal, Building Performance Research Unit, University of Strathclyde.

Maver, T.W. (1979) "A Time-Space Odyssey", in T. Markus (ed.), *Building Conversion and Rehabilitation: Designing for Change in Building Use*, Butterworths, London, pp. 12-30.

Moharram, L.A. (1980) A Method for Evaluating Flexibility of Floor Plans in Multi-Storey Housing, Ph.D. Dissertation, University of Pennsylvania.

Oddie, G. (1974) A Strategy for Flexibility of Building for Education, *in Proceedings of UIA Second Seminar on Flexibility of Building for Education*, Working Group on Educational Spaces, Berlin, pp. 112-126.

Oddie, G.B. (1975) *Industrialised Building for Schools*, OECD, Paris.

Oxmon, H.S. (1978) Flexibility of Support: An Analysis of the Selected Design Variables Upon the Flexibility of Support Type Housing, D.Sc. Thesis Research, Technion Institute of Technology.

Pearson, E. (1975) *School Building and Educational Changes*, OECD, Paris.

Pena, W. et al. (1977) *Problem Seeking, An Architectural Programming Primer*, Cahners, Boston.

Rawlinson, C. (1978) "Space Utilization in Hospitals", *Journal of Architectural Research* **7** , 4-12.

Steadman, P. (1983) *Architectural Morphology: An Introduction to the Geometry of Building Plans*, Pion Limited, London.

Van Delan, D.B. (1979) *Understanding Educational Research: An Introduction*, McGraw-Hill, New York.

Vickery, E.C. (1979) The Measurement of Organisational Flexibility, Ph.D. Thesis, Dept. of Architecture, University of Bristol.

Weeks, J. (1964) "The Indeterminate Architecture", *Transactions of Bartlett Society* **2**, 85-106.

Weeks, J. (1969) "Multi-Strategy of Buildings", *Architectural Design* **39**, 530-540.

A DECISION-MAKING PROCESS FOR CHOICE OF A FLEXIBLE INTERNAL PARTITION OPTION IN MULTI-UNIT HOUSING USING DECISION THEORY TECHNIQUES

A. FRIEDMAN
Affordable Homes Program
School of Architecture
McGill University, 815 Sherbrooke St. West, Montreal, Quebec
Canada, H3A 2K6

ABSTRACT. Recent demographic changes have increased the heterogeneity of user groups in the North American housing market. Smaller households (e.g. elderly, single parent) have non-traditional spatial requirements that cannot be accommodated within the conventional house layout. This has created renewed interest in Demountable/Flexible internal partition systems. However, the process by which designers decide which project or user groups are most suited for the use of these systems is quite often complex, non-linear, uncertain and dynamic, since the decisions involve natural processes and human values that are apparently random. The anonymity of users when mass housing projects are conceptualized, and the uncertainty as to the alternative to be selected by the user, given his/her constantly changing needs, are some contributing factors to this effect. Decision Theory techniques, not commonly used by architects, can facilitate the decision-making process through a systematic evaluation of alternatives by means of quantitative methods in order to reduce uncertainty in probabilistic events or in cases when data is insufficient. The author used Decision Theory in the selection of flexible partition systems. The study involved a multi-unit, privately initiated housing project in Montreal, Canada, where real site conditions and costs were used. In this paper, the author outlines the fundamentals of Decision Theory and demonstrates the use of Expected Monetary Value and Weighted Objective Analysis methods and their outcomes in the design of a Montreal housing project. The study showed that Decision Theory can be used as an effective tool in housing design once the designer knows how to collect basic data.

1. Introduction

In the selection of design alternatives, decisions are often required, irrespective of the state of completeness and the quality of information. Decisions therefore have to be made under conditions of uncertainty in the sense that the consequences of a given decision cannot be determined with complete confidence. Housing design problems are very often complex, non-linear, uncertain, and dynamic, involving natural processes and human values that are often apparently random (Lifson and Shaifer 1982). Such phenomena are naturally indeterminate and cannot be definitively described. Alternatives for future design flexibility to be given to a user involve uncertainty due to insufficient data (i.e. anonymity of user when a mass housing project is conceptualized) and probabilistic events (i.e. uncertainty as to the alternatives to be selected by the user, given his or her constantly changing needs). For quantitative purposes probability can be considered as a numerical measure of the likelihood of occurrence of an event relative to a set of possibilities (e.g. the probability that the user will implement alternative A is higher than implementing B, therefore A is recommended).

There are many methods which can be used to assist the decision-maker in the decision-making process through systematic evaluations of alternatives by means of quantitative methods. These techniques are most often used in engineering applications where exact data is available and the uncertainty of the variables is quantifiable. In most cases, the methods are very complex and involve long calculation, thereby not lending themselves readily to architectural applications. In

H. Timmermans (ed.), Design and Decision Support Systems in Architecture, 179–188.

architecture, there are several indeterminate factors due to the nature of housing design and human nature, as mentioned above; quantifiable methods for decision-making are not necessarily useful or appropriate. However, the author found that the Decision Theory method allowed for a great flexibility of variables, and also allowed for the isolation of elements which could be modified for architectural applications using a common-sense, intuitive approach.

Decision Theory is not commonly applied to architectural problems in general nor to housing in particular. Very often architects prefer to use visual models when decisions are made and an alternative has to be selected. Visual models have been far easier to set and manipulate than available mathematical models. Gero (Gero and Radford 1980) suggests that recent technological developments, particularly in computer science, have helped probabilistic methods to take their place alongside (but in no way to replace) visual models. Most applications of Decision Theory deal with technical aspects of housing design; however, it will be useful to review some of these cases and study the criteria applied and their assistance in the problem-solving process.

Smith and Green (1975) used probabilistic methods for quantifiable determination of subsystems in building design. They concentrated on their operational behaviour where uncertainty in predicted performance is caused by variability of demand or loading. Cost of failure was used as a performance attribute. Bayazit (1985) used a multi-variable decision-making method for the determination of priorities of building performance attributes. However, he cites "some applicability difficulties in practice because of the possibilities of large numbers of performance attributes required from designers and initiators." Davidson (1971) used decision criteria with multiple objectives to solve a problem of regional development. The decision was made based on weighing of objectives, confronting them with alternatives and ranking them using a scale of 0 to 1. This method can serve as an efficient tool for the measurement of user satisfaction in the present study. Mattar et al. (1978) used the decision-by-exclusion rule in the design of the external wall of a single-family dwelling. The application of the theory was demonstrated in three stages, and the sensitivity of decision to changes in the designer input was also considered. Initial cost was an attribute in the process as well as fire resistance, condensation and sound. In spite of the fact that the studies presented above do not demonstrate the application of Decision Theory in housing layout selection, they show its potential as a useful tool in the selection of flexibility alternatives. Prior to demonstrating the use of the technique the author wishes to outline some of its basic principles which are relevant to this study.

2. Decision Theory: Relevant Principles

In general decision problems, a framework for systematic analysis is needed. It should include the following components (Ang and Tang 1982): a list of all feasible alternatives; a list of all possible outcomes associated with each alternative; an estimation of the probability associated with each possible outcome; the criterion for decision; and a systematic evaluation of all alternatives. Those components will be elaborated below; examples will be given using the case of designing for flexibility.

Listing the feasible alternatives is a phase in the design process preceded by collection of relevant information about user needs, evaluation and synthesis. The alternatives set by the architect are several options given to the user. These concern possible changes of internal subsystems (e.g. wet or dry partitions, mechanical systems). Listing alternatives is a 'creative' stage in the design process and, for the benefit of the design solution, a large number of alternatives is preferred. The possible outcomes will be valued based on the attributes that the architect has designated to the performance variables. They can be, for example: time needed for the erection of the system in days/hours/minutes; savings to the user when compared with conventional projects and size of material needed for implementation of design alternatives. Each alternative can be associated with more than one outcome according to participants' most important variables.

A diagrammatic tool that displays the sequence of alternatives and their associated outcome is the Decision Tree (Figure 1). The branches of the tree represent either decision alternatives or chance events. Decision alternatives start at decision nodes (represented by squares); chance events start at chance nodes (represented by circles). In a 'real life' decision-making process, chance nodes will be forecasted by the architect. All paths through the tree originate at the left of the diagram with a square representing the major decision milestone that is the reason for the subsequent decisions represented in the tree and for the activities of the decision process (Raiffa 1968).

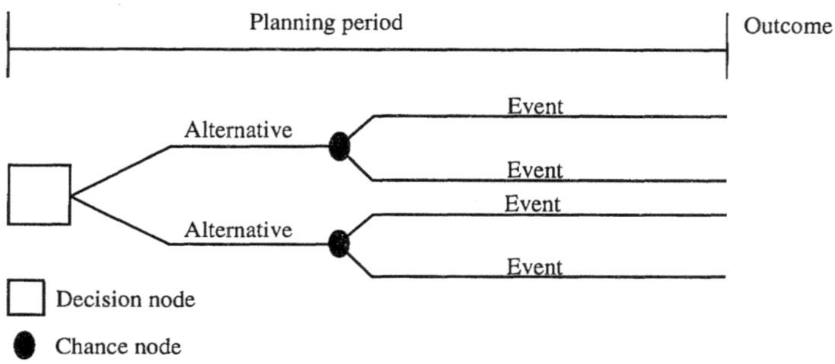

Figure 1. A decision tree

The probability that a particular alternative will be exercised is measured on a continuous scale ranging between 0 and 1. An alternative that has absolutely no chance of occurring has a probability value of 0, but on the other hand, an event that is sure to occur has a probability value of 1. The more likely an event is to occur, the higher will be its probability value which can be expressed in the following equation:

$$\text{probability that alternative A will occur} = \frac{\text{number of times alternative A occurred}}{\text{number of times counted}}$$

Ang and Tang (1975) suggested three methods to determine the probability value of an alternative: observed information, estimated from the frequency of observed events; subjective judgement of the decision-maker; and the Bayesian approach. In the last approach, the judgemental information of the decision-maker is combined with the observed information for an updated estimation of the probabilities. For flexibility problems, observed information and subjective judgement are more relevant due to the existence of studies and surveys mapping user behaviour (e.g. Teasdale 1977; Evendon 1983).

Most decisions are made under conditions of uncertainty in which performance variables of a given alternative may be stated only in terms of respective probabilities. Also, depending on the temperament, experience and degree of risk aversion of the decision-maker, the 'best decision' may mean different things to different people at different times. Nevertheless, some criteria which are relevant for a decision-making process for flexibility in housing are presented. One of these is Expected Monetary Value, a completely objective measure of the value of money. It implies that every dollar within a sum of money provides that same amount of satisfaction.

$$\text{EXPECTED MONETARY VALUE} = \sum_{j=1}^{n} X_j P(x_j) \tag{1}$$

where:

X_j = monetary value associated with alternative X
$P(x_j)$ = probability that alternative X_j will occur.

The other criterion is a Weighted Objective Decision Analysis with a determination of multi-attribute utility functions instead of single utility function. These criteria were used in environmental design primarily for the non-money attributes (Davidson 1971). The selection process is performed through summation of the contribution of alternatives to pre-determined objectives using a table (Figure 2).

Relative weights	w_1	w_2 ...	w_j ...	w_n		Overall relative utility
Alternatives	Objectives					
	O_1	O_2 ...	O_j ...	O_n		
A_1	p_{11}	p_{12} ...	p_{1j} ...	p_{1n}	U_1	$= \sum_j p_{1j}w_j$
A_2	p_{21}	p_{22} ...	p_{2j} ...	p_{2n}	U_2	$= \sum_j p_{2j}w_j$
:						:
A_i	p_{i1}	p_{i2} ...	p_{ij} ...	p_{in}	U_i	$= \sum_j p_{ij}w_j$
:						:
A_m	p_{m1}	p_{m2} ...	p_{mj} ...	p_{mn}	U_m	$= \sum_j p_{mj}w_j$

A_i = The ith alternative, $i = 1$ to m

O_j = The jth objective, $j = 1$ to n

w_j = The relative weight of importance to objective O_j, $j = 1$ to n

p_{ij} = The probability of attaining objective O_j through alternative a_i

U_i = The overall relative utility of alternative a_i

Figure 2. Principle of weighted objective decision analysis
Source: Ang and Tang (1982)

The decision criteria are instrumental in the selection of alternatives. In housing problems, where monetary and human values are often mixed, the lack of a tool that combines both is evident.

It may require the use of several criteria that cover both sets of values. Following the selection of the relevant criteria a systematic evaluation of all alternatives is made. The preferred alternative, or sequence of all alternatives, is selected on the basis that it represents the needs and objectives of all the participants.

3. Demonstration

The author selected a dwelling in a typical low-rise three-storey wood-frame walk-up apartment building in Montreal (Friedman 1987). Five flexibility design options (called alternatives) were suggested for the unit. The plan (Figure 3) and the description of their use is presented below:

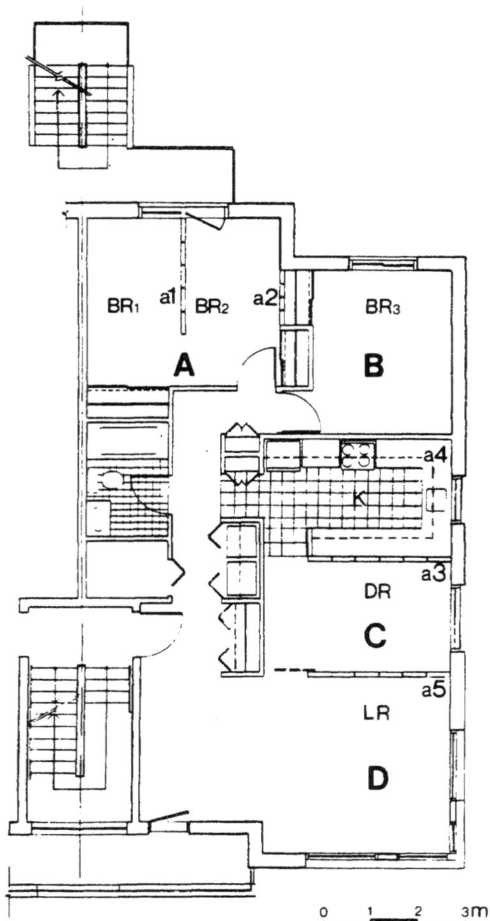

Figure 3. Plan showing the five flexibility options

ALTERNATIVE 1: *Demountable wall dividing room A.*
Description of Use: Upon occupancy dual-headed households without children can a) use room A as a master bedroom and store the partition, b) use room A with the dividers for their personal use (studies) or c) when children are born room A can serve as two bedrooms.

ALTERNATIVE 2: *Demountable wall section, divider between rooms A and B.*
Description of Use: If room B is used as a master bedroom (dual-headed households without children), wall 2 can be folded and stored to make more space.
When a child is born, the crib can be in room A and easy access can be provided at night when care is needed.
When a second child is born and/or when privacy is required when they attend school, the partition can be closed and more closet space can be added.

ALTERNATIVE 3: *Enclosure of kitchen space.*
Description of Use: Dual-headed families without children can have the partition removed and stored if wanted to allow open space between the LR and the kitchen. When children are older and/or more consumer goods are purchased, the partition can be re-installed and more kitchen cabinets added.

ALTERNATIVE 4: *Installation of mechanical and electrical service lines to allow future dismantling of kitchen.*
Description of Use: Pending their economic priorities, buyers may upgrade kitchen cabinets, appliances and finishes. To facilitate this, mechanical and electrical service lines have to be installed in a flexible manner.

ALTERNATIVE 5: *Demountable divider between the LR and the DR.*
Description of Use: When children are born and play space is needed, the wall can be removed. The same wall could also have been removed upon occupancy and stored and re-installed when a definite space for DR is required.

The author then evaluated the above design alternatives. Prior to the evaluation process he assessed each alternative and summed up the outcome in Table 1. It is important to mention that these alternatives were selected based on their contribution to the long-term economic objectives (i.e. return on initial economic investment). The architect displays the collected data on a Decision Tree (Figure 4) and uses the Expected Monetary Value criterion in the selection process.

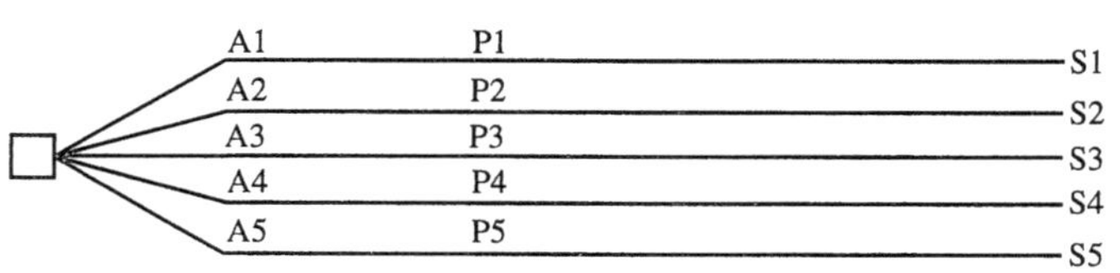

Figure 4. The decision tree

$$EMV = \sum_{j=1}^{n} S_j \, P(S_j) \qquad (2)$$

$EMV\ 1 = 500 \times 0.9 = 450$

$EMV\ 2 = 250 \times 0.5 = 125$
$EMV\ 3 = 700 \times 0.2 = 140$
$EMV\ 4 = 400 \times 0.7 = 280$
$EMV\ 5 = 250 \times 0.75 = 150$

Based on Expected Monetary Value, alternatives A1 and A4 will maximize the contribution to the user's objectives. In spite of the fact that alternative A3 promises the highest future savings potential, its probability of execution is the lowest. The author therefore selected those alternatives that yielded the highest values of S.

TABLE 1. Evaluation of design alternatives

Alternatives		*S*	*P*
A_1	Demountable wall dividing room A	500	0.9
A_2	Demountable wall section, divider between rooms A and B	250	0.5
A_3	Enclosure of kitchen space	700	0.2
A_4	Installation of mechanical and electrical service lines to allow future dismantling of kitchen	400	0.7
A_5	Demountable divider between the LR and the DR	250	0.75

where:

S = assumed Life Cycle Savings potential in dollars (minimum residency period of eight years). All figures were calculated based on contractors' estimates and the autor's field research for the cost of labour and materials in Canadian dollars

P = the probability that the user's investment will result in economic benefits in the residency period. The probability values are based on observed information (i.e. survey of occupied projects) and on the architect's subjective judgement

4. Weighted Objective Decision Analysis

When one of the participants (in this case an architect and a builder) has more than one objective, which is most often the case, a Weighted Objective Decision Criterion can be used. In the following examples, the builder listed his objectives to the architect who processes and uses them in the selection of alternatives. The objectives are:

01 - The alternatives proposed to the user will be those which can promote sale (i.e., are known to be a marketing device)

02 - The implementation of the alternative will not require more than two meetings with the user.

03 - Not more than two different trades will be involved in the installation of each alternative.

04 - The installation of the alternative will not last more than two extra working days per apartment.

4.1. RELATIVE WEIGHTS OF OBJECTIVES

The author used ordinal ranking in which he lists the alternatives in decreasing order of importance (Table 2).

TABLE 2. Alternatives in decreasing order of importance

Weights	Rationale
01 > 02	The builder sees sale potential as a prime objective due to stiff competition in the area and loan constraints.
01 > 03, 01 > 04	The builder does not mind having an extra meeting with some of the users. However additional trades will raise the cost of flexibility and will increase construction time.
04 = 03	Based on the builder's calculations, the use of trades is equivalent to the addition of two working days.
01 > 03 + 04	High sale potential is more important to the initiator than objectives 03 and 04 combined.

4.2. RELATIVE RANKING OF OBJECTIVES

Subsequently a cardinal ranking of each of the objectives is established; the relative importance of each of the objectives with respect to the others is evaluated by assigning each of them a numerical weight. Starting with the most important objective, an arbitrary weight (e.g. 100) is assigned. Next, numerical weights are assigned to each of the other objectives. The set of initial weight assignments is then cross-checked for consistency with the preference statement in the ordinal ranking (Table 3).

4.3. RELATIVE UTILITY OF ALTERNATIVES

The next step in the decision analysis included a listing of the above alternatives and the assignment of probabilities (p_{ij}) that were formed earlier, based on objective information and the architect's subjective judgement. Finally, the overall relative utility of each alternative is computed as in Figure 5 (Ang and Tang 1982):

$$U_i = \sum \; p_{ij} \, W_j \tag{3}$$

TABLE 3. Cardinal ranking of objectives

Objectives		Rank		Rationale
		1st attempt	re-evaluation	
01	Marketing device	100	100	
02	Maximum 2 pre-occupancy meetings	25	25	
03	Maximum 2 trades	75	50	Both 03 and 04 seem to be equal to the builder in their importance in comparison to 01
04	Maximum 2 working days	75	50	

The recommended alternative is the one which has the maximum relative utility. In the above example, alternative A5 seems to be the one which will maximize the contribution to the builder's objectives. Weighted Objective Decision Analysis provides a useful means by which many objectives of a participant can be evaluated through the assignment of probability values.

Relative weights	$W_1=100$	$W_2=25$	$W_3=50$	$W_4=50$	
	Objectives				Overall relative utility
Alternatives	01 Marketing device	02 Meetings	03 Trades	04 Working days	
A_1 Demountable wall dividing room A	0.7	0.2	0.8	0.6	145
A_2 Demountable section divider between room A and B	0.9	1.00	0.3	0.5	155
A_3 Enclosure of kitchen space	0.2	1.00	0.7	0.6	110
A_4 Mechanical and electrical system for dismantling	0.8	0.3	0.5	0.3	127.5
A_5 Demountable divider between LR and DR	1.00	0.95	0.4	0.3	158.75

Figure 5. Relative utility of alternatives

5. Conclusions

Architects in North America tend to rely on a variety of models in their decision-making process,but rarely on scientific ones. In housing, the need to incorporate the builder's objectives into the design provides an opportunity to experiment with other models since most of his decisions are based on quantifiable attributes (e.g. economic). In the above demonstration the author showed that Decision Theory can be a useful tool in combining a range of attributes. Expected Monetary Value proved to be a simple technique for deciding what the probability value in the application is going to be. That determination counts for most of the successes and the reliability of the applications. In examining the results in the above demonstration, the author believes that they make sense with regard to family space needs through its life cycle.

The Weighted Objective Decision Analysis technique, although more complicated, provides margin for adjustment to the needs of the decision-maker. The primary advantage of the technique, in the author's opinion, is the inclusion of a subjective approach in the process. The architect can list all of his/her options and assess them. To conclude, Decision Theory can be a useful tool when quantifiable attributes are combined with other factors in the decision-making process.

References

Ang, A.H. and Tang, W.H. (1975) *Probability Concepts in Engineering Planning and Design Vol I: Basic Principles*, John Wiley and Sons, New York.

Ang, A.H. and Tang, W.H. (1982) *Probability Concepts in Engineering Planning and Design Vol II: Decision, Risk and Reliability*, John Wiley and Sons, New York.

Bayazit, N. (1985) "Multi-Variable Decision-Making Methods for the Determination of Properties of Building Performance Attributes", *Design Methods and Theories* **19**, 324-336.

Davidson, R. (1971) "An Application of Utility and Decision Theory in a Planning Environment", paper presented at the 39th National Meeting, Operations Research Society of America, Dallas, Texas.

Evendon, L.J. (1983) Single Family House Expansion, Report submitted to CMHC, Simon Fraser University, B.C.

Friedman, A. (1987) A Proposed Decision Making Model for Initiators of Flexibility in Multi-Unit Housing, Ph.D. Dissertation, Université de Montréal.

Gero, J. and Radford, A. (1980) "On Optimization in Computer-Aided Architectural Design", *Building and the Environment* **15**, 73-80.

Lifson, M.W. and Shaifer, E. (1982) *Decision and Risk Analysis for Construction Management*, John Wiley and Sons, New York.

Mattar, S., Manning P., Fazio, P., and Bitterlich, W. (1978) "Application of the Decision By Exclusion Rule: The Design of External Walls", *Building and the Environment* **13**, 217-232.

Raiffa, H. (1968) *Decision Analysis: Introductory Lectures on Choices Under Uncertainty*, Addison Wesley, Reading, Massachusetts.

Smith, S.B. and Green, M.F. (1975) "Probability Theory as a Basis for Building Design", *Building Science* **10**, 177-183.

Teasdale, P. (1977) User-Generated Program for Low-Rise Multiple Dwelling Housing, University of Montreal.

A TRICHOTOMIC SEGMENTATION PROCEDURE TO EVALUATE PROJECTS IN ARCHITECTURE

E. MORTOLA and A. GIANGRANDE
Dipartimento de Progettazione Architettonica e Urbana
Universita di Roma "La Sapienza"
Via A. Gramsci, 53 00197 Roma
Italia

ABSTRACT. This paper illustrates a model used to construct the evaluation module for 'An Interface for Designing' (AID), a system to aid architectural design. The model can be used at the end of every cycle of analysis-synthesis-evaluation in the intermediate phases of design development. With the aid of the model it is possible to evaluate the quality of a project in overall terms to establish whether the project is acceptable, whether it should be elaborated ex-novo, or whether it is necessary to begin a new cycle to improve it. In this last case, it is also possible to evaluate the effectiveness of the possible actions and strategies for improvement. The model is based on a procedure of trichotomic segmentation, developed with MCDA (Multi-Criteria Decision Aid), which uses the outranking relation to compare the project with some evaluation profiles taken as projects of reference. An application of the model in the teaching field will also be described.

1. Introduction: Design Process and Evaluation

Under the pressure of social needs and the desire to address the complex problems of today's world, the 60s and the early 70s witnessed the birth of a number of methods and techniques intended as aids in the design process. Many of these methods sought to break down the barriers between disciplines and to demonstrate the applicability of the scientific method to the fields of artistic research and design (Alexander 1964; Archer 1965, 1968; Broadbent 1970; Gregory 1970; Jones 1963, 1970; Luckman 1967). The intrinsic limits of the methods proposed, together with the declining interest in a more systematic way of projecting, led architects in the mid-70s to concentrate on the language and form of architecture.

In recent years, however, the widespread diffusion of computer technologies, the emergence of the so-called environmental crisis, and the growing complexity of the decision-making context have led architects to reconsider method. They came to think of it as an important tool to manage the design process, from a technical point of view as well as from an organizational one. In fact, the architects' tasks seem to be growing more and more difficult as time passes. Architects are finding themselves in situations where they have to coordinate multidisciplinary working groups, where they have to identify design actions to protect human and environmental values and, at the same time, keep in mind the limits imposed by the decision-making context.

A number of methods consider designing to be a process that consists of a sequence of cycles of analysis-synthesis-evaluation (Archer 1965; Jones 1963; Luckman 1967; Maver 1977, 1988). Such methods usually differ from one another owing to the number of cycles in the process or the number of steps in each cycle. The basic idea is that every cycle must be repeated several times before completing the project. This way of conceiving the design process is useful in some kinds of projecting (for example, in mechanical engineering), but it is much less useful in architecture. In fact, the designing of architecture - especially in the early stages - appears as an ill-structured and ill-defined problem, one that's difficult to control. We can encounter alterations in the sequence as well as in the duration of the various steps in a cycle. A procedure based on

H. Timmermans (ed.), Design and Decision Support Systems in Architecture, 189–200.

successive cycles of analysis-synthesis-evaluation proves to be useful only in the more advanced phases of the design process, when the solution has reached relative stability.

Whereas methods of analysis are fairly numerous and efficient, methods of synthesis are virtually absent. In fact, this step in the cycle depends entirely on the capacity for synthesizing on the part of the individual architect. As far as evaluation is concerned, the methods available refer to very specific aspects of the process. What we are talking about, in general, are methods regarding technical control. It would be inexact, in our opinion, to consider them as real tools of evaluation, because they do not allow us to evaluate a project from an overall point of view.

This paper describes an evaluation method that is based on MCDA (Multi-Criteria Decision Aid). This method is particularly suitable to a process with different 'actors' (designers, consultants, users, etc.). But MCDA can also be used by the individual designer who intends to simulate the decision-making context. The method allows direct comparison of the performances related to building efficiency in terms of environmental comfort, costs, etc., with other intrinsically qualitative aspects, such as the relationship between the project and the landscape or the historical and cultural context. The model helps the designer evaluate the project in an explicit and global manner without overlooking qualitative aspects. It aids him in making choices which imply complex trade-offs between performances of different types. The method has been used to construct a module of evaluation that forms part of a design support system named 'An Interface for Designing' (AID), which is in the working stage (Mortola 1988, 1989a, 1989b, 1990; Giangrande et al. 1991). This system, which can be applied to different phases of the design process, was developed with the aid of Hypercard. It is made up of a number of modules, some of which can be interfaced with other commercial systems and programs frequently used in CAAD or in scientific fields.

2. Evaluation and MCDA

In many disciplines, including economics, statistical theory of decisions, operational research, games theory, etc., the rationality of a choice is guaranteed when the decision-maker, knowing all the feasible alternatives and their consequences, selects the best alternative on the basis of a single criterion (e.g., profit maximization). This paradigm of decision-making went into crisis in the 50s as a result of studies carried out in the social choice theory (Arrow 1951), in the behavioural sciences, and in organization theory (March and Simon 1958). These studies have progressively led to a lack of conviction about the possibility of explaining and forecasting decision-maker's behaviour according to codes of rationality (Simon 1957). These studies questioned the validity of many classical evaluation models and favoured, from the mid-60s, the development of a new field of research which put to good use the theories of the previous decade: the Multi-Criteria Decision Aid (MCDA) (cf. Jacquet-Lagrèze et al. 1978; Roy 1968, 1985). The methods and techniques developed in MCDA furnish the actor with the tools that help him to make a decision in the light of multiple and often conflicting points of view.

According to this approach, the term 'optimization' loses its meaning. In contrast with some evaluation techniques developed in the field of operational research, such as MOLP (Multiple Objective Linear Programming), the MCDA methods do not lead to the 'objectively' best solution, which, in fact, does not exist. In MCDA, the decision is seen not so much as a result but as a process with various steps in which one or more dominant points of view prevail. The rationality of the decision process is evaluated as 'coherence with the objectives of the decision-makers' (Sfez 1973). In this point of view, it is important that a method accommodates the complex game of interaction between the different elements of the decision-making process.

The 'hard' models of operational research are substituted by 'soft' ones (less mathematical but equally rigorous). This allows the actors to interact directly and more easily with the evaluation tool. In MCDA, concepts such as actor, action and criterion assume particular importance. The actor is an individual (or a group of individuals who participate in a homogeneous and coherent way in the process) who can directly or indirectly influence the decision. By action, we mean the representation that an actor constructs of a solution (or of an element which helps to obtain a solution). In the architectural design, the action is the entire project or a design decision that modifies it. By criterion, we mean a 'model' (indicator, objective function, etc.) that allows us to compare more actions and to evaluate their effectiveness in an explicit way.

In MCDA, the criteria are usually multiple and conflicting. In any case, the criteria used to evaluate the action (or the actions) must make up a coherent 'family', i.e., a set of exhaustive and not redundant criteria (Roy 1985, p. 310). Another fundamental concept is problem formulation. Four problem formulations are considered in MCDA: α, β, γ, δ, (Roy 1985) The objective of the problem formulation is to help in choosing one action considered the 'best among those studies (selection procedure). In the problem formulation β, the objective is to help to sort out all those actions that are "good' (segmentation procedure). In the problem formulation γ, the actions are ranked in order of decreasing preference (ordering procedure). The objective of the problem formulation is to analyze the actions and their consequences to help the decision-maker make a choice. This problem formulation does not necessarily lead to ranking the actions, nor to choosing the best action(s). The applications of MCDA are numerous in the planning and management fields but are less diffuse in architectural fields (Calderaro et al. 1986). In the following sections, we will illustrate the method we used to develop the evaluation module for the AID system. We will then give an example of its application to an experiment in architectural design in a didactic context.

3. The Evaluation Method

The evaluation method is based on problem formulation β. In this problem formulation, the objective is to help to sort out all those actions that are 'good' (segmentation procedure). This is the only MCDA usable problem formulation when we need to evaluate an action not according to a set of alternative actions but with respect to absolute references. In this specific case, we used a trichotomic segmentation procedure (Moscarola and Roy 1977; Roy 1981) to establish if the action is good, bad or uncertain. The procedure assigns the action to one of three categories: K^+ (action accepted), K^- (action rejected) and $K^?$(the action merits an additional examination). The action is evaluated with respect to n criteria by l actors ('judges'). The n criteria are represented by the same number of semantic differentials (Osgood et al. 1957) which correspond to all those aspects, and only those which are meaningful to evaluate the project with reference to the design theme, the context, and the level of definition of the project itself. A semantic differential consists of one or more pairs of adjectives or statements of antonyms. These are the polar terms or factors which have to be used in judging the objects presented. In order to increase the precision of the responses, a range of choices is allowed between each extreme.

In our case, every semantic differential is defined on a scale of 7 levels. The two extreme levels of the scale, 1 and 7, represent respectively the negative and the positive polar terms of the differential. The meaning of every polar term is defined by a sentence and some images, maps, or diagrams of exemplary projects. Every judge assigns the project to a level on the scale which will be all the more close to the positive (or negative) polar term as the project is judged better (or worse) in respect to the considered criterion. In this manner, the judges define l vectors with n components $a_j^k(j=1,2,\dots n;\ k=1,2,\dots l)$. The jth component of this vector represents the level of the project on the criterion j in the opinion of the judge k.

In the applications which interest us, the judges can be the designers and the consultants of a professional studio, the jury of a competition, the members of a building commission, or a board of examining professors. To guarantee a correct evaluation of the project with respect to all the criteria, it is necessary that the jury be composed of experts in different sectors. In a no-conflict situation, the evaluation of the project can be carried out by the judges in collaboration. In the didactic context, a student can simulate the judgments of a board of examiners to identify the aspects (criteria) where his project is weaker. Every judge assigns afterwards the weights w_j^k ($j=1,2,...n$; $k=1,2,...l$) to the criteria. These weights are coefficients which reflect the importance that each judge, on the basis of his own system of values, assigns to the different criteria, considering the design theme and the context. The sum of weights assigned by a judge to all the criteria must be equal to 1.

The procedure asks, moreover, that each judge define a pair of vectors with n components: b_j^k, c_j^k, named respectively the upper and the lower profile. These profiles represent two sets of reference projects which discriminate between acceptable and uncertain projects (b_j^k) and between uncertain and unacceptable projects (c_j^k). Therefore, for every criterion, the level of the upper profile must be greater than that of the corresponding lower profile. If the project is over all upper profiles (or under the lower ones) the project can be assigned, without hesitation, to the category K^+(or K^-). If these conditions are not met, as often happens, it is necessary to compare the project with all the profiles by the outranking relation S defined as follows. Given two actions, a_1 and a_2, we observe that a_1 S a_2 (a_1 outranks a_2) if a_1 is at least as good as a_2 (Roy 1985). To establish if project a outranks a specific profile p (or vice versa), the indexes of concordance $C^k(a,p)$ (Roy 1968) have to be calculated. This is done as follows:

$$C^{k}(a,p)= \Sigma_j w_j^k c_j^k (a,p) \qquad (k=1,2,...l) \qquad (1)$$

where,

$$c_j^k (a,p) = \begin{cases} 1 \text{ if } g_j^k (a) >= g_j^k (p) \\ 0 \text{ if } g_j^k (a) < g_j^k (p) \end{cases} \qquad (j = 1,2,...n;\, k = 1,2,...l) \qquad (2)$$

$g_j^k(a)$ and $g_j^k(p)$ are respectively the levels of a and p on criterion j. The condition $g_j^k(a) >= g_j^k(p)$ is fulfilled if the project is assigned to a level of the semantic differential equal to or higher than the level of the reference profile. The index of concordance can vary between 0 and 1: the closer this value is to 1, the greater is the comprehensive weight of criteria which agree upon the sentence 'a is at least as good as p.' To calculate the symmetrical indexes $C^k(p,a)$, it is sufficient substitute a with p in (1) and (2). In general, $C^k(p, a) \neq C^k(a,p)$.

To be able to assert that a outranks p, it is not sufficient that $C^k(a,p)$ assumes values near 1. If the difference $g_j^k(p)$ - $g_j^k(a)$ is considerable, even for only one criterion, the outranking could not occur, or it would not appear very credible, aside from the value of $C^k(a,p)$. The procedure asks that every judge assigns a veto threshold v_j^k to each criterion (Roy 1978). v_j^k represents the maximum value of difference $g_j^k(p)$ - $g_j^k(a)$ that judge k could accept for criterion j in order to attribute a high credibility to the outranking aSp. The veto thresholds are used to calculate the 'fuzzy' indexes $S^k(a,p)$, which represent the credibility degrees of outranking asp. To calculate these indexes, we must calculate the indexes of discordance $D_j^k(a,p)$.

These are defined as follows:

$$D_j^k(a,p) = \begin{cases} 1 & \text{if } g_j^k(p) > g_j^k(a) + v_j \\ 0.5 & \text{if } g_j^k(p) = g_j^k(a) + v_j \\ 0 & \text{if } g_j^k(p) < g_j^k(a) + v_j \end{cases} \quad (j = 1,2,...n;\ k = 1,2,...l) \quad (3)$$

The indexes $S^k(a,p)$ are calculated in the following way:

$$S^k(a,p) = \begin{cases} C^k(a,p) \text{ if } D_j^k(a,p) \leq C^k(a,p), & (j = 1,2,...n) \\ & (k = 1,2,...l) \\ C^k(a,p) \prod_j \{[1- D_j^k(a,p)] / [1- C^k(a,p)]\}, & \end{cases} \quad (4)$$

where, $\prod_j$ is extended to all the values of j for which $D_j^k(a,p) > C^k(a,p)$.

In general, the smaller the thresholds of veto v_j, the lesser the degree of credibility of the outranking. Comparing the project with all the profiles, upper and lower, for each judge we obtain four indexes: $S^k(a,b)$, $S^k(b,a)$, $S^k(a,c)$, $S^k(c,a)$.

To assign the project to the category K^+ (project accepted), K^- (project rejected) or $K^?$ (the project needs improvement), we use the decision tree shown in Figure 1 (Scharlig 1985; Vansnick 1979), where B^+, B^-, C^+ and C^- represent respectively the number of judges for which the indexes $S^k(a,b)$, $S^k(b,a)$, $S^k(a,c)$ and $S^k(c,a)$ exceed a threshold T. This threshold is chosen empirically. The more it grows in value, the more severe the 'filtering' of the corresponding index and the fewer the number of judges for which the corresponding outranking relation assumes credibility.

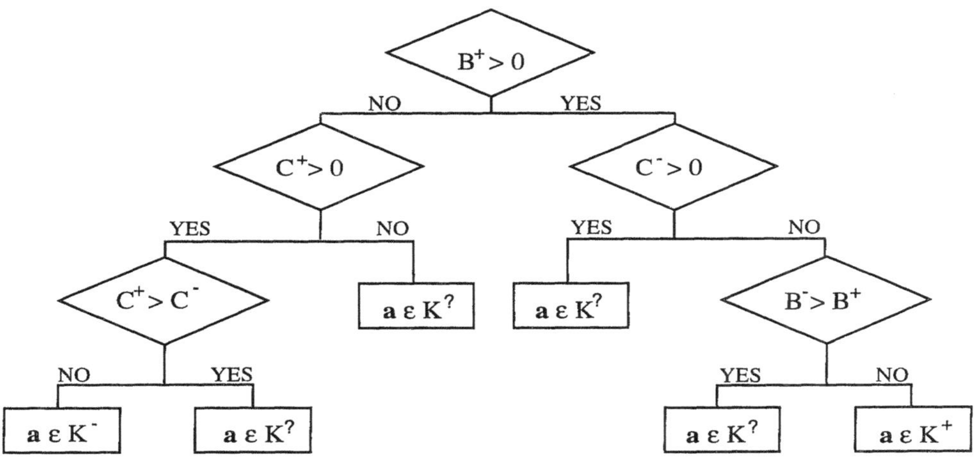

Figure 1. The Decision Tree

Considering that the choice of the threshold is arbitrary, it is necessary to conduct a sensitivity analysis, varying its values in a suitable range (e.g. 0.6 - 0.9) in order to verify the stability of the results of evaluation.

4. An Application of the Evaluation Model

The procedure of evaluation has been applied, as a didactic exercise, to one of the projects of the

Italian students who participated in an inter-school competition sponsored by Strathclyde University in Glasgow for European students of architecture. The theme of the competition was the design of a Media Lover House, a home for the 21st century. This competition intended to celebrate an important event in English architecture. This was the competition proposed in December 1900 by the German magazine 'Zeitschrift für Innendekoration' for the design of an Art Lover House ('Ein herrschaftliches Wohnhaus eines Kunstfreundes'). One of the participants was the architect C.R. Mackintosh. His project is now taking shape in BellaHouston Park, for the occasion of the celebration of Glasgow architecture, which marked the European City of Culture for 1990. The building is intended for habitation. Some of its characteristics are connected with the kind of user (a media lover). Specifically, it should provide for collateral activities to satisfy professional, social, and cultural needs, to promote hobbies, and to encourage communication and meeting at the professional and family level, with the aid of advanced technologies. In addition, the

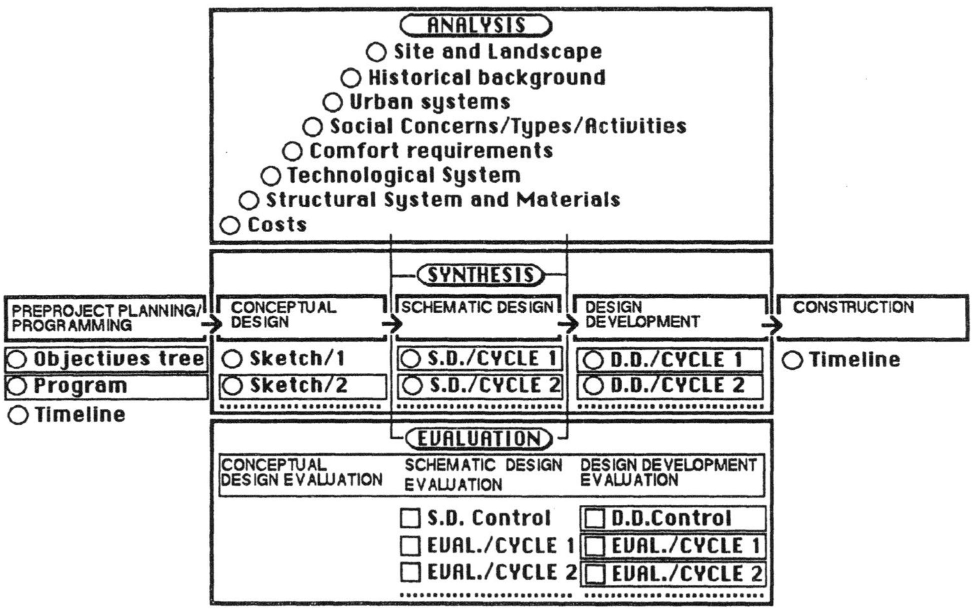

Figure 2. The menu of an interface for designing (AID)

project should propose innovative solutions for the management of the building, in order to obtain the energy to be used from the natural surroundings. The resident group consists of six permanent inhabitants, not necessarily linked by family ties, and two guests. The participants in the competition could also determine the localization of the house.

In the following, we will describe the application of the evaluation procedure to the project developed for this competition and successively for the 'thesis of laurea' by the students Francesca R. Castelli and Anna Gadola (Chairman: Prof.Arch. Elena Mortola; Co-chairman: Arch. Maurizio Clarotti). In developing their project, the students used the AID system. The project was emblematically located in BellaHouston Park, in front of the Art Lover House of Mackintosh.

The evaluation procedure has been applied by the students to the last two cycles of analysis-synthesis-evaluation of the development design phase (Figure 2). With the aid of the procedure, in

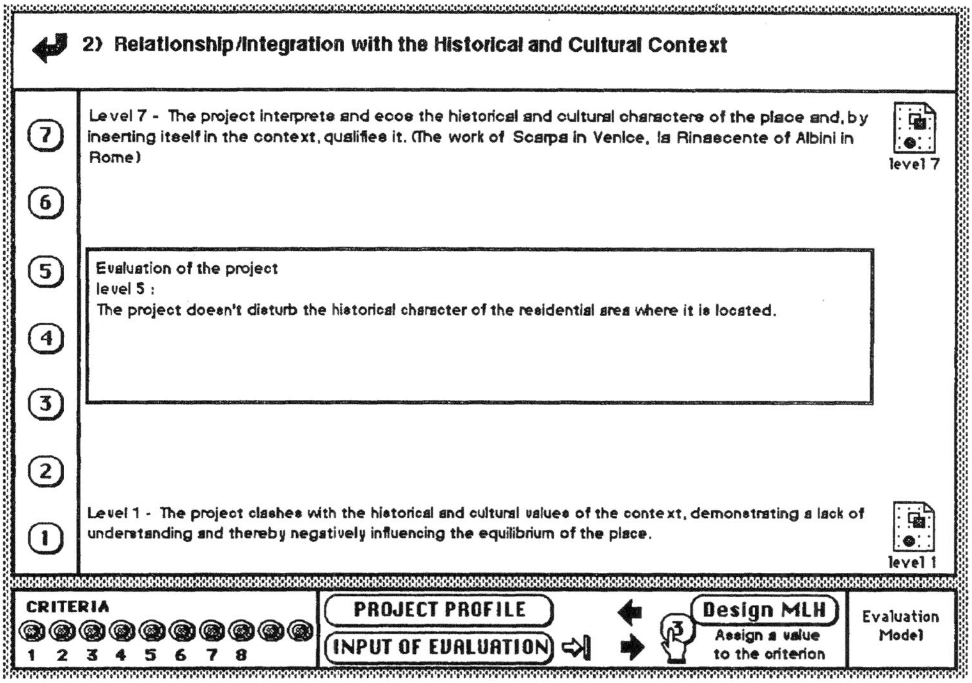

Figure 3. An example of a semantic differential (criterion 2)

the next-to-last cycle, the students defined a strategy to modify the project. A strategy is the prefiguration of a set of design actions which could improve the project with respect to one or more criteria. To evaluate the efficacy of the strategy, it is sufficient to apply the evaluation model to the vectors which represent the estimated performances of the modified project.

To choose the best strategy, the students considered only the actions that do not lead to the total transformation of the project and that allow the minimum design effort. The choice of the evaluation criteria was made jointly by the students and six professors (see Annex). The lack of a criterion on the quality of the formal aspects is justified by the fact that every criterion takes this requirement into account (see, for example, criterion 5). In order to choose the examples that would clarify the meaning of the polar terms of the semantic differentials, many projects were analyzed and evaluated in relation to the different criteria (Figure 3). This helped the students develop a multicriteria mentality, i.e. the capacity to consider together many different aspects of designing. The weights of the criteria were assigned separately by every judge (Figure 4), who successively evaluated the project in collaboration. This evaluation is a result of a preliminary discussion which allows the judges to reach agreement. The judges decided to use the same reference profiles, i.e. b^k = (5,5,5,5,5,5,5,5), and c^k = (3,3,3,3,3,3,3,3), for k=1,2,...l. The profiles are 'flat', that is, the levels are the same for every criterion. At the end of the considered cycle, the project is evaluated as follows: a = (5,5,4,5,4,4,3,4). Considering the results of evaluation, we deduce that the technological systems are not innovative (level 3). Moreover, we deduce that the relationship with the urban context (in functional terms), the comfort in the

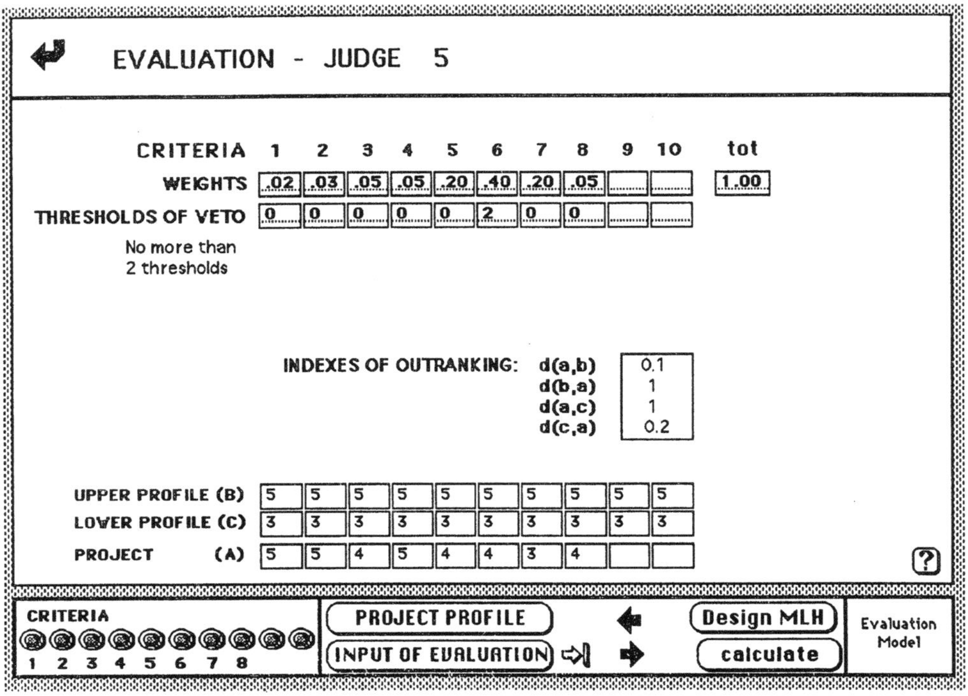

Figure 4. The evaluation of a judge

building, the static solution, the construction, and the maintenance costs cannot be considered satisfactory (level 4). The performances referred to by the other criteria are sufficient (level 5). The project, as evaluated with the trichotomic segmentation procedure, is assigned to the category $K^?$. The sensitivity analysis showed that the evaluation result was stable.

With the aid of the model, the students explored some strategies to improve the project. A strategy is the prefiguration of a set of design actions which could improve the project in respect to one or more criteria. To choose the best strategy, the students considered only the actions that did not imply a revolutionary transformation of the project and allowed the minimum design effort. A first strategy was intended to improve the originality of the technological solutions, especially concerning techniques to obtain the energy to be used for the management of the building from the natural surroundings (from level 3 to level 4), the relationship with the historical pre-conditions, through a study of English houses and the architectural language of Mackintosh (from level 5 to level 6), and the layout problems (from level 5 to level 6).This strategy should improve the project as follows: $a = (5, \mathbf{6},4,\mathbf{6},4,4,\mathbf{4},4)$. These modifications are not sufficient to change the evaluation of project from category $K^?$ to category K^+. The students decided to explore a strategy which considered other improvements, in addition to those considered previously (Figure 5). The new improvements are related to the static solution (from level 4 to level 5) and the comfort conditions inside the building (from level 4 to level 6), i.e. $a = (5,6,4,6,\mathbf{5},\mathbf{6},4,4)$. The evaluation model, with these performances, assigned the project to category K^+ (Figure 6).

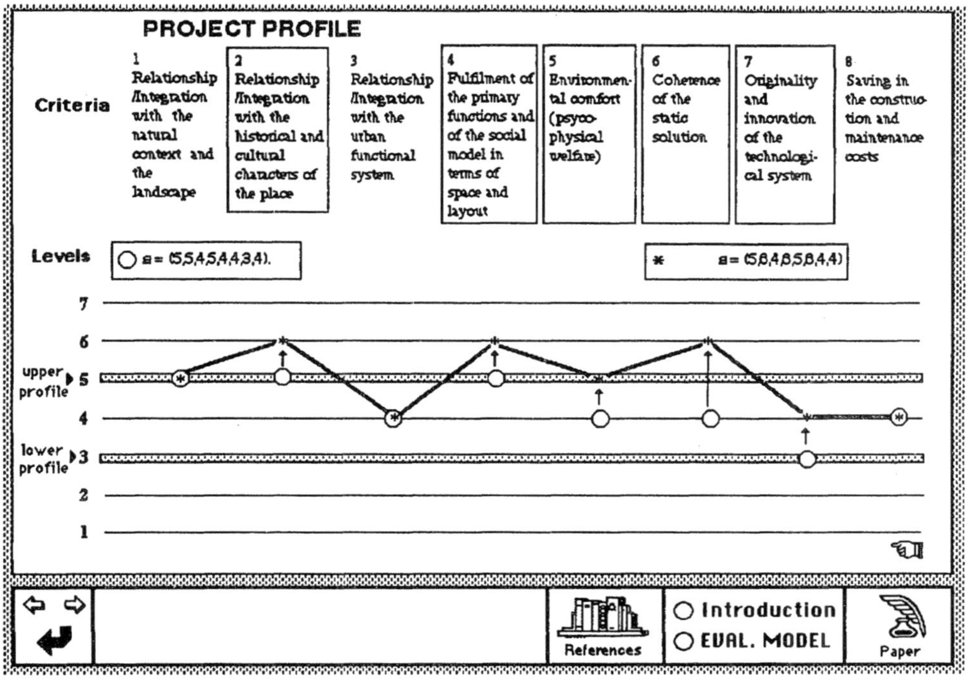

Figure 5. Searching for strategies to improve the project

5. Conclusion

The AID system can be used interactively in a very easy way. It can help the designer, not only in the last but also in the first phases of the design process. It allows the designer to make a direct comparison between the performances related to building efficiency (in terms of environmental comfort, costs, etc.) and other intrinsically qualitative aspects (such as the relationship between the project and the landscape or the historical and cultural context). Therefore, the AID system helps the designer to evaluate the project in an explicit and global manner without overlooking qualitative aspects. It assists him in making choices which imply complex trade-offs between performances of different types.

With the help of the evaluation module of AID, it is possible to simulate how some design decisions will affect the project before deciding to carry out these design transformations. These simulations, achieved with the aid of this module, allow the designer to identify the paths to be taken in order to improve the global quality of the project. In other words, the simulations help him to elaborate strategies, i.e. prefigurations of a set of design actions, which can improve the project with respect to one or more criteria. To choose the best strategy, it is better to consider only those actions that did not imply a revolutionary transformation of the project and at the same time produce the best results with the minimum design effort.

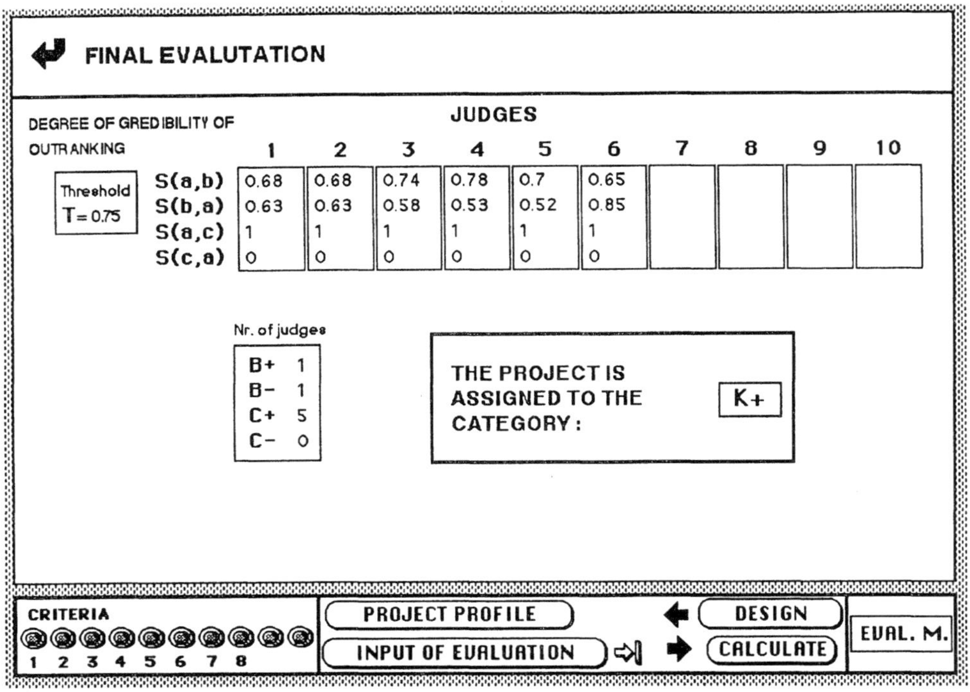

Figure 6. The final evolution

References

Alexander, C. (1964) *Notes on Synthesis of Form*, Harvard University Press, Cambridge, Mass.

Archer, L.B. (1965) Systematic Method for Designers, Council of Industrial Design, London.

Archer, L.B. (1968) The Structure of Design Processes, Ph.D. Thesis, Royal College of Art, London.

Arrow, K.J. (1951) *Social Choice and Individual Values*, John Wiley and Sons, New York.

Broadbent, G.H. (1970) "La creatività", in S.A. Gregory (ed.), *Progettazione Razionale*, Marsilio, Milano, pp. 140-153.

Calderaro, V., Giangrande, A., Mirabelli, P., and Mortola, E. (1986) "Decision Support Systems (DSS) in Computer Aided Design (CAAD)", in *Proceedings ECAADE Conference*, Roma, pp. 20-38.

Giangrande, A., Magrini, A., and Mortola E. (1991) "Uno Strumento di Aiuto Alla Valutazione nel Processo Progettuale", Atti del Convegno Aicographics, Mondadori Informatica, Milano, pp. 370-382.

Gregory, S.A. (ed.) (1970) *Progettazione Razionale*, Marsilio, Milano.

Jacquet-Lagrèze, J., Moscarola, J., Roy, B., and Hirsch, G. (1978) "Description d'un Processus de Décision", Cahiers du Lamsade 13, Université Paris IX Dauphine.

Jones, J.C. (1963) "A Method of Systematic Design", in J.C. Jones and D.G. Thornley (eds.), *Conference on Design Methods*, Pergamon Press, Oxford.

Jones, J.C. (1970) *Design Methods*, John Wiley and Sons, New York.
Luckman, J. (1967) "An Approach to the Management of Design", *Operational Research Quarterly* **18**, 345-358.
Negroponte, N. (1970) *The Architecture Machine*, MIT Press, Cambridge, Massachusetts.
Maver, T.W. (1977) "Building Appraisal", in J.S. Gero (ed.), *Computer Application in Architecture*, ASP, London, pp. 63-69.
Maver, T.W. (1988), "Software Tools for the Technical Evaluation of Design Alternatives", in T.W. Maver and H. Wagter (eds.), *CAAD Futures 87*, Elsevier, Amsterdam.
March, J.G. and Simon, H.A. (1958) *Organization*, John Wiley and Sons, New York.
Mortola, E. (1983) "Una Esperienza Didattica nel Settore del CAAD", in *Computer Graphics: Elaborazione di Immagini*, Jackson, Milano, pp. 147-163.
Mortola, E. (1988) "Computer Grafica e Progettazione: Una Esperienza Didattica", *Pixel* **11**, 21-23.
Mortola, E. (1989a) "Un'Interfaccia per Progettare", *Ricerca e Progetto* **12**, 51-60.
Mortola, E. (1989b) "An Interface for Designing", in *Proceedings CAAD: Education-Research and Practice ECAADE Conference*, School of Architecture, Aarhus, Denmark, 8.5.1-8.5.15.
Mortola, E. (1990) "Computer Grafica e Progettazione", 10 Lezioni di Architettura, Cangemi, Roma, pp. 212-219.
Mortola, E. (1991) "Concorso di Architettura per Studenti Europei: Progettare a House for a Media Lover (A home for the 21st century)", *Progetto e Ricerca* **13**, 73-75.
Mortola, E. and Sansoni, C. (1990) "L'Insegnamento della Computer Grafica e del CAD nella Facoltà di Architettura di Roma", in D. Marini and L. Moltedo (eds.), *Conoscenza per Immagini*, Il Rostro, Milano, pp. 133-160.
Moscarola, J. and Roy, B. (1977) "Procédure Automatique d'Examen de Dossiers Fondée sur une Segmentation Trichotomique en Présence de Critères Multiples", *Revue Française d'Informatique et de Recherche Operationelle* **17**, 145-173.
Osgood, C.E., Suci, C.J. and Tannenbau, P.H. (1957) *The Measurement of Meaning*, University of Illinois Press, Urbana.
Roy, B. (1968) "Classement et Choix en Présence de Points de Vue Multiples: La Méthode ELECTRE", *Revue Française d'Informatique et de Recherche Operationelle* **8**, 57-75.
Roy, B. (1978) "ELECTRE III: Algorithm de Déclassament Basée sur une Représentation Flue de Préference en Présence de Critères Multiples", Cahiers du CERO **20**, pp. 3-24.
Roy, B. (1981) "A Multicriteria Analysis for Trichotomic Segmentation Problems", in P. Nijkamp and J. Spronk (eds.), *Multiple Criteria Analysis*, Gower, Aldershot, pp. 245-257.
Roy, B. (1985) "Méthodologie Multicritère d'Aide à la Décision", *Economica* ***21***, 75-122.
Scharlig, A. (1985) *Décider sur Plusieurs Critéres*, Presses Polytechniques Romandes, Lausanne.
Sfez, L. (1973) *Critique de la Décision*, A. Colin, Paris.
Simon, H. (1957) *Models of Man*, John Wiley and Sons, New York.
Vansnick, J.C. (1979) Une Approche Nouvelle des Problèmes de Décision: L'Aide à la Décision, Faculté des Sciences Economiques et Sociales, Université de l'Etat, Mons.

Annex: Criteria

1. Relationship/integration with the natural context and the landscape

Level 7: The project takes into account the characteristics of the place and its natural aspects,itself becoming an important element of the place and/or improving its quality.

Level 1: The project contrasts with the landscape, altering its characteristics.

2. Relationship/integration with the historical and cultural character of the place

Level 7: The project interprets and recalls the historical and cultural character of the place and, inserting itself in the context, qualifies it.

Level 1: The project puts itself in contrast with the historical and cultural values of the context.

3. Relationship/integration with the urban functional system

Level 7: The project matches closely the hierarchical structure of the road network and urban services. In addition, this integration is plainly rendered in the project's solution.

Level 1: The project neglects the problems of the urban functional system, hindering its function.

4. Fulfilment of the primary functions and of the social model in terms of space and layout

Level 7: The layout and the dimensions of the spaces take into consideration the needs of the users. Different activities and possible changes with time of the needs are alsoforecasted. In addition, the project represents an innovative solution.

Level 1: The layout and the spaces do not take into consideration the needs of the users.

5. Environmental comfort (psycho-physical welfare)

Level 7: The environmental comfort performances are satisfied and do not contrast in any way with the formal solution.

Level 1: The project does not take into consideration the comfort requirements, necessitating extraordinary interventions or costly installations that, nevertheless, do not guaranteea complete solution to the problem.

6. Coherence of structural solution

Level 7: The project points out a precise correspondence between the structural typology and the materials utilized. The static solution shows its structural clarity and coherence.

Level 1: The structural typology is not coherent with the formal solution and contrasts with the materials.

7. Originality and innovation of the technological system

Level 7: The project includes the most advanced technological systems, meeting the highest level of quality neds without altering the architectural image itself.

Level 1: The project does not take into consideration any innovation concerning technological systems. The systems used are out of date and do not integrate with the formal solution.

8. Saving on the construction and maintenance costs

Level 7: The cost level is contained in this respect and takes into consideration the obsolescence problems, security and environmental comfort. Costs fit the resources of the users.

Level 1: The project presents elevated costs and low quality because of erroneous solutions from the financial point of view.

ROP: AN INTERACTIVE SPATIAL OPTIMIZATION AND GROUPING COMPUTER APPLICATION

P.W.M. WINTERAEKEN-BRULS
Department of Architecture, Urban Planning and Management
Eindhoven University of Technology
P.O. Box 513, Postvak 4, 5600 MB Eindhoven
The Netherlands

ABSTRACT. As a part of a research project at Eindhoven University of Technology, the computer application ROP for space-planning problems was tested in practice. The use of the application in a real-world project was evaluated. The decision-making process for the development of alternatives for a courthouse was observed to see how the computer application could support decision-making. The aim of this paper is to describe the performance of ROP in a real-world setting. ROP appears to be a useful instrument in decision-making for space-planning problems. Especially in the early stages of the design process, it enhances insight among all participants in a project team. It can also be used in situations where little information is available. To conclude, ROP appears to enhance communication between members of a design team.

1. Introduction

Computer applications are not new to space-planning problems. A new kind of approach for combinatorial problems has led to the development of the computer application ROP. This spatial layout planning application was developed by the Government Building Agency (GBA) and the Faculty of Architecture and Urban Planning, Eindhoven University of Technology. It consists of two modules: ROP-TRI and ROP-CAD. ROP-TRI shows a relationship matrix and a relationship graph, and ROP-CAD is a module connected to a commercial CAD application: MiniCad™ (1990).

The technique underlying ROP-TRI is defined as an interactive graphic heuristic procedure (Daru and Adams 1989). This technique is based on the work of Bertin (1981). It is argued that a person can recognize patterns of data easily when these are represented by symbols. This means that one can use ROP-TRI to organize the functional demands for space and physical proximity in a relationship matrix and graph by rearranging the symbols. The matrix and graph offer features to construct groups of functional units with strong relationships. What kind of alternative grouping is seen as optimal depends on the user. This is an important point, because other optimization criteria can be considered in addition to the criterion of physical proximity. For example, the number of floors in a building constitutes a criterion for the number of functional groups one wants to differentiate. In general, the human ability of recognizing patterns is used for operating ROP-TRI. It is argued that by manipulating matrix rows, a user can try to get what he or she considers the optimal functional 'clustering'. This means that the user can group functional units with strong relationships in the matrix. Visually, these groups appear as triangles.

The ROP-CAD part provides the user with features to construct and depict spatial grouping. It is the user of the program who decides what kind of optimization actions should be performed. ROP offers tools to do this in an easy and efficient way. At the same time, it is possible to see what alternative layouts mean in space use. One can also see how relationships are rewarded in the spatial layout. How this fits one's demands can be seen in the ROP-TRI matrix. For the GBA, ROP is especially useful as a support in the definition and design stages of building

H. Timmermans (ed.), Design and Decision Support Systems in Architecture, 201–212.

projects. It offers tools by which users can prepare and develop parts of their brief. Furthermore, it assists in matching the building to the organization. ROP performs best when there are several parties to the decision and when the criteria for optimization depending on the individual situation.

To elucidate the above advantages, this paper will explain how ROP can operate from two points of view. First, ROP is explained in terms of the internal tools that are offered. In fact, ROP is described as if a person is performing a task without concern for external influences. Secondly, the external environment in which ROP can act is illustrated in a project description, resembling reality for space-planning problems. Out of the two points of view, the discussion will deal with the benefits of ROP for space planning undertaken by the GBA. The next section deals with features of ROP to give the reader more insight in the program. The subsequent section describes a real-world case in which ROP was used. On the basis of the practical use of ROP, Section 4 discusses how ROP supports decision-making.

2. Space planning and ROP

ROP is an application for the Apple Macintosh computer (recommended level Macintosh II). As already mentioned, it consists of two modules: ROP-TRI and ROP-CAD. When using ROP, four general stages in a procedure can be distinguished. Figure 1 depicts these four general stages.

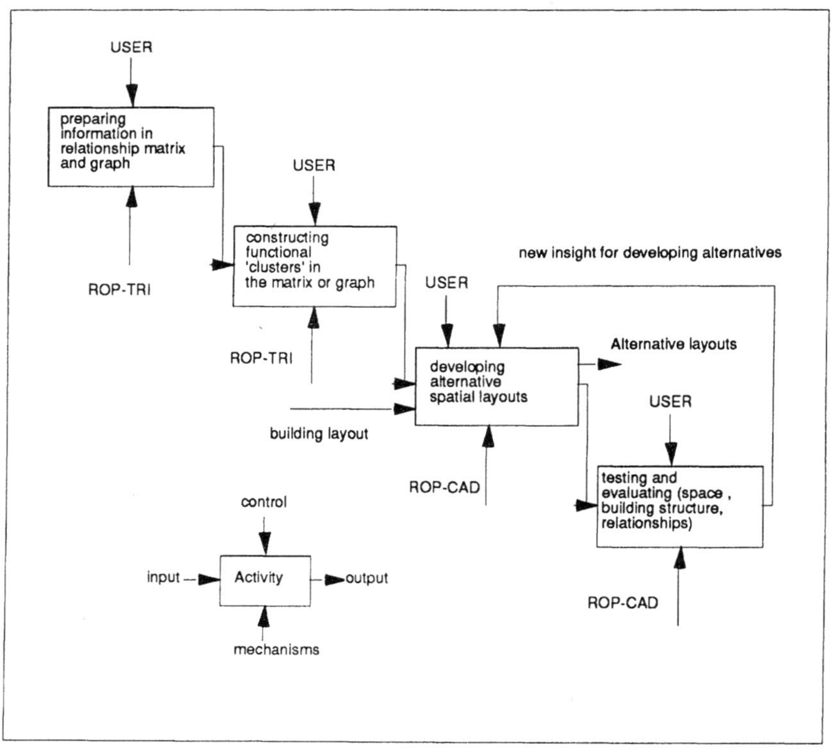

Figure 1. Process description of how ROP cab be used

First the data has to be prepared for ROP-TRI. In space-planning problems, this data is dependent on the request of the organization for which the new accommodation is going to be planned. Secondly, this data is transformed into new information for the planning procedure. The user of the program is the one who performs the task. These two stages concern the handling of the requirements ascribed to the functional units of an organization. In the next two stages, the building is also in the picture. This is where the actual allocation of functional units in a building takes place. Often there is no best solution at hand. Therefore, alternative spatial layouts are tested and evaluated to see which is best. This leads to new insights, and the spatial layouts will be changed accordingly. The rest of this section explains in more detail how ROP can be used.

2.1. ROP-TRI AND ITS FEATURES

In ROP-TRI, demands for space, reference names, and a reference number are shown in the left border of the matrix or along the perimeter of the graph. The matrix and the graph both represent the same data. In the top border of the matrix, the same reference numbers are shown. The relationships are defined as an expression of physical proximity. They should be rated as primary, secondary or unwanted, representing some degree of demand for proximity. What this actually means depends on the interpretation of the user. Mostly, primary relationships indicate a demand for rooms close to each other. Secondary relationships are less important than primary ones. Unwanted relationships indicate that the organizational units should be located separately in the building. Empty cells in the relationship matrix indicate that there is no demand for proximity among the organizational units. The relationships should be described by the members of the organization that will occupy the new accommodation. The scored relationships are represented by symbols in half of the matrix. On the left side of the diagonal, the primary and secondary relationships are depicted; on the right side, the unwanted relationships will be visible. The information can be imported in TEXT format. However, it is also possible to insert the information directly.

When the matrix is filled, it will probably reveal a scattering of dots (see Figure 2a). The user can then initiate the actions for clustering or diagonalization. He or she can do this by picking up rows with the mouse and move them to another position in the column. The functional units, with primary and secondary relationships, should be positioned near each other in the left column. The relationship dots will also shift positions when a functional unit is moved. This action can also be undertaken in the relationship graph. The matrix and graph are connected to each other. Changes in one are seen in the other. By performing the described actions, a user can construct clusters along the diagonal (see Figure 2b). This is optimizing on the criterion of physical proximity. The unwanted relationships should be located as far as possible from the diagonal. Other important criteria for a certain situation can inspire the user to undertake certain optimization actions. On the computer screen, something might seem not as optimal as possible, but it will appear to be more optimal when other aspects are taken in consideration. When a user considers a certain optimization as the 'optimum', then the information can be exported to the commercial CAD application MiniCad™.

2.2. ROP-CAD AND ITS FEATURES

In the MiniCad™ application, features of the ROP-CAD module are used. These support the spatial grouping process. The imported information from ROP-TRI appears in a spreadsheet in the order in which the optimization was made (see Figure 3a). With the features of the ROP-CAD module, one can manipulate the data. First, the functional units can be placed in an existing floorplan. It is also possible to construct a new floorplan. Secondly, physical units can be assigned to functional units in the spreadsheet (see Figure 2b). The difference between the quest for space and the actual assigned space is recorded in the spreadsheet.

These two actions can be performed by the user by manipulating the mouse and by choosing what he or she wants to do.

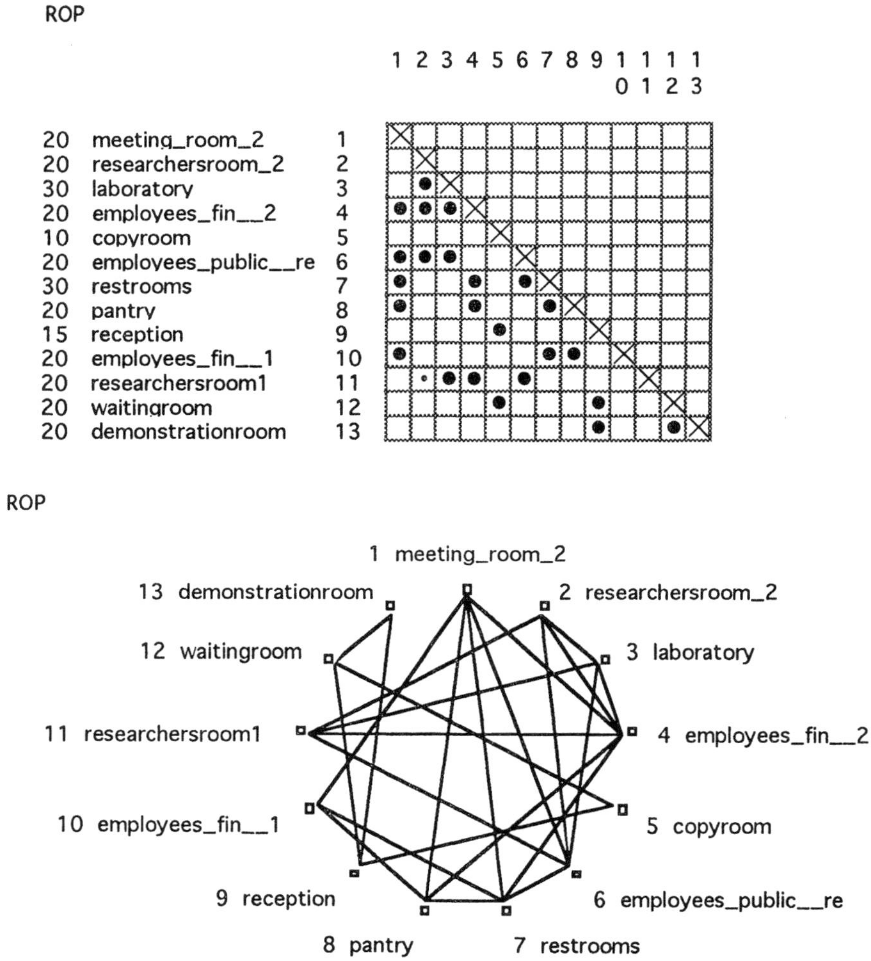

Figure 2a. Relationship TRI-graph and TRI-matrix

Whenever a unit is located, it is also possible to change that location by deleting the unit in that position and assigning it to another position. The spreadsheet will immediately record the actual situation. In this way the user can search for better space use in a very easy manner.

It is also possible to evaluate the relationships. That information can be sent back to the ROP-TRI module to compare the demand for relationships with the relationship achieved in a floorplan. The operations with the commands give the user the opportunity to construct, in an easy way, several alternative layouts. One can immediately see the consequences on the space budget or on the way requested relationships are realized. It is important to note that the variables of the spatial grouping process include functional units, requested and assigned floor area, and relationships. On the other hand, from his or her own experience as a space planner, the user can introduce all kinds of grouping criteria; e.g. structural restrictions of the floorplan or criteria that are dependent on the particular situation.

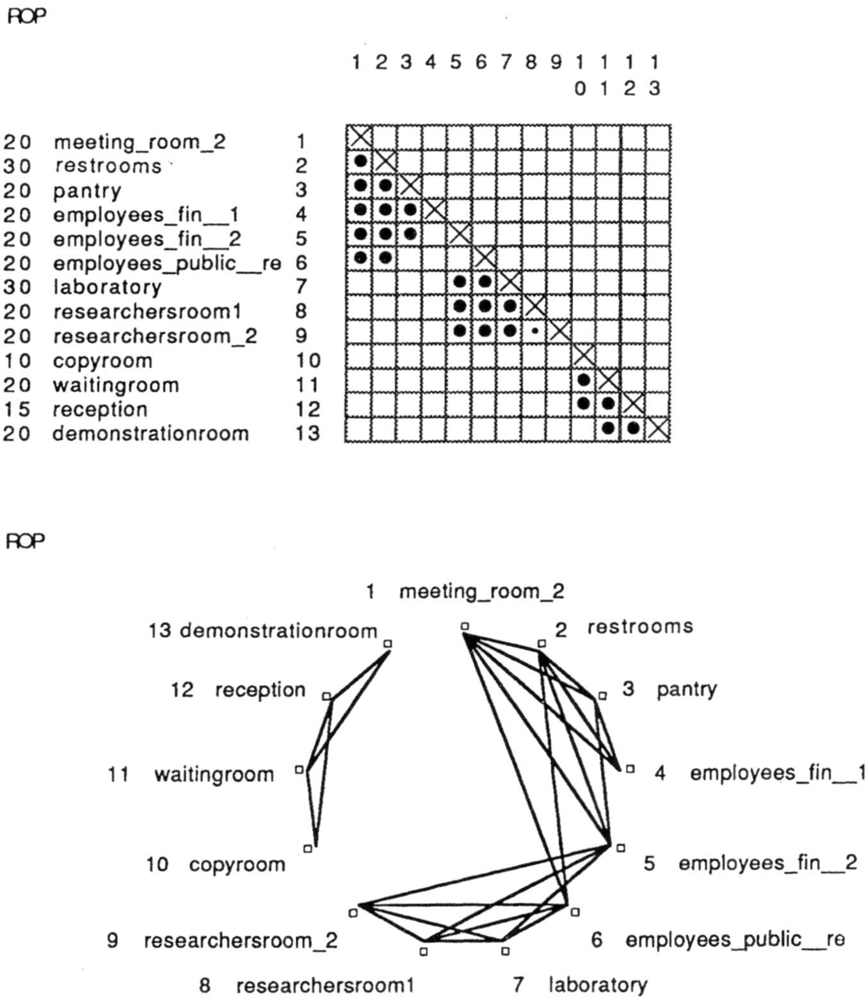

Figure 2b. A TRI-matrix and TRI-graph with functional clusters

In the ROP-CAD module, the 'optimum' is also what a user thinks it is. For difficult situations, an experienced space planner often has very creative solutions. The ROP applications allow these creative moves, since there are no restrictions. The features of MiniCad™ allow the planner to change the structure of the building if one thinks a better layout can be constructed.

FEATURES OF ROP-TRI

1. Possibility to import TEXT information, e.g. out of a spreadsheet
2. Inserting information directly in the matrix
3. "Clustering" by picking up the rows
4. Export possibility as TEXT or for MiniCad ™

Figure 2c. General features of ROP-TRI

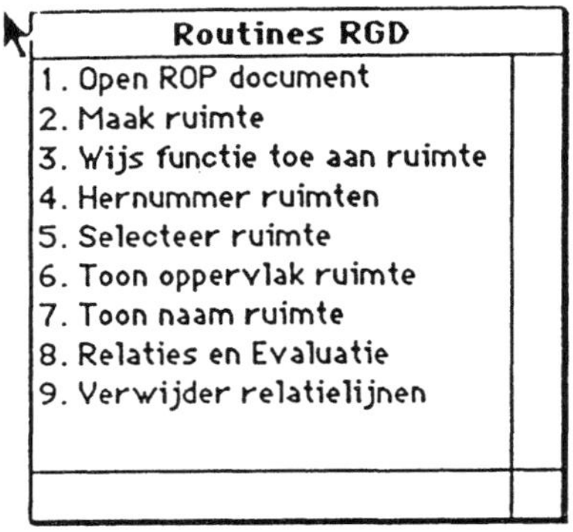

Translation:
1. Open ROP document
2. Make a room
3. Allocate functional unit to a room
4. Give rooms a new number
5. Select a room
6. Show area of room
7. Show name of room
8. Relationship and evaluation

aWorksheet

	A	B	C	D	E
1	m2 FNO	Functie	m2 NO	NO-FNO	NO/FNO
2	20	meeting_room_2			
3	30	restrooms			
4	20	pantry			
5	20	employees_fin__1			
6	20	employees_fin__2			
7	20	employees_public__r			
8	30	laboratory			
9	20	researchersroom1			
10	20	researchersroom_2			
11	10	copyroom			
12	20	waitingroom			
13	15	reception			
14	20	demonstrationroom			
15	265.0		0.0	0.0	0

Figure 3a. ROP-CAD connection with routines and spreadsheet

3. ROP in Practice: A Real-World Case

A general overview of the actions supported by ROP is useful, but even more interesting is to see how such an application actually works. As a part of a post-graduate research project, projects were analyzed to see how ROP had been used in actual situations. In this paper the results of one case study will be discussed. The application ROP was used as an experiment in a real-world project for a design for a courthouse. It was possible to attend (as an observer) the meetings in which ROP was used. During observation and in interviews with members of project groups, activities were recorded, information input and output of every action was analyzed, and criteria used to control the activities were explored. This was done to make predictions about possible decision processes in which ROP could operate. On the other hand this benefits the development of ROP, because special demands coming out of the actual decision processes were revealed.

In the following, we will show how ROP can be used and what kind of external information influences the decision-making. An example of a decision process concerning space planning will be given in a description of a project. Some generalizations have been made which bring it close to reality. The information is based on the observations of a real-world project.

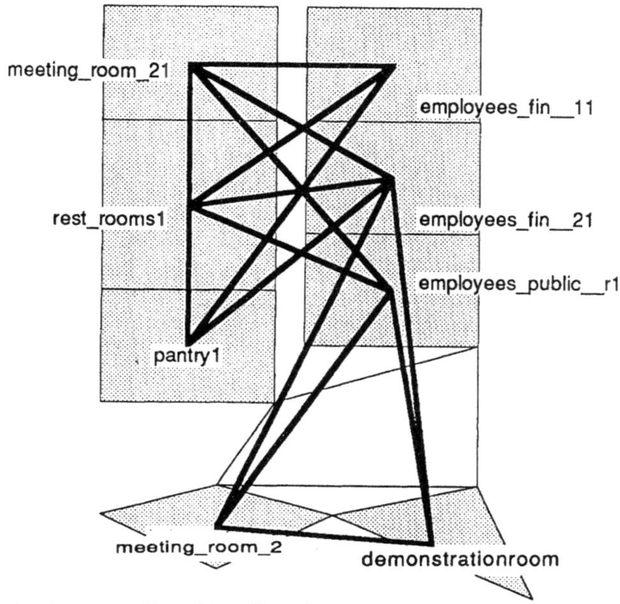

Figure 3b. Functional units are positioned in a floorplan

FEATURES OF ROP-CAD

1. Possibility to import ROP-TRI information
2. Assigning functional units to 'rooms' in a drawing
3. Construction new 'rooms' out of the spreadsheet
4. The spreadsheet shows space assigment: demanded space actually assigned space, and the difference between both
5. Evaluation features:
   - Presenting relationship lines in drawing
   - Exporting information back to ROP-TRI

Figure 3c. General features of ROP-CAD

3.1. DEVELOPING AND TESTING THE STRUCTURE OF A DESIGN FOR A COURTHOUSE

3.1.1. *Background information: the status of the project.* ROP was tested in a project for the design of a new building. Before ROP was actually used, some actions were already undertaken. A location had been chosen, and an architect had been selected. In the total procedure of the public building authority, the project can be situated partly in the definition stage and partly in the design stage. Participants in the project team included the architect, representatives of the GBA and the Department of Justice, future users, and a security expert. In this part of the design process the members of this project team were the decision-makers. In this project, decisions had to be made about the structure of the new building. The architect had already developed some structural sketches from a preliminary design.

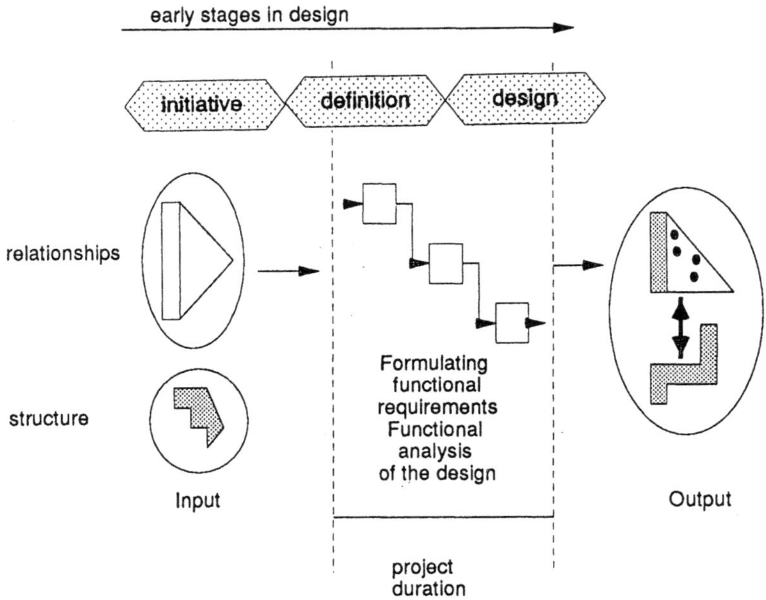

Figure 4. Situating the case

3.1.2. *The aim of this part of the design process.* In this part of the design process, a sketch design had to be developed. ROP was used in two kinds of actions. First, functional requirements had to be specified more precisely. There had been some changes in the organizational structure, so the relationships would have to be updated. Secondly, the developed structure of the design would have to be tested. As a result, more insight should be gained in the potential of the building structure for the functional units.

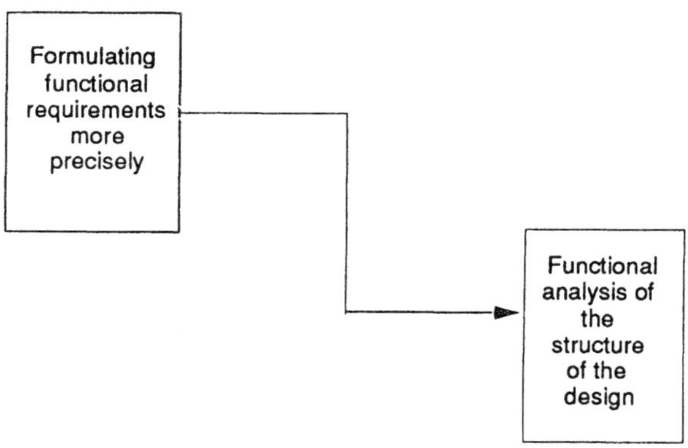

Figure 5. Two general stages in the project

3.1.3. *Stage 1: Formulating functional requirements.* In this stage of the project, three activities are identified. The building was planned to accommodate two organizations. There were thus two different briefs, describing on paper the functional units and the relationships between these units. This paper information had to be translated to be useful for ROP. The relationships that were mentioned in the briefs needed to be evaluated and updated to reflect the organizational change. During a meeting of the project team, a ROP-TRI relationship matrix was used to determine which relationships were still valid and which should be enhanced. The user group was able to identify the actual situation, and the security expert pointed out the unwanted ones.

After reaching agreement about the relationships, one could start to group the functional units. The designer took the ROP-TRI document back to his office. There he optimized the matrix from his point of view. He discovered some contradictions; he also found some situations that are difficult to realize when all relationships should be maximized. At this point, the relationships were reviewed again by the project team. The designer used the matrix to point out the expected problems. The clusters were also evaluated by the project team from their point of view. An important change was required by the decision that the two organizations should have an integrated accommodation. This led, for example, to the request that the conference rooms of the two organizations should be located near each other. New relations were identified and other clusters were formed. ROP-TRI was the tool that visualized the consequences of these decisions.

There also appeared to be some contradictions in demand. A secretary had a strong relationship with her boss, and an unwanted relationship with the main courtroom. Her boss also had a strong relationship with this courtroom. This situation is very difficult to deal with. Therefore the architect asked the members of the project team to think about how this unwanted relationship should be reflected in the accommodation. At the end of this activity, functional groups were constructed. These were represented in the relationship matrix and graph of ROP-TRI.

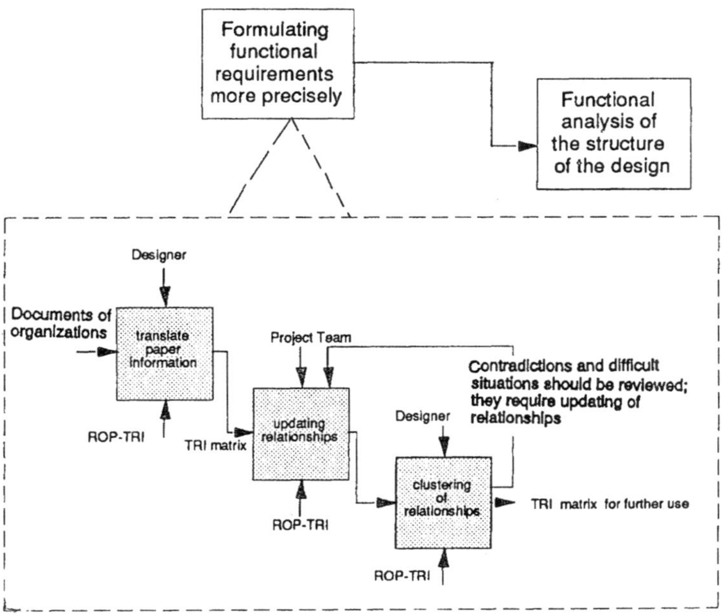

Figure 6. Activities to specify functional demands more precisely

3.1.4. *Functional analysis of the structure of the design.* In this stage two activities were identified. First, the functional units were positioned in the drawing. A computer drawing of the structure of the design was available. It was no problem to import the information into

MiniCad™. The ROP-CAD commands were activated, and the designer started to place the functional units in the drawing. He developed three alternative layouts. In doing this he also made minor changes in the structure to fit spatial requirements. The principal group also asked for an exercise whereby a certain part of the building could be tested for flexibility in regard to different kinds of functional units. Consequences for the space use that was actually assigned to a unit was recorded in the spreadsheet. In the second activity, the alternative layouts were discussed during a meeting of the project team. The designer had taken the ROP-CAD document, so that minor changes could be introduced on the spot. The consequences could be seen immediately and taken into account in the decisions that had to be made. In this process, external information was important to the decision-making. Personal opinions influenced the construction of the layout. In fact, one can say that this kind of decision-making is more or less intuitive.

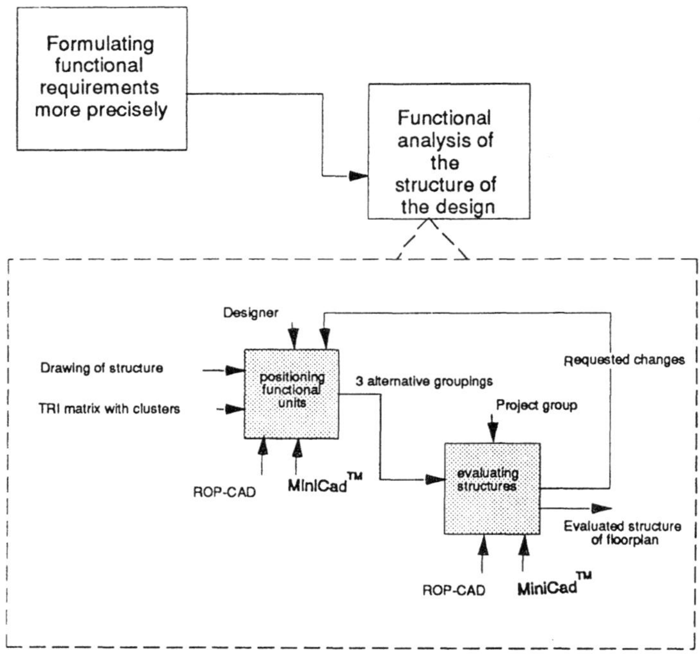

Figure 7. Activities to test and develop the structure

3.1.5. *Results of the project.* The proposed structure of the building was thoroughly evaluated in regard to the possibility of granting the requests for proximity of functional units. Also, the structure was tested on the space ratio. This means that the structure was looked at to see if it created lots of space that was not suitable for use or if the rooms for the individual functions were much too large. If this were the case, the building would be expensive. In this part of the design process, one can change such a structure very easily so that the space ratio will be better. The functional requirements toward the building were elaborated. But these requirements change over time. To be sure to adapt the accommodation according to the users' wishes, these changes should be monitored carefully. ROP can be used in subsequent stages, whenever the design is supposed to be tested in regard to its functional space use. However, this relationship matrix should also be updated according to expected changes in future functional use.

4. Discussion

Performance aspects derived from the internal functioning of ROP contribute to decision-making in real space-planning projects in an important way. Therefore it is important to distinguish these internal performance aspects. These include the following: (i) all the information that is imported in the two modules of the application is usually visible in the module for the user, (ii) the two modules follow a procedure that is common for space planning and can be followed easily, (iii) the user of the program keeps full control of the actions that are performed, (iv) both modules offer features to change the relationship matrix or a layout plan in regard to details, (v) several features are offered for evaluating the realized spatial layout. When these five performance aspects are translated to external conditions, decision-making in space-planning projects can be supported from two viewpoints. One can look at the kind of space-planning project where ROP performs best. Moreover, you can also look at managerial aspects of a space-planning procedure that are supported by ROP.

In regard to the kind of space-planning problem, ROP operates excellently where extraordinary organizations and buildings should be matched. ROP has no trouble assigning functional units to 'rooms' with strange shapes. There are also no restrictions on the kind of functional units that can be handled. ROP can assign office units to rooms, but it also performs well in a space-planning project for a hospital. One of the advantages of the application is that the user of the program remains in full control. So, in situations that are not common, unusual optimization actions can be performed.

As mentioned above, ROP supports decision-making from a managerial point of view. First, ROP can support decision-making in the early stages of a design process. At that point, the knowledge of the consequences of choices is limited. However, these choices may have an impact that is expensive or undesirable. This can also be said about the early stages in which decisions on space planning are made. ROP, on the other hand, needs little information about the building and potential users to determine whether a certain building layout will have a good ratio for space use. ROP allows parts of a design to be tested without putting much effort into gathering uncertain information; that information can change dramatically during the design stages. A principal can evaluate the first sketch alternatives on space budget without much effort.

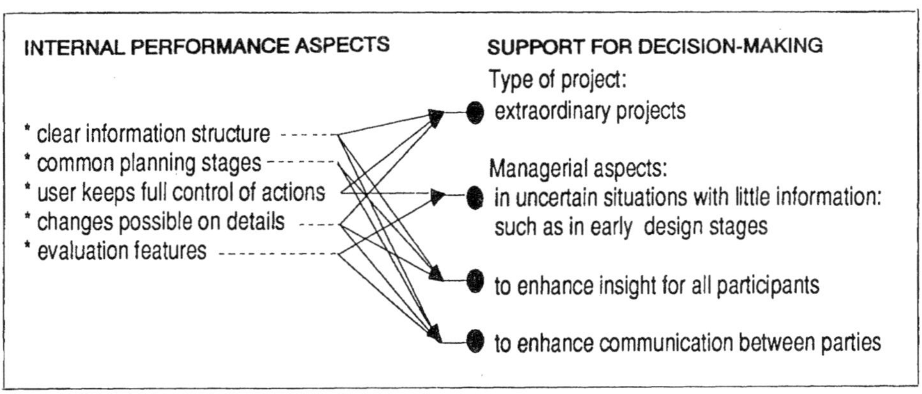

Figure 8. Internal performance aspects connected to decision-making

Furthermore, ROP offers tools to transform functional requirements into spatial layouts based on the criterion of physical proximity. People who formulate the functional relationships can see this transformation in stages through ROP-TRI and ROP-CAD. They can also intervene in the separate stages if they think the design is heading in the wrong direction. People involved in such a process who have no architectural background can more easily follow how their functional demands are transformed in an actual floorplan. It can be argued that ROP enhances insight. Thus, ROP can make an important contribution to the goal that principals and potential users are supposed to have more influence in the design of their accommodation. In this way, ROP can support quality, in the sense of better fit between primary processes and new accommodation. Last but not least, the process of developing a new accommodation for a governmental agency is a matter of communication and negotiation between all kinds of parties, e.g. working units of the GBA, members of governmental departments, designers, and users. ROP acts as communication medium between parties. In the stages where the functional demands should be defined, the ROP-TRI tool can be used to point out contradictions in the demands when possible options are considered. These contradictions can be easily illustrated in the relationship matrix. This is also true for the ROP-CAD module. A space planner can easily change a layout of a floorplan on small details, according to the wishes of others, and then show them the consequences immediately.

References

Bertin, J. (1981) *Graphic and Graphic Information Processing*, De Gruyter, Berlin.

Daru, R. and Adams, W.T.C.F. (1989) "Interactive Graphic Heuristic Procedure", in F. Kimura and A. Rolstadås (eds.), *Computer Applications in Production and Engineering*, Elsevier Science Publishers, Amsterdam, pp. 783-806.

MiniCad™ (1990) Diehl Graphsoft Inc., Ellicot City.

AN APPROACH TO EVALUATING EXHIBITION SPACES IN ART GALLERIES

Y.S. KIM and M. BRAWNE
School of Architecture and Building Engineering
University of Bath, Claverton Down
Bath, BA2 7AY
United Kingdom

ABSTRACT. There are certain building types in which movement of people is the most significant evaluation factor. Among these are art galleries and museums. Unlike other building types, which are often explicated by investigating the relationship between people and people, and between people and the built environment, art galleries and museums are a building type in which the social relationship between people hardly exists and people's movement through space, that is, the functional relationship between people and space, is one of the most significant factors for their description. The typical museum experience is through direct, sequential, and visual contact with static objects on display as the visitor moves. Therefore, the movement pattern of the visitors must exert a significant influence on achieving the specific goal of a museum. There is a critical need for predicting the consequences of particular spatial configurations with respect to visitors' movement. In this sense, it is the intention of this paper to find out the relationship between the spatial configuration of exhibition space and the visitors' movement pattern.

1. Introduction

Museum professionals today tend to emphasise the importance of the quality of visitors' experiences more than ever. In order to accomplish this goal, they have mainly concentrated on development of versatile programmes such as lectures, publications, video, explanatory labels and so on in order to provide visitors with a worthwhile experience. Apart from the various programmes, they have extensively dealt with the effectiveness of the arrangement of objects in already existing gallery spaces (Robinson 1928; Melton 1935, 1972; Bechtel 1967; Gardner 1986).

However, it is argued that this goal can be facilitated more fundamentally by manipulating the spatial organisation before all the decisions stated above are made. The typical museum experience is above all through direct and sequential contact with static objects as people move. In this sense, it is believed that the way in which objects are arranged in sequence, that is, how we see one after the other, does affect our understanding. Therefore how a visitor follows a route is an important factor of the layout. Accordingly, the first consideration should be to define the pattern of visitors' movement through the manipulation of spatial configuration, hopefully in accordance with the arrangement of the objects. Furthermore, it could also be said that it is absurd to force the visitor's viewing sequence by orientation signs, since moving through a space is not passive perception, but a dynamic process in which all the elements present around are compared both to each other and to others which are not present except in the memory of the visitor. Despite such an importance of the exhibition space, however, it is true that architects have been more concerned with the style of the building, interior decoration and an effective lighting system. When they consider the spatial organisation, they often borrow answers from various precedents in history, without explicit knowledge of the consequence of space with respect to the visitors' experience. It is therefore a critical task for architects to offer useful information by providing an

H. Timmermans (ed.), Design and Decision Support Systems in Architecture, 213–225.

empirical study of existing museum buildings, aimed at what the consequences of a particular spatial organisation with relation to the visitors' experience would be.

The main objective of this study is to find out how art galleries could effectively encourage visitors to arrive, understand and to take away a worthwhile experience through the manipulation of the spatial organisation. It is widely believed that to achieve this worthwhile experience requires both time and proper sequence to look at art objects. In this respect, this study investigates the effect of spatial organisation of exhibition spaces on visitors so that a useful architectural statement can be made. It should be noted that this paper limits itself to the investigation of individual spaces. The investigation of overall exhibition space will be dealt with at the next stage. By being able to predict the effect of the spatial organisation on visitors' behaviour, museum professionals and architects can offer visitors a pleasant and enjoyable place as well as an effective educational institution. That is to say, the information to be offered as the result of the study should be crucial at an early stage of planning, when influential decisions are often made without explicit presupposition of the effect, and at later stage when the addition or redesign of the exhibition space is reconsidered.

2. Methodology

The trailing technique is employed in order to investigate visitors' behaviour in the exhibition space by using stopwatches, pencil and paper to record the time spent either in a gallery as a whole or looking at a painting. This technique has been prevalent among museum professionals and behavioural researchers to evaluate mainly the effectiveness of display (Robinson 1928; Melton 1972). Although there has been some criticism, such as reliability of the data obtained from it and lack of efficiency, this method is ideal for the present study, which does not need precise readings of short time like one or two seconds, usually spent before an object. This study deals with the relationship between space and visitors' behaviour (especially, their movement pattern), not the relationship between the display and visitors' behaviour (especially, the time spent before an object), which has been the main task of museum professionals. The data to be obtained from this observation were the time spent in each gallery and visitors' movement line in the individual gallery. While staying in a gallery, the observer drew visitors' movement lines in the space and, at the same time, recorded the time they spent in it. The observation was carried out between March 1990 and September 1990 in the Tate Gallery and the National Gallery which are both located in London.

The data obtained from the trailing technique is correlated to the physical variables (independent variables), which are the geometrical dimension and geometrical proportion of individual galleries, in order to find out how the gallery space affects visitors' behaviour.

3. Variables

By carrying out the preliminary observation in several museums and art galleries located in London, criteria to evaluate the visitors' behaviour and variables to assess the criteria were identified. Dependent variables are the criteria used to assess the effect of space on the visitors' behaviour. Independent variables are linked to exhibition space, which is supposed to influence the visitors' behaviour.

3.1. DEPENDENT VARIABLES

In order to assess the effect of spatial configuration on the visitors' behaviour, we have chosen two factors, namely, the visitors' movement pattern and movement rate. It was found that there are two explicitly different movement patterns in a gallery, that is, a linear movement pattern and a circular movement pattern. As far as a long shape of gallery is concerned, the visitor tends to walk towards the other end of the gallery and hardly turn around to look at paintings on the other side of the wall. In this sense, this kind of movement pattern can be called linear movement towards the other end of the gallery. On the other hand, in a square-like gallery, the visitor tends to look at most of the paintings hung on both sides of the gallery. That is to say, the visitor turns around at the end of one side to look at paintings of the other side and then finally walks towards the exit. In this sense, this kind of movement pattern can be called circular movement. The visitors' movement pattern is the most significant factor in art galleries since it can be said that an art gallery's success depends on it in the sense that the objects on display are not isolated events but have some meaning which depends on an understanding of their relation to each other.

The movement rate means the time that visitors spend in a gallery rather than the speed which they move. Therefore, the movement rate represents both the time spent in walking around the gallery space and the time spent in appreciating the works of art. However, it is a characteristic of an art gallery or museum that there is no clear demarcation between moving and looking at the works of art, because in many cases, visitors look at the works of art as they keep moving around. In this sense, the rate at which visitors move around in a gallery is represented by the time that they spend in the gallery whether they look at the works of art or move around. Accordingly, it can be said that the movement rate is closely related to the visitors' movement pattern, and thus related to the spatial configuration.

The first variable, movement rate 1, is simply the time that visitors spend in a gallery from start to finish. As we pointed out earlier, it is assumed that the longer the movement rate 1 is, the more successful the gallery is, since one of the main intentions of art galleries is to attract more visitors and to hold them longer in a gallery. The second variable, movement rate 2, is the time that visitors spend actually looking at or appreciating paintings. The movement rate 1 might not be a significant criterion because it includes walking time as well as looking and appreciating time. In this respect, movement rate 2 is the time which subtracts the walking time from movement rate 1.

At this point, to calculate the walking time, the walking speed should be defined. From an experiment executed by the researcher and with the data obtained by observing the visitors, we established that the average walking speed in museums or art galleries is 1.2 m/sec. Therefore, we can formulate an equation:

$$\text{walking time} = \text{walking distance} / 1.2$$

Thus,

$$\text{movement rate 2} = \text{movement rate 1} - \text{walking time}$$

Here, it should be noted that walking distance is recorded by measuring the length of each visitor's movement line drawn on the plan during the observation.

3.2. INDEPENDENT VARIABLES

Having done the preliminary observation, some physical variables, which were significantly influential on museum visitors' behaviour in terms of their movement pattern and movement rate, were taken into account for actual observation. Those independent variables concerned geometrical proportion (ratio of length to width) and perimeter. It was generally noticed during the preliminary visits that the higher the length to width ratio was, the more people tended to make a linear movement line. Although there are some exceptions in very small rooms, or long and narrow rooms, this variable is closely related to how people move around in a gallery room. Thus, it could be also related to the movement rate. The perimeter of a gallery room is important on the grounds that the visitors move along the walls on which objects or paintings are actually displayed. It is quite certain that if the perimeter is long, visitors spend more time to look at all the paintings. However, this is closely interrelated with the rule concerning the geometrical proportion. In addition to these two main variables, length and area have been taken into consideration on the same grounds as the perimeter. The length of a gallery room seems to have a considerable effect on the visitors' movement pattern and movement rate. It was seen that during the preliminary observation in a particular room, which has long length and high ratio, this variable plays a significant role in defining the movement pattern and the movement rate.

4. Samples of Case Studies

We chose two galleries as case study, namely, the Tate Gallery (TG) and the National Gallery (NG). They were further divided into five wings: the West-Wing (WWT) and the East-Wing (EWT) of TG; and the West-Wing (WWN), the East-Wing (EWN) and the North-Wing (NWN) of NG.

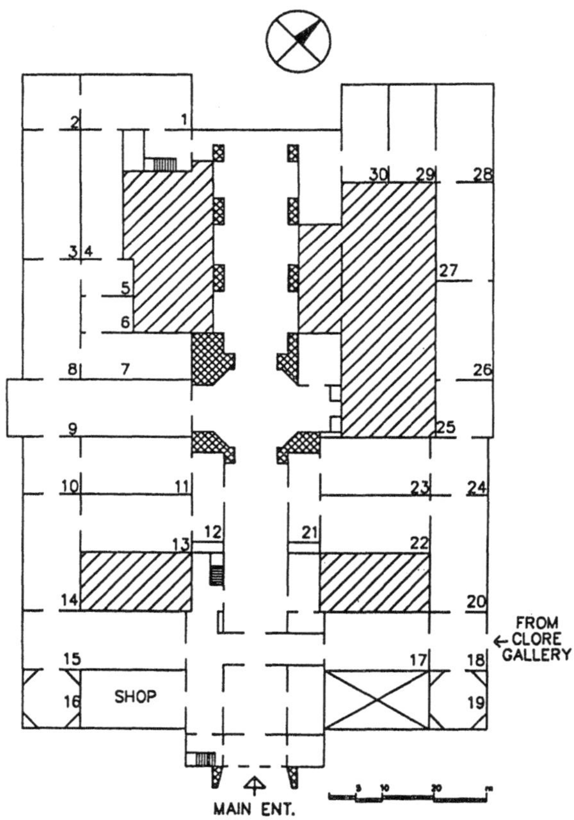

Figure 1. The floor plan of the Tate Gallery, London

As seen in plans, TG shows a symmetrical disposition of two separable exhibition spaces, spaces, which are WWT and EWT. The spatial configuration of the two wings is almost identical (see Fig. 1). NG can also clearly be divided into three exhibition spaces, that is, WWN, EWN and NGN (see Fig. 2). Those five exhibition spaces selected for the observation show quite different aspects in their spatial disposition. In particular, NWN is different from the rest in the sense that the galleries in NWN are almost identical in terms of their shape and size and encourage the visitors to move through one gallery after another, whereas the other wings have alternative routes diverging from a major traffic line. The other wings also display some differences. The two wings of TG have a long, straight traffic line on which main galleries are located, and from the major traffic line there are some minor galleries. On the other hand, WWN and EWN differ from the wings of TG in the sense that there is no one single major traffic line which is straight like in TG.

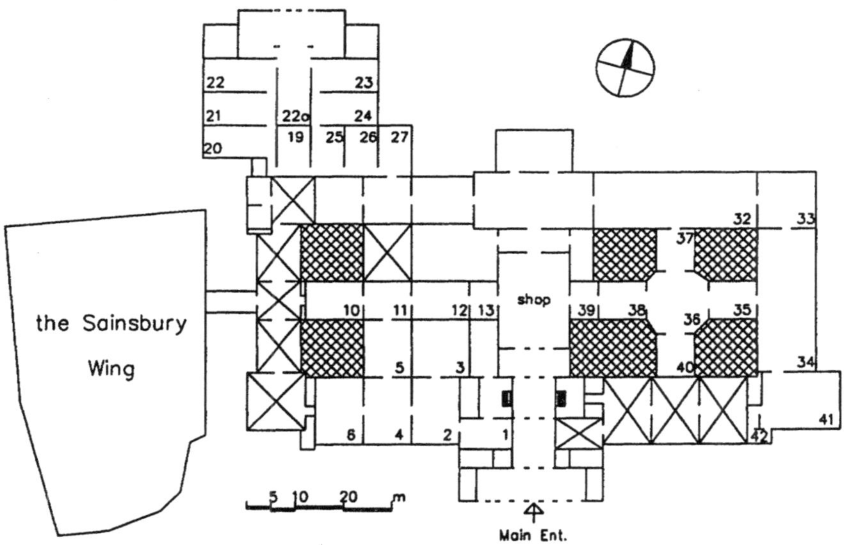

Figure 2. The floor plan of the National Gallery, London

5. Individual Gallery Spaces

5.1. ATTRIBUTES OF THE PHYSICAL VARIABLES IN BOTH GALLERIES

Table 1 shows the average values of every physical variable of the five wings to be analyzed. Those physical variables presented in a table give an overall view of the five exhibition spaces as well as the differences and similarities between them. Here, it is important to make a comparison between the five wings. It can be seen at a glance that the number of paintings in a gallery shows considerable difference, in particular between WWT and EWT. In spite of the similar spatial configuration of the two wings, WWT has 25.8 paintings or sculptures in a gallery, while there are only 10.6 in EWT. This could be because the average size of the paintings of EWT is bigger. Such difference in the number of paintings naturally results in another considerable difference in Pe/Pa (perimeter/the number of paintings). That is to say, a painting occupies 6.25 m of the

As for the geometrical ratio, the average is 1.67 which is derived from 9.1 m x 15.1 m. WWT and NWN show a little higher ratio, that is, 2.06 and 1.80 respectively. While the geometrical ratio does not show a significant difference, the area of each wing shows a considerable difference between the wings. In general, the galleries of TG are larger than the ones of NG. The average area of WWT is more than double that of the NWN. At this point, what is interesting is that although the average area of NWN shows the smallest figure, the geometrical ratio (1.80) is comparatively high. Except for the NWN, therefore, it can be noticed that the higher the geometrical ratio is, the larger the gallery space is.

As far as the average number of paintings of a gallery is concerned, EWT stands apart from the rest, that is, the number of paintings displayed there is far smaller in comparison with other wings, because of the large paintings and big sculptures there. As far as the relationship between the geometrical ratio and the area is concerned, NWN is exceptional in the sense that despite the fact that the geometrical ratio is quite high, the average area of the galleries is the smallest among these wings. Other wings have common physical characteristics.

TABLE 1. The physical variables of the five wings

Variables	WWT	EWT	WWN	EWN	NWN	MEAN
No. of Gal.	16	14	10	11	9	12
No. of Pa/G	25.8	10.6	16.8	20.4	18.7	18.5
Width (m)	9.8	10.4	8.8	9.3	7.0	9.1
Length (m)	19.7	17.2	11.7	14.2	12.8	15.1
Ratio (L/W)	2.06	1.64	1.34	1.49	1.8	1.67
Area (m2)	196.3	177.9	106.2	149.1	89.4	143.8
Perimeter (m)	58.9	55.1	41.0	46.9	39.6	48.3
Pe/Pa (m)	2.75	6.29	3.01	3.55	2.30	3.58

5.2. INDIVIDUAL GALLERY SPACE AND VISITOR'S BEHAVIOUR

5.2.1. *WWT*. In WWT, there are 16 gallery rooms, of which eight are off the main circulation line. Eleven were observed and analyzed. Among the observed gallery rooms, room 7, 11 and 12 are situated off the major traffic line. In the correlation analysis between the independent and dependent variables (Table 2), it can be seen that there is a significant correlation between the geometrical ratio and the movement rate 1, and between the geometrical ratio and the movement rate 2. Although it is not so strong, the length and the movement rates also show reasonably good correlation. On the other hand, the movement line and the other physical variables do not show any significant correlation. Again the time spent in front of a painting does not make a good correlation with other independent variables. In other words, apart from the geometrical ratio and the length, other independent variables, namely the area, perimeter and Pe/Pa, do not appear to play a role in influencing the visitors' behaviour in this wing.

At this point, it can be seen that room 9 and room 12 have an exceptional character. In spite of the fact that room 9 is the largest (385 m2) and has high geometrical ratio (3.18), the movement line shows 80% of the circular movement, which is supposed to be characteristic of gallery rooms of low geometrical ratio. It can be speculated that because this room is at an angle of 90° to the main circulation line rather than parallel to it, visitors are encouraged to go further inside. Once they get inside, they tend to return to the beginning point of the room and then proceed in spite of the fact that there is another exit on the other side which leads to the main corridor of the gallery. This indicates that sense of orientation is an important factor in defining the visitor's movement pattern. The big door of the exit is always kept closed, which plays a role in preventing people from using it. By naturally enforcing circular movement on the visitors, therefore, the paintings hung on both sides are effectively shown. The character of this room may suggest therefore that the way in which a space is placed in the overall configuration is very important in terms of the visitor's movement pattern. Room 12 measures 6 x 17 metres so that the geometrical ratio is considerably high in comparison with other galleries in this wing.

TABLE 2. Correlation coefficient between variables (WWT)

Variables	Ratio	Length (m)	Area (m2)	Peri. (m)	Pe/Pa (m)
Move-	0.785	0.629	0.400	0.508	0.060
Rate.1	(0.787)	(0.786)	(0.786)	(0.786)	(0.407)
Move-	0.753	0.554	0.316	0.427	0.094
Rate.2	(0.730)	(0.729)	(0.729)	(0.729)	(0.406)
Linear	-0.148	0.195	0.362	0.294	0.216
Move.	(0.886)	(0.886)	(0.886)	(0.592)	(0.177)
Rate 1	0.190	-0.245	-0.450	-0.371	0.292
/Painting	(0.022)	(0.018)	(0.018)	(0.018)	(0.401)

( ) correlation coefficient after eliminating room 7, 9 and 12

Despite the high ratio, in which visitors are expected to make a linear movement, this gallery shows a relatively high proportion of circular movement line, i.e. 55%. The rather narrow gallery space of 6 metres enables visitors to look at both sides of the room almost simultaneously without any great effort. That is to say, the paintings hung on both sides are easily accessible from any point of the room, which tends to encourage visitors to make a circular movement. In addition, even if there is an opening leading to the main corridor of the building, most visitors tend to turn back toward the major traffic line.

Room 7 has 95% of circular movement line and the greatest movement rate in spite of the high geometrical ratio (3.00). Because this space is located off the major traffic line and there is no opening on the other side, it can be presumed that people in this room are more serious visitors so that they spend more time to appreciate the works of art. In fact, this room seems the only space in which the serious or motivated visitor can look at the paintings without being interrupted by people who just pass through.

At this point, if we look at the correlation again after eliminating three galleries, namely room 7, 9 and 12, it can be seen that the correlations among the variables are significantly improved (Table 2). The best correlation occurs between the geometrical ratio and the movement line. In addition, what is more interesting is that the correlation between the length and the linear movement line has improved considerably. These facts indicate that in general, the spatial configuration of a gallery space significantly affects the visitors' behaviour, in particular their movement pattern. The more detailed results from the analysis of the West-Wing indicate that (i) the higher the geometrical ratio is, the less the visitors make circular movement in a gallery room, (ii) the higher the geometrical ratio is, the more time the visitors spend in a gallery room, and (iii) the length of a gallery space also plays a significant role in defining the visitors' movement. In fact, this variable is obviously related to the geometrical ratio. Here, it should be indicated that perfect positive correlation (1.00) is shown between the geometrical ratio and the rest of physical variables. That is to say, the higher the geometrical ratio is, the larger the gallery space is. Therefore, the second and third results seem to be obvious.

5.2.2. *EWT*. This wing has 14 gallery rooms, and the observation was carried out in nine of those. This wing is quite similar to WWT in terms of the allocation and distribution of gallery rooms. The visitors' movement pattern and their movement rate in EWT, however, shows considerable difference. First of all, the average time spent in a gallery room is 77.9 seconds, whereas WWT shows 171.2 seconds, which is more than twice. Another variable, T/P (movement rate 1/perimeter: sec/m), also shows the difference: 2.75 sec/m in WWT and 1.36 sec/m in this part. In fact, in room 28, in which paintings by one painter, Mark Rothko, are displayed, the spend time is 31.9 seconds, which is far less than average. Furthermore, the visitors' movement pattern in this wing indicates 78.33% of linear movement, whereas it is 35.91% in WWT.

TABLE 3. Correlation coefficient between variables (EWT)

Variables	Ratio	Length (m)	Area (m2)	Peri. (m)	Pe/Pa (m)
Move-	0.276	0.415	0.482	0.462	-0.436
Rate.1	(0.382)	(0.570)	(0.610)	(0.610)	(-0.436)
Move-	0.273	0.395	0.450	0.434	-0.449
Rate.2	(0.388)	(0.555)	(0.578)	(0.584)	(-0.449)
Linear	-0.083	0.132	0.319	0.245	0.195
Move.	(-0.750)	(-0.430)	(-0.042)	(-0.203)	(0.232)
Rate 1	0.117	0.088	0.045	0.064	0.322
/Painting	(0.499)	(0.479)	(0.349)	(0.412)	(0.359)

( ) correlation coefficient after eliminating room 24

As for the correlation between variables (Table 3), we can hardly see any significance. The best correlation occurs between the area and the movement rate 1, and between the length and the movement rate 1. The correlation between the physical variables and the movement line is insignificant. It can be argued that these results are determined by the special character of the paintings displayed in these galleries. As stated before, most of these are modern or post-modern paintings, which are generally extremely abstract or unfamiliar and the general public do not seem to enjoy them, let alone fully understand them. Or they are very easy to understand, and thus only need a very brief appreciation. Therefore, visitors tend to pass through this gallery space looking briefly at the objects rather than stopping to contemplate the paintings. In this sense, visitors tend to pass through making mostly linear movement lines regardless of the geometrical ratio. This tendency results in relatively better correlation between the physical variables and the movement rate. In other words, the visitors' spending time is directly related to the area, perimeter and length of the gallery rooms.

At this point, let us look at some rooms more closely. Room 28 displays seven huge paintings which were painted by Mark Rothko and all look quite similar in terms of colour and pattern. It was therefore observed that most visitors looked at a painting by the entrance and then did not go inside to look at the others because of their similarity. Rooms 29 and 30 also display abstract paintings and sometimes sculptures. Because of the allocation of the openings, like in room 9 in WWT, however, some visitors tended to make a circular, though relatively fast, movement in spite of the fact that the works of art were abstract or less well-known. In contrast, although room 26 has a very similar character in terms of the properties of object size and geometrical ratio, the movement pattern shows 100% linear lines.

As a result, it can be seen that the properties of the objects on display play a significant role in defining the visitors' movement pattern and movement rate. However, although the properties of the paintings on display influence people, the physical settings such as openings, size and length still play a considerable part in defining the visitors' behaviour in this part of the gallery. In particular, room 29 and 30 show that the visitor's movement pattern and movement rate can be greatly influenced by manipulating the location of openings in connection with other variables such as geometrical ratio and length in a gallery space.

5.2.3. *WWN.* Table 4 shows that the best correlation occurs between the movement rate 2 and Pe/Pa (Perimeter/No. of Paintings) (-0.888) and between the movement rate 1 and Pe/Pa (-0.861), which are negative. Again, Pe/Pa makes a significant correlation with the linear movement line (0.707) and the time spent before a painting, that is, R/Pa (0.810). In addition, significant negative correlations are shown between the MR1/Pa (Movement Rate 1/No. of Paintings) and other physical variables, such as the length, the area and the perimeter ($r = -0.686$, -0.662 and -0.675 respectively). This implies that the larger a gallery space is, the less time visitors spend before a painting.

As for the physical variables, there is a good correlation between the perimeter and the movement rate 1 (0.672), and between the area and the movement rate 1 (0.654). On the other hand, the correlations between the geometrical ratio and other dependent variables are almost insignificant.

At this point, it can be noticed that in spite of the lack of difference of the geometrical ratio inother galleries, rooms 6 and 10 show a remarkably low percentage of the linear movement line. This fact has resulted from having only one opening, that is, being located off the major traffic line, whereas other galleries have two or three. Therefore, once the visitors get into the room, their returning route is supposed to be the other side of the room. In addition, the fact that visitors in a gallery or museum hardly ever repeat the same route also seems to encourage a circular movement.

In this respect, if we consider another correlation after eliminating those two galleries, it can be seen that the overall correlation has been considerably improved. In particular, the correlation between the geometrical ratio and the linear movement line becomes quite significant (0.822). Meanwhile, the correlations between the Pe/Pa and other dependent variables are still highly significant. All in all, some important behavioural aspects can be drawn from Table 4. In general, the larger the gallery space is, the less time the visitor spends before a painting. The higher the geometrical ratio is, the more linear movement line is made in the gallery. Finally, the larger the space a painting occupies on the gallery wall, the less time the visitor spends in the gallery. In addition, it can also be seen from Table 4 that the relatively low ratio (1.38) of the gallery spaces results in the high percentage of circular movement line (70.7%).

Therefore, it can be said that the geometrical ratio and the movement line are closely related to each other. The variable MR1/Pe (Movement Rate1/Perimeter: sec/m), indicates that the visitors spend more time in this part: 3.16 sec/m, whereas they spend only 2.75 sec/m in WWT and 1.36 sec/m in EWT.

TABLE 4. Correlation coefficient between variables (WWN)

Variables	Ratio	Length (m)	Area (m2)	Peri. (m)	Pe/Pa (m)
Move-	-0.308	0.551	0.654	0.672	-0.861
Rate.1	(-0.616)	(0.484)	(0.675)	(0.678)	(-0.823)
Move-	-0.327	0.498	0.606	0.618	-0.888
Rate.2	(-0.683)	(0.429)	(0.644)	(0.634)	(-0.844)
Linear	0.246	0.273	0.228	0.219	0.707
Move.	(0.822)	(0.677)	(0.489)	(0.505)	(0.506)
Rate 1	-0.200	-0.686	-0.662	-0.675	0.810
/Painting	(-0.062)	(-0.661)	(-0.688)	(-0.683)	(0.764)

( ) correlation coefficient after eliminating room 6 and 10

5.2.4. *EWN*. Table 5 shows correlations between the variables. The most significant correlation occurs between MR1/Pa and all other independent variables, which show a negative correlation except with Pe/Pa (0.758). This implies that the larger the gallery space is, the higher the geometrical ratio is, the less time the visitor spends before a painting.

There are also significant correlations between Pe/Pa and other dependent variables which are negative except with MR1/Pa. Like in WWN, the greater the space a painting occupies on the gallery wall, the less time the visitor spends in the gallery. However, what is different from WWN is that the correlation between Pe/Pa and MR1/Pa is negative. At this point, it can be said that this good negative correlation is more convincing than the positive one of WWN. The reason is that the ratio does not have any correlation with other physical variables in WWN, whereas in this wing it shows a highly significant correlation.

As for the movement rate and the movement pattern, and other independent variables, there is no significant correlation at this stage. Room 38 has an exceptional character. In spite of the relatively low ratio (1.25), it shows 55% of linear movement line, and thus the time spent in a gallery is very short. The relatively high percentage of linear movement line of room 38 is probably caused by the way the paintings were arranged on the walls. By hanging well-known paintings on one side only, the balance between the walls has been lost so that people do not readily approach the other side of the room even though the geometrical ratio is low. In fact, most gallery rooms observed in this study showed great consideration in balancing both sides in terms of the paintings' size and quality. Therefore, if the balance was seriously broken, the effect on people's behaviour was enormous. At this point, if we eliminate room 38, it can be seen that the correlations between the variables are considerably improved. In particular, the correlation between the geometrical ratio and the linear movement line shows great significance (0.821). In addition, the correlation between the geometrical ratio and the movement rate, and between physical variables and the movement rate have improved remarkably.

The low ratio of the gallery spaces in this wing results in high percentage of circular movement line (71.7%) and thus much spending of time. This aspect is well illustrated in room 34, which is the largest gallery space. This room shows far less time spent than others in terms of MR1/P (1.89 sec/m). As a result, some behavioural characteristics can be summarised.

TABLE 5. Correlation coefficient between variables (EWN)

Variables	Ratio	Length (m)	Area (m2)	Peri. (m)	Pe/Pa (m)
Move-	-0.625	0.580	0.578	0.555	-0.689
Rate.1	(0.853)	(0.793)	(0.782)	(0.765)	(-0.888)
Move-	0.431	0.374	0.373	0.346	-0.593
Rate.2	(0.590)	(0.498)	(0.484)	(0.460)	(-0.826)
Linear	0.561	0.502	0.481	0.483	-0.604
Move.	(0.821)	(0.734)	(0.713)	(0.701)	(-0.930)
Rate 1	-0.817	-0.781	-0.765	-0.767	0.758
/Painting	(-0.920)	(-0.875)	(-0.861)	(-0.854)	(0.888)

( ) correlation coefficient after eliminating room 38

The higher the geometrical ratio is, the more time visitors spend in the gallery. This obviously results from the fact that the higher the ratio is, the larger the gallery space is. The higher the geometrical ratio is, the less time visitors spend on each painting. This is also related to the fact that the higher the ratio is, the larger the gallery space is. Therefore, it is natural that other physical variables are closely correlated to the time spent for a painting. There is a good negative correlation between Pe/Pa, and the movement rate and the linear movement line.

5.2.5. NWN. At a glance, it is obvious that there cannot be any significant correlation between the variables because of the resemblance of the spaces. However, the movement lines of the visitors indicate 86.1% of circular movement. If room 25, which has an exceptional character, is removed, most visitors tend to make circular movement (93.7%). As a matter of fact, room 25 is a transition space from which visitors have a choice as to whether to go to room 24 or room 19. This seems to encourage visitors to make a linear movement line. The figure of MR1/P is the highest among the cases (3.89 sec/m), even though the geometrical ratio is relatively high compared to the remaining parts of the gallery. In these respects, the characteristics of this wing are the high percentage of the circular movement line and thus the long time spent in a gallery.

5.3. CONCLUSIONS

Five exhibition spaces (wings) have so far been analyzed in terms of the spatial characteristics of the individual gallery space. It has been noted that museum visitors' behaviour is significantly influenced by the physical dimension and geometrical shape of a gallery space and the location of openings in the space.

Openings in a gallery

First of all, it was noticed that the location of the opening(s) in a space plays a great role in defining the visitors' movement pattern and thus their movement rate. When the opening is situated on one side of the gallery space, like in the North-Wing of the National Gallery, the visitors are naturally forced to make a circular movement line along the wall to finish the space. Therefore, visitors tend to spend more time in this kind of space than in other gallery spaces. When the openings are situated on both sides of the gallery space, like in the Tate Gallery and the rest of the National Gallery, the higher the geometrical ratio is, the more visitors make a linear movement line.

Spatial Configuration
The higher the geometrical ratio is, the less time the visitors spend, provided that the physical dimensions are identical. In this sense, geometrical ratio, movement rate and movement pattern are closely interrelated. As for the movement rate, when the visitors make a circular movement, it is natural that they spend more time per metre measured on the perimeter of the gallery. One of the interesting facts was that the gallery spaces with high geometrical ratio tend to be also large spaces, in particular in the West-Wing of the Tate Gallery and the East-Wing of the National Gallery.

It can be said that the geometrical ratio is closely related to the size of the space, that is, the proportion of a space is generally likely be constant in an integrated part of an art gallery building, though there are differences between art galleries.

Character of paintings
It was seen that in the East-Wing of TG, the perimeter per painting shows 6.29 m, whereas the overall average is 3.58 m. This implies that either the paintings in this wing are generally larger than the ones in other wings or they are hung at much less density. Furthermore, the time spent per painting is 1.36 sec., which is much less than in other wings (the overall average is 2.86 sec.). In fact, more recent paintings are characterised by larger size and abstract aspects. Apart from the large size of the paintings and lower density of hanging, there is another reason why the time spent in front of a painting was less. This seems to lie in the fact that paintings in this wing were less well-known or more difficult to understand for the general public, or that the visitors are very quick to understand, and thus need a very brief appreciation. Therefore, the variables fail to make good correlations.

6. Discussion

Although recent museums show a considerable similarity in their spatial organisation with traditional museums, which were mainly built during 19th century, it is suggested that architects and museum professionals could benefit from a more systematic investigation of how spatial organisation affects the visitors' behaviour. Such an investigation could be useful in view of the growing concern about the quality of the visitors' experiences. This paper represents an attempt from the architectural point of view to find out the way in which museum people and architects can enhance the quality of the visitors' experiences.

As a result, it can now be said that there is a significant relationship between spatial organisation and the movement pattern and rate of the visitors. Throughout the analysis, it was noticed that other factors also played a role in defining the visitors' behaviour. The character of the paintings such as colour, size and the subject-matter certainly influenced the visitors' behaviour. However, the spatial structure of the exhibition space is the first and most important factor in defining the visitors' movement pattern and movement rate. In this sense, even if the character of the paintings is likely to discourage the visitors from spending enough time to look at them and from moving along the intended route, as we have seen in the East-Wing of the Tate Gallery, it may be possible to encourage them to do so naturally by manipulating the spatial structure.

References

Bechtel, R.B. (1967) "Human Movement and Architecture", *Trans-action* **4**, 53-56.

Gardner, T. (1986) "Learning from Listening - Museums Improve Their Effectiveness through Visitor Studies", *Museum News* **2**, 40-44.

Melton, A.W. (1935) Problems of Installation in Museums of Art. Washington, D.C., Publication of the American Association of Museums, New Series, No. 14.

Melton, A.W. (1972) "Visitor Behavior in Museums: Some Early Research in Environmental Design", *Human Factors* **14**, 393-403.

Robinson, E.S. (1928) The Behavior of the Museum Visitor. Washington, D.C., Publication of the American Association of Museums, New Series, No. 5.

DESIGN SUPPORT VIA SIMULATION OF BUILDING AND PLANT THERMAL INTERACTION

J.L.M. HENSEN
Eindhoven University of Technology
Department FAGO/HG 11.77
P.O. Box 513, 5600 MB Eindhoven
The Netherlands

ABSTRACT. Design decision support related to building energy consumption and/or indoor climate should be based on an integral approach to the environment, the building, heating, ventilating and air-conditioning (HVAC) system, and the occupants. The tools to achieve this are now available in the form of computer simulation systems which treat the building and plant as an integrated dynamic system. Although its potentials reach beyond the area of Computer Aided Building Design, the paper describes building and plant energy simulation within the context of CABD, design decision support and design evaluation. Currently, computer simulation is only used indirectly as a design decision support mechanism; that is, its power is not delivered very efficiently to the design profession. This paper suggests some future research directions. These are aimed at providing a mechanism to overcome this problem by developing an 'intelligent front end' which bridges the gap between sophisticated computer simulation tools and the design profession.

1. Introduction

The dynamic thermal interaction, under the influence of occupant behaviour and outdoor climate, between the building and its heating, ventilating and air conditioning (HVAC) system is still difficult to predict. In practice, this often results in non-optimal, malfunctioning, or even 'wrong' building/system combinations. Other topics belonging to the same problem domain are (in no particular order): the Sick Building Syndrome, Building Energy Management Systems, application of passive solar energy, HVAC system and control development and testing, integrated systems (e.g. floor heating, ice rinks, swimming pools), and unusual building/system combinations which may occur, for instance, when a historical building finds a new use (e.g. a church being converted into a multi-purpose centre) or in case of relatively new developments like atria.

The above mentioned problems and the need for an integral design approach to building, HVAC system and occupants, are becoming more and more important. Therefore, a research project was initiated on development/enhancement of building performance evaluation tools which treat the building and plant as an integrated dynamic system (Hensen 1991). One of the techniques available is modelling and simulation. Modelling is the art of developing a model which faithfully represents a complex system. Simulation is the process of using the model to analyze and predict the behaviour of the real system. Currently the most powerful tool available for the analysis and design of complex systems is computer simulation.

Modelling and simulation have become indispensable engineering techniques in the fields of design (e.g. of buildings, plant configurations, and on the component level) and operation (system control, understanding, and interaction). The main reason for this is that these techniques offer vast advantages over, for example experimentation, with respect to economy (in an increasing number of cases, simulation is faster, better and cheaper than experimentation), prediction (allows analysis of a system which does not yet exist), and education (models are

H. Timmermans (ed.), Design and Decision Support Systems in Architecture, 227–238.

easily adapted, inexpensive to operate, able to simulate adverse conditions and may also serve as an aid in communication). Of course, simulation and experimentation are effectively complementary: experimentation to discover new unknown phenomena or for validation purposes, and simulation to understand interactions of the known components of a system. In the current context, modelling and simulation is thus used for predictions in order to support design decisions on real-world problems regarding buildings and the HVAC systems which service them. The building in question may be an existing structure, a proposed modification of an existing structure, or a new design. Although the potentials of building and plant energy simulation reach beyond the area of Computer Aided Building Design, we will describe simulation within the context of CABD, decision support and design evaluation.

2. The CABD Context

Since the early 1960s, the use of computers in the field of building design - i.e. CAD (Computer Aided Design) which was only later specified to CABD (Computer Aided Building Design) - has been increasing steadily (Rooney and Steadman 1987). Nevertheless, according to Gero (1983), its potential has and is taking longer to realize than was first thought. This statement is still true, judging from comparison of Gero's predictions for the then immediate future of 1983 with the actual situation at present (see Radford 1991). Having mentioned this, CAD in the field of building design has received more and more attention, both from the research and commercial communities. Due to economic factors, the drafting function has received the most - commercial - attention and is now becoming well established in building design practices. The design process itself is very complicated, as may be concluded from the vast amount of work aimed at establishing models of the process of design. As Butera (1990) points out, the architectural design process may even be approached using principles from the so-called 'deterministic chaos' theory. The complexity and the diversity of parameters to be taken into account leave large opportunities to chance in identifying the design optimum. The optimum may be regarded as a 'strange attractor' in this context.

Due to its complexity, general software to aid in the design process is much less developed and has received much less attention than drafting and design process management tools. In recent years, promising research activities have been - or are about to be - initiated to relieve this deficiency. These studies often involve pluri-disciplinary research teams and employ very sophisticated research techniques (e.g. Dubois 1990; Clarke and Duffy et al. 1991). There is, however, one activity throughout the design process which has received much attention: building performance appraisal. Powerful, computer-based models were created to assess cost, performance and visual impact issues in design: from life-cycle cost estimates at the design proposal stage, through realistic visualizations of the design, to comprehensive evaluations of building energy and environmental performance. A demand for systems which possess both drafting and appraisal functions is, however, steadily growing. In response, appraisal programs were appended to drafting packages, thus creating what we may call early CABD systems.

As elaborated by Clarke (1989), CABD, and the sub-systems which it comprises, are affected by continuous changes in: power and cost of hardware, quality of software, elegance and effectiveness of human-computer interfaces (HCI), and user interface management systems (UIMS), and in (computer aided) software engineering (CASE) methods with which a greater degree of sub-system integration is possible. Table 1 (Clarke 1989), which is self-explanatory, summarizes several of the important issues in this respect. When this Table is projected on the actual situation at present, it seems that we are already in the short-term or perhaps even mid-term columns as far as the technology is concerned. This is due to recent technological and economic developments: i.e. relatively inexpensive, high-performance, graphics workstations; strong reduction of data storage costs; and emergence of early expert systems (see MacRandal 1988).

Table 1. Issues underlying CABD evolution (Clarke 1989)

Issue	Time Scale			
	Immediate (now)	Short-term (now-5 years)	Mid-term (5-10 years)	Long-term (10^+ years)
Technology	· Micros · Drafting · Early performance prediction	· Supermicros · Partial CAD integration · Early expert systems	· Networking (worldwide) · Expert systems · Full CAD integration	· Computer ubiquity · Artificial intelligence · Natural language · Full IKBS
Applications	· Drafting · Information technology · Performance prediction	· 'Accredited' performance prediction · 3-D Visualization · Regulations	· Performance specification · Solid modelling · Integrated functions	· Participation · Client-orientated CABD · Post occupancy applications
Education and Training	· Applications exploration · Hardware familiarization	· In-depth postgraduate training	· Advanced undergraduate & mid-career training	· Full computer assisted design systems
Research	· Application knowledge · Validation · System evaluation	· Human-orientated CAD · UIMS, Shells & Environments	· Systems for computer-naive designers · Implementation environments	· Non-traditional communication · Design optimization
Impact	· Expensive & time consuming · Job shifts	· Skills shortage	· De-skilling · Breakdown of professional boundaries	· De-professionalization
Net result				· Improved product performance

To illustrate this further, at the start of the present work (late 1986), a high-resolution, bit-mapped, graphics workstation, offering a performance of 1.5 Mips (million instructions per second) and 70 Mbyte data storage, was purchased for approximately 20 kECU (≈ NLG 50,000). Now, five years later, two new workstations have been ordered, one of which is only half the price and offers a 15 Mips performance, and another which still costs 20 kECU but offers 28 Mips performance and 1 Gbyte of disk storage capacity.

As another example of fast-developing technology, consider the following quote from Hartman (1988), which in addition illustrates usage of worldwide networking as may be deduced from the reference: "Technology which seems unearthly and ethereal today will be reality and commonplace tomorrow. Fiberoptics and electronic imaging will allow us instant access to networks and information systems throughout the world. CD-ROM, WORM, and huge optical disks will provide mass storage capabilities with speedy searching and retrieval. Animation, simulation, video and voice input and output, and supercomputer power will be focused on the desktop. And while computer processing will become decentralized as workstations continue to proliferate, networks and shared systems will weave strength into our interdependence. Not technology for technology's sake, but more people performing more computing and incorporating computers in new and innovative ways in the pursuit of excellence in teaching and research." With respect to applications, in 1992 we still seem to be in the 'immediate' column, except perhaps for 3-D visualization, which appears to be the next commercial goal (following drafting). The same is true for education and training; as yet, only a few educational institutions offer in-depth postgraduate training.

Regardless of technological developments, CABD will not become commonplace unless there is a high standard of user training. Although, as pointed out by Clarke (1989), it could be that ultimately user training becomes less important due to high-level assistance by the computer. This does not imply, however, that setting-up of education and training schemes is not of the utmost importance now. With respect to research, up to now most activity has been directed towards proving the system and towards the acquisition of application knowledge. In the field of building energy simulation, for example, a number of projects concerned with model validation have been or are being carried out.

Currently there are indications that the building energy simulation research activity is broadening in its scope (see Augenbroe and Laret 1989; Clarke and Maver 1991). More effort is being expended on human-orientated CABD through expert systems, HCI research and the like (Clarke and Rutherford et al. 1989). There is also a greater tendency to approach the problems underlying CABD in a multi-disciplinary, inter-institutional manner (Clarke and Hirsch et al. 1986; Clarke and Irving et al. 1988; Augenbroe and Winkelmann 1990). This is also reflected in the recent formation of building analysis clubs: International Building Performance Simulation Association (IBPSA) based in the United States; Building Environmental Performance Analysis Club (BEPAC) in the United Kingdom; and Building Analysis Groups (BAG) in the Benelux. CABD is not a cure-all. At worst, it is an automation of much of the mechanics of design. At best it allows an evaluation of the relationships inherent in a given design hypothesis. At present, the application of CABD is expensive in terms of required human resource and it is time-consuming. In the near future, the profession will probably experience a skills shortage. In the longer term, however, with further advances in technology, application knowledge and education and training, CABD might bring important changes in the design process, involving de-skilling and the breakdown of professional boundaries. CABD could well become the common denominator of all parties involved in the design process through some future IIBDS (Integrated Intelligent Building Design System). This will lower or even remove inter-professional barriers and improve the quality of the end product, the building.

As indicated, CABD is an evolutionary process which is characterized by several strong interrelations between quite different issues. For example, the level of application of energy simulation is as much a function of education and training as it is of hardware and HCI. Of course, CABD must also be regarded in the light of other technological and other developments which are taking place around us. That is, CABD will certainly become integrated in the 'office of the future', which might offer multi-media personal work environments, incorporating integrated CAD/CAE (Computer Aided Engineering) features.

One important issue not yet mentioned is that of politics. There are trends towards requiring that specified conditions must be achieved during the operation of buildings as well as in the design of buildings (as in ASHRAE's 1989 standard on ventilation and air quality, and towards setting up 'responsibility chains', ultimately making a design team liable for the performance of the end product (as implied in, for instance, ASHRAE's (1989a) guideline for commissioning of HVAC systems). It could well be that if these trends are followed and accepted by the building industry, such issues will become major catalysts in the evolution of CABD. Building energy simulation must be placed within this evolutionary context. Contemporary energy models are an important improvement compared with the traditional methods they replace. However, there are still several important developments which must be undertaken before valid, easy-to-use models can be delivered to the design profession.

3. Building Energy Simulation

As indicated in the introduction, modelling and simulation have become popular engineering tools since they permit us to predict the behaviour of a system before the conditions we are interested in occur, and indeed even without the system actually existing. In fact, modelling and simulation are the only techniques available that allow us to analyze arbitrarily non-linear systems accurately and under varying experimental conditions. Simulation is used in many areas of science and engineering. It is used in different senses to study a variety of systems which may be classified as continuous vs discrete, deterministic vs stochastic, or dynamic vs steady-state. It may be clear that building energy simulation addresses very complicated, highly interacting, continuous, deterministic, dynamic systems. Building energy modelling and simulation is part of an evolutionary process in the field of building design tools. Table 2 (from Clarke 1988) summarizes one view of the evolution of these design tools, from the traditional via the present-day simulation approach to the fourth generation tools by the late 90s. More information on the state of the art in building energy simulation can be found in reviews by Winkelmann (1988) and Wiltshire and Wright (1988).

Table 2. The evolution of building energy models (Clarke 1988)

1st generation	Handbook orientated		Indicative
	Analytical in formulation	<---	Application limited
	As simplified as possible		Difficult to use
	Piecemeal in approach		
2nd generation	Dynamics important	Feedback loop	
	Still analytical		
	Still piecemeal	>---	Increasing integrity
	Suitable for low-order projects		vis-à-vis the real
	with time invariance		world
3rd generation	Field problem approach		
(current generation)	requiring numerical methods		
	Integrated view of energy		
	sub-system		
	Suitable for high-order projects	>---	Leading to
	with time variation		
	Heat and mass transfer		
	considered		
	Better user interface and		
	partial CABD integration		V
Next generation	Full CABD integration		Predictive
	More advanced numerical methods	Feedback loop	Generalized
	Intelligent knowledge-based	>---	Easy to use
	Object-orientated software		
	architecture		

Early work in current generation approaches, focused on the relation between building design and energy consumption (Clarke 1977; Bruggen 1978), or between building design and thermal comfort (Lammers 1978). In this and in later work (Hoen 1987), the auxiliary system was still more or less regarded as a given boundary condition instead of as a variable. These workers emphasized the building side of the overall problem domain, while others (McLean 1982; Murray 1984; Tang 1985; Lebrun and Liebecq 1988) focused more on the plant side. In the former approach, the influence of the plant system is more or less neglected by over-simplification; estimation of energy consumption is based on some presumed, imposed indoor air temperature profile. In the latter approach, the complex building energy flow paths are usually grossly simplified, and the building (or each building zone) is commonly regarded as just another plant component which, in this case, imposes a thermal load on the system.

Although justifiable at that time, it is now felt that neither approach is preferable for the majority of problems which are affected by the thermal interaction of building structure and auxiliary system. We started from the principle that both building and plant have to be approached on equal levels of complexity and detail while taking into account all major fluid flow and heat transfer couplings. For the present work, we started from an established building energy simulation environment and enhanced this on the plant simulation side of the overall problem domain: the ESP^R (Environmental Systems Performance, Research version) energy simulation environment (Clarke and Hand et al. 1991), a system which is currently under development at various research centres throughout Europe, among which the Universities of Strathclyde and Eindhoven.

Reporting this work, Hensen (1991) describes a 'modular-simultaneous' technique for the simulation of combined heat and fluid flow in a building/plant context. The present performance of the system indicates that it is practical to solve the building/plant heat and mass flow network in detail. Moreover, the solution of complex building/plant/fluid flow networks in the transient state is now feasible on inexpensive computers. This enables an integral approach to the thermal interaction of building structure and heating and ventilating systems, and this also provides the basis/power for design decision support in this area.

4. Design Support via Simulation

In the field of building energy-related issues, there is a certain tradition of using computer simulation for design support, involving the generation of knowledge which is subsequently transferred to the design profession. This kind of design support is thus based on knowledge transfer from 'specialists' in a certain part of the overall problem domain towards the design profession. With respect to the transfer process itself, there are may different approaches. At one end of the spectrum of possibilities one finds the so-called design-aids, while consultancy work for a specific design could be located at the other end of the spectrum. The following sub-sections demonstrate both these approaches by an example.

4.1. DESIGN-AIDS

This is a form of knowledge transfer in which the 'specialists' try to generate generic knowledge which is supposed to be suitable for a range of buildings and which is usually aimed at being used by the designers. This kind of knowledge is commonly based on regression techniques applied to the results of multiple parametric runs of more powerful modelling systems. The results to emerge can often be reduced to simple relationships or presented in tabular or graphic form. Figure 1 is a typical example (from CEC 1986) showing summer overheating assessment graphs for medium-weight houses. It is obvious, however, that there are a number of drawbacks to such an approach. The most important of these are the following: (i) a particular aspect is regarded in an isolated manner; (ii) this approach is only possible for a limited number of variables; and (iii) the results are only valid for a certain combination of environment, building, installation, and occupancy pattern which is quite similar to the one used to generate the results.

4.2. CONSULTANCY WORK

In the current context, this involves the generation of specific knowledge for a particular design by a specialist. Work we did on predicting air flow through proposed shopping arcades; (the Heuvel Galerie in Eindhoven; Figure 2) is a typical example of such an approach. The Heuvel Galerie includes an extensive shopping mall. This four-level complex incorporates a 220-metre-long shopping arcade, interspersed with atria and dome-shaped roofs, approximately 20,000 m^2 of shops, an 8,600 m^2 concert hall, 3,000 m^2 of restaurants, a 1,200-unit car park, offices, and apartments.

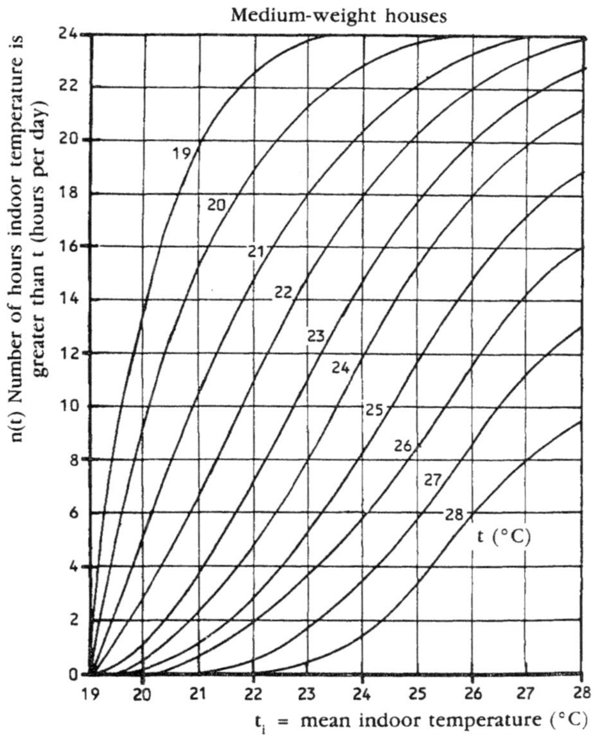

Figure 1. Summer overheating assessment graphs for medium-weight houses
Source: CEC 1986

It should be apparent that such a building is a highly complicated system. For instance, the manner in which air will flow depends on the external pressures on entrances and domes, temperature differences inside and with respect to outdoors, and impulses by the ventilation system. For this building ESPR was used to make various predictions with respect to the indoor environment (Pernot and Hensen 1990). As an example, consider Figure 3. This graph shows the air velocities which may be expected in the pedestrian entrance area. For commercial reasons, the architects and the developers want the entrance areas to be as open as possible. From the first results it was clear, however, that the original design proposal (incorporating air curtains for the main entrance) would lead to unacceptably high air velocities. One of the main conclusions was that additional air flow restrictions were necessary. For this it was suggested to apply double sets of sliding doors (wind sluices) at the 'cafe' and 'west' entrances, and to incorporate extra sliding doors as well as side-hung doors at the main entrance. In case of severe wind, it should be possible to further restrict the (open) cross-section of the main entrance.

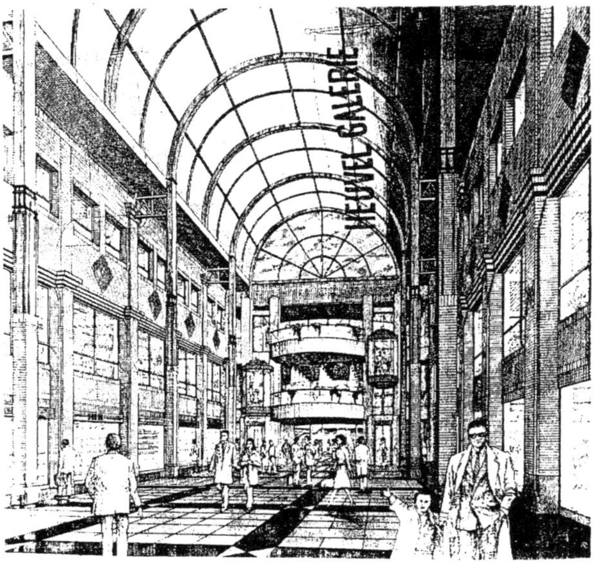

Figure 2. Shopping arcade in the Heuvel Galerie in Eindhoven

These are merely two examples, more or less demonstrating both ends of the spectrum of knowledge transfer/design decision support possibilities in the area of building-energy use. Obviously, there are various intermediate approaches; these often take the form of some simplified calculation method. The envisaged user profile usually shifts from more designer-like towards more specialist-like as the method shifts from design-aids towards the full simulation

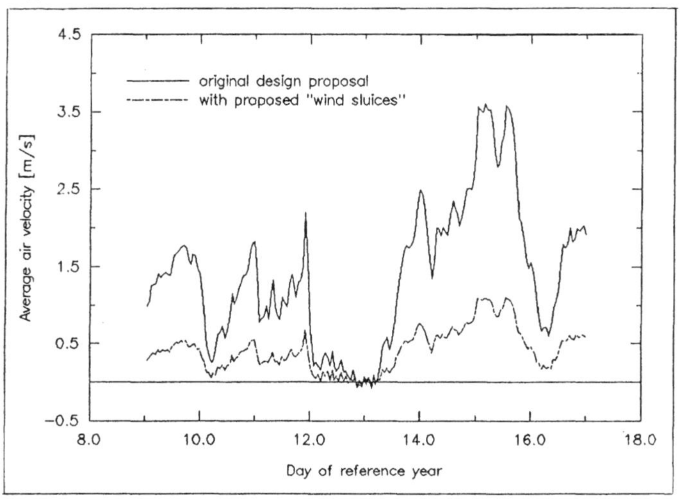

Figure 3. Predicted effect of proposed wind sluices to decrease the air velocities in the pedestrian entrance area (the passage connecting to the main square)

model. Since buildings are complex mechanisms, involving phenomena such as transient conduction and air movement, there is a growing realization that traditional design tools cannot cope with this complexity. Particularly at the earlier stages of the design process, there is a need for rapid feedback on the cost and performance consequences of alternative design scenarios. The present system of specialist consultants, while adequate for the detailed design and final specification phases, fails to provide this immediate 'ad hoc' advice.

5. Conclusions and Future Directions

While development of sophisticated building performance evaluation tools will add to the building engineer's toolkit, they also create new problems. These problems emanate from the conflict between two sets of requirements. On the one hand, the tools have to be powerful and comprehensive and, according to first thermodynamic law principles, to adequately represent the real world complexity. On the other hand, they have to be simple, straightforward, and intuitive in order to facilitate user interaction. Such problems are not restricted to novice users; they apply to experienced users as well (Van Nes 1991).

As Clarke (1991) points out, the conflict between power and ease of use is further exaggerated by the divergence of the conceptual outlook of the design-orientated program users and the technically orientated program developers. And to add to the confusion, the various engineering professions use subtly different terminology. One - very promising - way to tackle these problems is by utilization of Knowledge Based System (KBS) and Human-Computer Interaction (HCI) techniques to create an Intelligent Front End (IFE). Using these techniques it is possible to

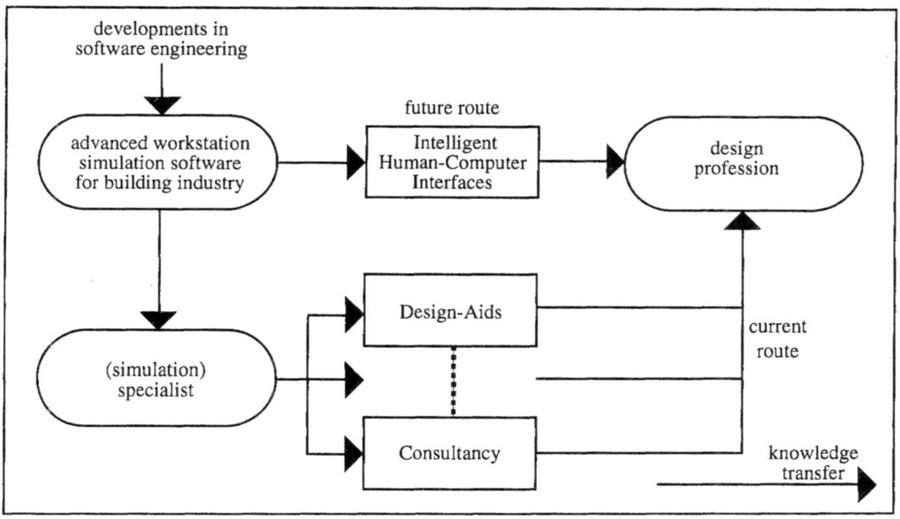

Figure 4. Current and future route of knowledge transfer/design decision support, starting from simulation tools which treat the building and plant as an integrated dynamic system

construct a user interface which incorporates a significant level of knowledge in relation to building description. This is possible in the face of real-world uncertainty and realistic performance assessment methodologies. Such a system would direct a user's line of enquiry, allowing 'What do you suggest?' and 'Why do you ask?' type responses. It would also be expert enough to devise an appropriate performance assessment methodology and to coordinate model operation with this. Using an IFE, the powerful simulation core may be invoked much earlier in the design process, because it is readily available to the designer. Obviously, specialist consultancy will still be necessary, but this can be limited to the more common questions/problems.

This shift away from the more traditional approach using design-aids and specialist consultancy, towards future direct application of powerful simulation tools by the design profession is indicated in Figure 4. This future kind of design decision support in the area of building energy and indoor climate thus derives its power from its simulation core while deriving its ease of use from some intelligent interface. It is in this direction that we are currently orienting our research activities.

Acknowledgement:
The author is deeply indebted to Professor Joe Clarke of the Energy Simulation Research Unit at the University of Strathclyde in Glasgow, whose support was essential to this work.

References

ASHRAE (1989a) Guideline for Commissioning of HVAC Systems, ASHRAE Guideline 1-1989, American Society of Heating, Refrigerating and Air-Conditioning Engineers, Atlanta, GA.

ASHRAE (1989b) Ventilation for Acceptable Indoor Air Quality, ASHRAE Standard 62-1989, American Society of Heating, Refrigerating and Air-Conditioning Engineers, Atlanta, GA.

Augenbroe, G.L.M. and Laret, L. (1989) COMBINE: Computer Models for the Building Industry in Europe, Information Package for the DG XII Programme 'JOULE', Commission of the European Communities DG XVII, Brussels.

Augenbroe, G. and Winkelmann, F. (1990) "Integration of Simulation into Building Design: The Need for a Joint Approach", in *Proceedings of the ASME International Solar Energy Conference on Design Tools for Passive Solar and Building Energy Conservation*, Miami.

Bruggen, R.J.A. van der (1978) Energy Consumption for Heating and Cooling in Relation to Building Design, Ph.D. Dissertation, Eindhoven University of Technology.

Butera, F.M. (1990) "Conflicts between Physics and Architecture in Solar Systems Optimization", in T.C. Steemers and W. Palz (eds.), *Proceedings of the 2nd European Conference on Architecture*, Kluwer Academic Publishers, Dordrecht, pp. 346-352.

CEC (1986) *European Passive Solar Handbook: Basic Principles and Concepts for Passive Solar Architecture*, Commission of the European Communities DG XVII, Brussels.

Clarke, J.A. (1977) Environmental Systems Performance, Ph.D. Thesis University of Strathclyde, Glasgow.

Clarke, J.A. (1988) "The Energy Kernel System", *Energy and Buildings* **10**, 259-266.

Clarke, J.A. (1989) "Building Energy Simulation: The State-of-the-Art", *Solar and Wind Technology* **6**, 345-355.

Clarke, J.A. (1991) "Intelligent Front-ends for Engineering Applications", *International Journal of Artificial Intelligence in Engineering* **6**, 2-4.

Clarke, J.A., Duffy, A., Lees, B., Lewis, J., MacRandal, D., and Maver, T.W. (1991) Elicitation of the Knowledge Underlying Design Conceptualisation, Grant Proposal to the EDRC (Engineering Design Research Centre), (unpublished), University of Strathclyde, Glasgow.

Clarke, J.A., Hand, J., Strachan, P., Hensen, J.L.M., and Pernot, C.E.E. (1991) ESPR: A Building and Plant Energy Simulation Research Environment, Energy Simulation Research Unit, ESRU Manual U91/2, University of Strathclyde, Glasgow.

Clarke, J.A., Hirsch, J., Winkelmann, F., Buhl, F., and Erdem, E. (1986) "Planned Developments in Building Energy Simulation", in *Proceedings of the 5th International Symposium on Use of Computers for Environmental Engineering Related to Buildings*, Bath, UK, pp. 482-491.

Clarke, J.A., Irving, A.D., Lockley, S., and MacRandal, D. (1988) "An Object-Oriented Approach to Building Performance Modelling", in *Proceedings USER 1 Working Conference at Ostend*, Society for Computer Simulation International, Ghent, Belgium, pp. 144-151.

Clarke, J.A. and Maver, T.W. (1991) "Advanced Design Tools for Energy Conscious Building Design: Development and Dissemination", *Building and Environment* **26**, 25-34.

Clarke, J.A., Rutherford, J.H., and MacRandal, D. (1989) "An Intelligent Front-End for Computer-Aided Building Design", in *Proceedings 3rd European Simulation Congress, September 1989, Edinburgh*, Society for Computer Simulation International, Ghent, Belgium, pp. 362-368.

Dubois, A.M. (1990) "Presentation of CIBAO: A New Approach for Integrated Computer-Aided Design Systems for the Building Activities", in T.C. Steemers and W. Palz (eds.), *Proceedings 2nd European Conference on Architecture, Paris, December 1989*, Kluwer Academic Publishers, Dordrecht, pp. 507-510.

Gero, J.S. (1983) "Computer-Aided Architectural Design - Past, Present and Future", *Architectural Science Review* **26**, 2-5.

Hartman, A. (1988) Plus ça Change ..... 'Comment' from the Tulane Computing Services Newsletter, in Articles database of CCNEWS, the Electronic Forum for Campus Computing Newsletter Editors, EDUCOM, BITNET.

Hensen, J.L.M. (1991) On the Thermal Interaction of Building Structure and Heating and Ventilating System, Ph.D. Dissertation, Eindhoven University of Technology.

Hoen, P.J.J. (1987) Energy Consumption and Indoor Environment in Residences, Ph.D. Dissertation, Eindhoven University of Technology.

Lammers, J.T.H. (1978) Human Factors, Energy Conservation and Design Practice, Ph.D. Dissertation, Eindhoven University of Technology.

Lebrun, J. and Liebecq, G. (1988) System Simulation Synthesis Report, Energy Conservation in Buildings & Community Systems Programme, Final Report Annex X, International Energy Agency, Operating Agent: University of Liège, J. Lebrun, Report AN10 881020-RF.

MacRandal, D. (1988) "Some Trends in Computing: The Implications for Simulation", *Energy and Buildings* **10**, 249-258.

McLean, D.J. (1982) The Simulation of Solar Energy Systems, Ph.D. Thesis, University of Strathclyde, Glasgow.

Murray, M.A.P. (1984) Component Based Performance Simulation of HVAC Systems, Ph.D. Thesis, Loughborough University of Technology.

Nes, F.L. van (1991) "Boeken, Computers en Ergonomen: Gemak van Gebruik door Gebruiksgemak", Inaugural Speech, Eindhoven University of Technology.

Pernot, C.E.E. and Hensen, J.L.M. (1990) Adviezen inzake de Kwaliteit van het Binnenmilieu van de Heuvel Galerie te Eindhoven, TPD-FAGO-RPT-90-42, Eindhoven.

Radford, A. (1991) "Editorial: Developments in Computer-Aided Design", *Building and Environment* **26**, 25-34.

Rooney, J. and Steadman, P. (1987) *Principles of Computer-Aided Design*, Pitman, London.

Tang, D. (1985) Modelling of Heating and Air-Conditioning System, Ph.D. Thesis, University of Strathclyde, Glasgow.

Wiltshire, T.J. and Wright, A.J. (1988) "Advances in Building Energy Simulation in the UK - the Science and Engineering Research Council's Programme", *Energy and Buildings* **10**, 175-183.

Winkelmann, F. (1988) "Advances in Building Energy Simulation in North America", *Energy and Buildings* **10**, 161-173.

A MODEL APPRAISING THE PERFORMANCE OF STRUCTURAL SYSTEMS USED IN SPORTS HALL AND SWIMMING POOL BUILDINGS IN GREECE

V. BOURDAKIS and R.F. FELLOWS
School of Architecture and Building Engineering
University of Bath, Claverton Down
Bath, BA2 7AY
United Kingdom

ABSTRACT. The selection of the best performing structural system (among steel, timber laminated, concrete, fabric tents) for medium span (30-50m) sports halls and swimming pools in Greece formed the impetus for this research. Decision-making concerning selection of the structural system is difficult in this sector of construction, as was explained in the "Long Span Structures" conference (November 1990, Athens, Greece). From the literature it has been found that most building appraisals end up at the level of data analysis and draw conclusions on the individual aspects they investigate. These approaches usually focus on a fraction of the problem, examining it very deeply and theoretically. Their drawback is loss of comprehensiveness and ability to draw conclusions on an overall level and consequently being applicable to the existing conditions. Research on an inclusive level is sparse. In this particular research project, an inclusive appraisal approach was adopted, leading to the identification of three main variables: resources, human-user-satisfaction, and technical. Consequently, this led to a combination of purely quantitative and qualitative data. Case studies were conducted on existing buildings in order to assess the actual performance of the various alternative structural systems. This paper presents the procedure followed for the identification of the research variables and the focus on the development of the model of quantification. The latter is of vital importance if the problem of incompatibility of data is to be solved, overall relation of findings is to be achieved and holistic conclusions are to be drawn.

1. Appraising Building Performance

The problem of decision-making during building design, in terms of structural materials, subassemblies and constructional systems used, is an old one (CIB 1975) and, as Preiser (1988) argues, still has not been solved. This decision-making problem is closely related to the evaluation of the building elements in order to identify their properties and subsequently their applicability for certain needs. The same problem exists as far as evaluation in use is concerned; according to Shibley (1982) "Buildings need constant adjustments to the needs of their users", leading to a continuous need for assessment evaluation and fine tuning of the building objectives, following users' feedback-feedforward, to use Markus' (1972a, 1972b) terminology.

Building performance has been investigated and researched a lot in the past - especially during the 1960s and 1970s. W60 (1975), commissioned by the CIB on the performance concept, stated that: "As a term to characterize the fact that products must have certain properties to enable them to function when exposed to stresses the word performance has been chosen." The notions and approaches of the last 15 years can be identified through Preiser's (1988) definition: "In the performance concept, the behaviours, qualities and accomplishments of people and things are measured and evaluated."

The aims of the performance-related studies can be classified in three main categories. The first is the performance concept itself, in terms of identifying problems, investigating and setting performance requirements, criteria and performance specifications for particular building types. The second is evaluation of buildings in the design stage, employing the performance concept as

H. Timmermans (ed.), Design and Decision Support Systems in Architecture, 239–248.

a tool for the analysis and evaluation of proposed solutions. And finally, the third is the appraisal of buildings in use, evaluating existing buildings.

The evaluations of buildings are of two types: the partial, dealing with particular aspects of building performance; and the inclusive, incorporating all aspects of performance in a structured and systematic way. The partial ones may be cost-dependent, analyzing the economic factors and considerations; quality-dependent; or cost-time-dependent. The last is a relatively new trend due to implementation of more advanced economics and the changes in the financing methods followed.

There is great confusion as far as the terms evaluation and appraisal are concerned. It is widely believed that they both imply a process by which various solutions to a particular problem are tested and selected - that is prior to the actual implementation. Furthermore, evaluation is closely related to some sort of scores or values. According to Bishop (1978), "Appraisal is the process by which completed buildings are assessed, judged or evaluated by the client, the users (if distinct from the client), the designers or by a combination of any of these." Another name given to appraisals during the late 1970s, is post-occupancy evaluations (POE). Preiser (1988) points out that the philosophical and theoretical foundation of POE is the performance concept: "POE is an appraisal of the degree to which a designed setting satisfies and supports explicit and implicit human needs and values of those for whom a building is designed." In their more rigorous forms, POEs follow a systematic approach where all user groups are represented and all important design elements are examined. The concept of appraisal, as analyzed by the Building Performance Research Unit (BPRU, Markus 1972a, 1972b), is related to the establishment of the quality or performance of a solution and incorporates three basic steps: "Representation in which the solution is modelled in a suitable way. This representation may be verbal, mathematical, visual or even full-scale (a building in-use is a full model). Measurement where the performance of the model is obtained on as wide a variety of counts as necessary. Finally evaluation, when the measured results are processed."

The literature survey showed that the approaches followed in the evaluation of buildings can be broadly classified in two categories: theoretical and practical. Theoretical partial approaches examine rigorously a narrow section of the whole building performance issue, leaving many aspects and variables untouched. On the contrary, inclusive approaches being systematic analyze all aspects and present a holistic view. However, they tend to be "all breadth and no depth" (Bishop 1978) and unless used with great care do not produce significant results (SSBRT 1976). The majority of theoretical approaches are of the framework type, identifying some variables and measuring each one individually (partial). They consequently fail to develop an overall measure of building performance or a model for the quantification of the results.

On the other hand, the criteria set and the models developed in practical approaches are suitable for specific building types only. As these are mostly conducted by public sector departments, the criteria are mainly related to the use, function and maintenance costs. These are more technically oriented than the highly theoretical inclusive approaches. Only a few theoretical approaches, the so-called model appraisals, reviewed in the literature develop a model, BPRU being one of the better established and widely known. This model is inclusive, focusing on school buildings, although the team that developed this model argue that it can be modified and applied in various building types. The important characteristic of this approach is that the building is divided in four sub-systems (building, environmental, activity-behaviour and organizational) and related to each sub-system is a measure of cost. Consequently, all four sub-systems are interrelated and examined in total, enabling the drawing of inclusive conclusions.

Another model is that for the assessment of expected performance effectiveness developed by Crise (1975). This model focuses on industrialized buildings and examines them in terms of technological and economic performance capabilities. It is not an inclusive model since sociological, psychological and environmental issues are not considered; the two points that make it worth noting are its application and model. It is applied in the pre-selection stage, assessing various alternatives. An explicit classification of techno-economical variables is presented in three levels. Following the calculation of the relative effectiveness indices for the two first levels, the overall effectiveness index of each industrialized building is calculated.

2. The Problem

The already-mentioned global problem of evaluating and assessing the appropriateness, effectiveness and in general the performance of the various structural systems and materials used in buildings is equally applicable in Greece, though it is more apparent in long spanning structures. In buildings spanning up to 15 metres, in-situ reinforced concrete is the dominant structural material being produced locally. It is widely accepted by the engineers, labour force and public. In long spanning structures, over 25 metres, the situation is confusing. There are various opinions and arguments from architects, building engineers, contractors and theoreticians leading to a long-lasting debate. The importance of the debate, together with an increasing need for these structural types (exhibition, conference and sports halls, warehouses and swimming pools), forced the Technical Chamber of Greece to organize a Scientific Seminar on Long Span Structures (Athens, 1990). The main problems highlighted were the availability and suitability of structural materials, the need for new building standards, import of know-how on a permanent basis, quality control and educational needs of the engineers.

Considering the existing needs, future trends, number of buildings built, availability of data and variety of structural systems implemented, this research focuses on sports halls and swimming pools spanning 30 to 60 metres. Among the structural systems used are steel frames, ranging from space framed beams or 3D trusses up to space frames, all the forms of fabric and pneumatic structures, tents, inflated balloons, etc. and Glu-lam. Even reinforced concrete has been used, though rarely. Korbas (interview, 1990), architect and member of the Design Department of the General Sport's Secretariat (DDGSS), pointed out that the vast majority of these buildings are General Sport Secretariat (GSS) developments. As far as availability of information is concerned, the existence of this public board is critical.

The aim of the research is to find out which is the best performing structural system for sports halls and swimming pools in Greece. The methodological objectives set were to develop a model appraising the performance of sports halls and swimming pools in Greece and, following the analysis of a representative sample, to investigate and evaluate any impact of the structural system on the performance of sports halls and swimming pools in Greece. This paper is concerned with the initial analysis only, leading to the identification of the main variables. Additionally the problems related to the development of a model of quantification are presented.

3. Procedure of Analysis

The research method used is selected according to the type of research and information needed (Yin 1989). Following a classification of the sports hall and swimming pool buildings, an in-use evaluation of 15 buildings was conducted. This method combined some of the advantages of case studies (depth of the research) and surveys (variety - all building types examined), helping reach reliable conclusions as well as being more flexible. A model was constructed aiming at an overall measurement of building performance and facilitating the comparison between buildings of different structural systems. Building performance, according to SSBRT (1976) framework, is assessed in five systems: environmental, activities, technical, cost-resources, cultural-symbolic. Subsequently the effects of structural system changes on the environment, the activities, the resources and culture were examined, evaluated and correlated to produce an overall measure of building performance. The consequences of a change in the SSBRT's (1976) "technical system" (or according to Markus (1972a, 1972b) "building system") upon the building performance were investigated.

Analyzing the GSS briefs and particularly the Design and Build ones, the GSS model became more explicit. This model is based on the evaluation of proposals submitted by the contracting firms and incorporates two main variables: cost and quality. The first includes the building costs and overall cost of the building (excluding running costs), the costs of penalties for exceeding the construction deadlines set, and finally the overall construction period. Quality incorporates eleven parameters: aesthetics (external and internal appearance of the building), function (of the building in general, in relation to sports and to its surrounding environment), loadbearing structure (overall arrangement, selection, combination of the loadbearing members of the building and durability), lighting of the main hall, heating, plant room, machinery and e/m equipment, building materials, energy conservation systems, acoustics of the main hall and finally repairs and maintenance (selection of materials and constructional methods facilitating simplicity, flexibility and economy in repairing and maintaining the building). The relative importance of these parameters is not the same; analyzing this GSS model, their exact weighting was established, which is presented later in this paper.

4. Identification of the Variables

The model of appraisal developed for this particular research is based on the GSS model. The research focuses on aspects-issues influencing the performance of sports halls and swimming pools in Greece and at the same time affected by the selection of structural system. Consequently, few of the parameters listed were discarded as irrelevant to the research aims and objectives. The remaining parameters were analyzed from the different viewpoint of the research model and classified under three main variables: resources, human-user satisfaction and technical aspects. Resources, the first variable and the more clearly defined, is similar to the resources modifier of Hillier and Leaman (1972) and the SSBRT's (1976) cost-resources heading. This variable incorporates the cost and time variables of the GSS model. These are not only examined at the provision-construction level (as in construction costs and time spent of the GSS approach), but in-use (HVAC costs, energy conservation and maintenance) as well.

The second main dependent variable to be examined is the human-user satisfaction, including the physical environment, one of the two issues discussed in Markus' (1972a, p.10) environmental system: "aspects of the environmental system directly perceived as heat, light, sound, texture and smell" and psychological aspects. This second aspect that this variable deals with is presented in Zunde's (1982) analysis under the delight, visual aesthetics section and Preiser's (1988) behavioural elements of post-occupancy evaluation. Essentially, physical environment concerns heating, lighting and acoustics, whereas psychological aspects refer to the GSS aesthetics as well as the spatial configuration - similar to the functional parameter of the GSS model. The last main variable of the research is technical, similar to Preiser's (1988) category, where structural integrity, durability and general performance of the various building systems is assessed. This technical variable is identical to the SSBRT (1976) technical category and very close to the BPRU (1972) building system. However, only two out of BPRU's three sub-systems are examined: construction and services. The technical variable is divided in two: building structure and repair and maintenance. Building structure is assessed in terms of the two GSS model parameters of building materials and loadbearing structure, whereas repair and maintenance is reviewed as in the GSS analysis. The following table presents the three main variables, the aspects considered in each, and the relevant GSS parameters.

TABLE 1. Main research variables

Main variables	Aspects considered	GSS parameters
Resources	Cost Time	Cost Heating Repair and maintenance Time Energy conservation
Human-user satisfaction	Psychological aspects Physical environment	Function Aesthetics Heating Lighting Acoustics Machinery and equipment
Technical variables	Building structure Repair and maintenance	Building materials Loadbearing structure Repair and maintenance

Having structured the main research variables, the next step is to identify the independent and dependent ones, justify the selection and present the functions measuring them. As far as resources are concerned, two provision-construction cost indicators are considered in this particular research: the envelope and the services. The former includes sub-structure, superstructure and cladding. The latter incorporates heating, ventilating, air-conditioning and fire protection costs. In order to enhance comparability between various buildings with varying sizes, the floor area of the building is considered as well.

$$\text{Envelope} = \frac{\text{Substructure}+\text{Superstructure}+\text{Cladding}}{\text{Usable Floor Area}}, \quad \left(\frac{\text{drx}}{\text{m}^2}\right) \tag{1}$$

$$\text{Services} = \frac{\text{HVAC}+\text{Fireprotection}}{\text{Usable Floor Area}}, \quad \left(\frac{\text{drx}}{\text{m}^2}\right) \tag{2}$$

The only provision-construction time indicator the research examines is the rate of construction.

$$\text{Construction Rate} = \frac{\text{Usable Floor Area}}{\text{Actual Construction Period}}, \quad \left(\frac{\text{m}^2}{\text{days}}\right) \tag{3}$$

The in-use cost indicators that the research considers are running costs, regular maintenance and failure repair. Running costs include fuel, electricity and the wages of the permanent staff. The other two categories incorporate the labour cost involved as well as the cost of the necessary spare parts and materials.

$$\text{Running Cost} = \frac{\text{Fuel}+\text{Electricity}+\text{Wages}}{\text{Usable Floor Area}}, \quad \left(\frac{\text{drx}}{\text{m}^2}\right) \tag{4}$$

$$\text{Regular Maintenance} = \frac{\text{Labour}+\text{Materials}}{\text{Usable Floor Area}}, \quad \left(\frac{\text{drx}}{\text{m}^2}\right) \tag{5}$$

$$\text{Failure Repair} = \frac{\text{Labour} + \text{Spaces}}{\text{Usable Floor Area}}, \quad (\frac{\text{drx}}{\text{m}^2}) \tag{6}$$

Two of these are the in-use time indicators that this research considers: the regular maintenance and the failure repair.

$$\text{Regul.Mainten.Disruption} = \frac{\text{Days Lost Maintaining}}{\text{Usable Floor Area}}, \quad (\frac{\text{days}}{\text{m}^2}) \tag{7}$$

$$\text{Failure Repair Disruption} = \frac{\text{Days Lost Repairing}}{\text{Usable Floor Area}}, \quad (\frac{\text{days}}{\text{m}^2}) \tag{8}$$

It should be pointed out again that the in-use indicators are calculated annually. The following table draws the conclusion on the basis of previous analysis.

TABLE 2. Resources indicators

	Cost	Time
Provision	Envelope Services	Construction rate
In-use	Running Regular maintenance Failure repair	Regular maintenance Failure repair

Human-user satisfaction is a qualitative and highly subjective variable assessed through psychological and physical environment indicators. These indicators are classified in five main categories: aesthetics, HVAC, lighting, acoustics and functionality. As Markus (1972a, 1972b) explains, "as a consequence of the importance of the interaction of people and buildings, any examination of building performance should also take into account the quality of that interaction and the way in which the building influences it". Furthermore in order to proceed with the specific measurement of the above-mentioned interaction, it is necessary to collect empirical information about what people actually do and say in regard to their environment. Therefore, the qualitative parameters of the environment-activity interaction are measured through the subjective users' opinions. On the other hand physical environment indicators are assessed by the researchers, through objective or subjective measurements. Thus, in each category, psychological and physical indicators are related and compared in order to evaluate and assess the validity of the users' opinions and to enhance the capacity to draw conclusions.Aesthetics are further analyzed in appeal of the building as a whole and appeal of the hall alone. The relevant physical indicators are the shape and size of elevations, colours used for the first and main hall's properties in general and colours used for the second, recorded following personal observation of the buildings by the researchers.

In terms of HVAC, users' opinions are monitored. Additionally, the following physical indicators are considered: temperature considerations (proximity to the sea, open space or protected by buildings, strong winds, etc.), air quality (dust, pollution) and condensation. Users are also questioned on lighting issues, focusing on daylight intensity and glare problems. The relevant physical indicators are the window surface, texture and orientation, the properties (colour, texture, etc.) of the hall's walls and the number of hours that the artificial lighting is on during a day (time of day the lights are turned on during a certain month of the year).

Acoustics in general and the noise penetration from the surrounding environment are also considered. The physical indicators of the environment investigated by the researchers are traffic load, factories and other noise sources that are potential causes of disturbances. In terms of functionality, users are questioned on the safety and visibility of the buildings. The researchers are carrying out an observation of the whole building in order to identify areas of potential

TABLE 3. Human-user satisfaction indicators

	Psychological	Physical
Aesthetics	Appeal of the building Appeal of the main hall	Shape and size of elevations Colours used
HVAC	Heating Air quality-comfort Ventilation	Temperature considerations Air quality Condensation
Lighting	Natural light Glare	Window surface, texture and orientation Main hall's wall properties Usage of artificial lighting
Acoustics	Noise level in the hall Echo, clarity of sound	Proximity to noise sources
Function	Safety Visibility from seating	Wall's protrusions, smoothness Potential problem areas

problems such as the wall's protrusions and smoothness, etc. Technical aspects form the third main research variable. Repair and maintenance resources have already been discussed in previous sections, leaving aspects related to the building structure and maintenance to be presented and analyzed. Several indicators were identified: industrialization of structure, origin of structural materials, complexity of structure and maintainability. It was decided to develop cost-related functions for the measurement of these technical issues, as they offer the best indicators of the size of the relevant assemblies and are easily comparable.

$$\text{Industrialization} = \frac{\text{Prefabricated Members Cost}}{\text{Envelope Cost}}, \quad (\%) \tag{9}$$

$$\text{Origin} = \frac{\text{Native Members Cost}}{\text{Envelope Cost}}, \quad (\%) \tag{10}$$

$$\text{Complexity} = \frac{\text{Structural System Cost}}{\text{Envelope Cost}}, \quad (\%) \tag{11}$$

$$\text{Maintainability} \quad \begin{cases} \text{Maintenance Staff Capabilities Needed} \\ \text{Spares Availability} \end{cases} \tag{12}$$

5. Model of Quantification

Following the presentation and analysis of the three main variables, their relative importance must be established to enable an accurate evaluation of the performance of each structural system examined. The starting point for the development of a model of quantification was the GSS model and the relative importance of its parameters.

TABLE 4. Technical indicators

Industrialization	Extent of prefabrication
Origin of materials	Percentage of native materials
Complexity	Number and cost of sub-assemblies
Maintainability	Maintenance of staff capabilities Availability of spares

The next step was structuring the GSS parameters according to the main research variables and calculating their relative importance. As indicated previously, resources represented costs, time, part of repair and maintenance, energy conservation and part of heating. The percentages derived are therefore 42 + 4.2 + 3 + 1.9 + 1.6 = 52.7. Similarly, human-user satisfaction represents

TABLE 5. GSS parameter grading

Parameter	%
Cost	42.0
Time	4.2
Aesthetics	8.4
Function	11.2
Loadbearing structure	8.4
Lighting	2.5
Heating	3.6
Plant room	2.3
Machinery, e/m equipment	5.3
Building materials	3.4
Energy conservation	1.9
Acoustics	2.3
Repair and maintenance	4.5

aesthetics, function (only part of the whole), lighting, the rest of heating, machinery and equipment (part only) and acoustics, which leads to 8.4 + 5 + 2.5 + 2 + 3 + 2.3 = 23.2. Finally, technical indicators are related to loadbearing structure, building materials and the remaining part of repair and maintenance), which gives 8.4 + 3.4 + 1.5 = 13.3. These percentages were then normalised such as to sum to a hundred. Thus, resources are 59%, human-user satisfaction is 26% and technical indicators are 15%. However, this is not an inclusive approach to establishing the relative importance of the three main research variables. In order to cover all cases and furthermore compare and evaluate various opinions, Greek engineers were included in this process. Therefore, a small-scale survey was carried out in Greece, where building engineers and architects were asked to evaluate the three main variables and comment on the GSS percentages. The average of this survey was: resources 42.3%, human-user satisfaction 32% and technical indicators 25.7%. It is interesting to note the significant drop in regard to resources (by almost a third) and the increase in technical aspects. As a result, two sets of weights were developed: one according to the GSS and another according to the engineers.

Having established the coefficient factors of each main variable, as well as the coefficient of the indicators in each main variable, the problem is to relate the performance levels of the qualitative indicators of human-user satisfaction and technical indicators with the quantitative indicators of resources and technical aspects. Therefore, the quantitative resource and technical indicators are processed first, drawing some conclusions on which is the most economical building, the cheapest to run and the fastest to maintain. Next, the qualitative indicators of human-user satisfaction and technical aspects are analyzed and evaluated according to their coefficient. As a result another set of conclusions on the qualitative performance of the buildings is obtained, leading to the identification of both the best and worst structural system in qualitative terms. These sets of conclusions may indicate that a particular structural system is performing better in both cases than the other systems examined. In order to obtain scientific results, the correlation between the two sets of conclusions was examined, resulting in an overall evaluation of the structural systems analyzed. The relative importance of the three main variables facilitated the drawing of inclusive conclusions.

Concluding, it should be noted that each building is examined at three levels: overall, research-related, and envelope. In contrast, information is gathered from all four stages of the building life: briefing, construction design, construction and in-use. Furthermore, the methodology developed appraises the buildings systematically and practically: inclusively and down to earth, collecting information from the users, neighbours, building management team, construction supervisors, architects, engineers and clients as well as from personal observation.

6. Use of Findings

The most important contribution of this particular research project is the development of an inclusive appraising model for long spanning buildings, particularly sports halls and swimming pools. Bearing in mind the results of the literature review and the general lack of inclusiveness in the methodologies developed already, this is the main theoretical use of the findings. It is worth mentioning the work carried out on the actual data processing methodology concerning the problem of integrating qualitative and quantitative data sets for the evaluation of buildings, through development of coefficient factors and relative importance indicators.

The practical use of the findings is their immediate and straightforward application to sports hall and swimming pool buildings in Greece. As the GSS criteria, and consequently the research criteria, are realistic, it is anticipated that the conclusions are going to be of great help and importance for all the clients, designers and engineers concerned with long spanning buildings (especially for conference halls and exhibition centres, where the intrinsic environmental needs are similar). In other words, it is possible to generalize the findings, as the use of the various structural systems is actually the issue under investigation. However, there may be cases where the conclusions of this research constitute only an indicator of the expected performance levels to be obtained with various alternative structural systems.

References

Bishop, J. (1978) The Appraisal of Buildings: A Case-Study and Critique of Approaches, University of Bristol, School for Advanced Urban Studies, Occasional Paper No. 5.

C.I.B. (1975) "The Performance Concept and its Terminology", *Building Research and Practice* **3**, 18-22.

Crise, D.J. (1975) "A Procedure for Assessing Expected Performance Effectiveness of Industrialized Housing Systems", *Journal of Architectural Research* **3**, 17-25.

General Sports Secretariat (G.S.S.) (1985) Construction of Indoor Sports Halls in Koridallos and Petroupolis, Athens - Building brief.

General Sports Secretariat (G.S.S.) (1990) Construction of Indoor Swimming Pool in Argostoli,

Kefallonia - Building brief.
Hillier, B. and Leaman, A. (1972) "A New Approach to Architectural Research", *RIBA Journal* **79**, 517-521.
Markus, T.A. (1972a) *Building Performance,* Applied Science, London.
Markus, T.A. (1972b) "Building Performance Appraisal", in *Performance Concept in Buildings*, Joint RILEM-ASTM-CIB Symposium Proceedings, U.S. Department of Commerce, National Bureau of Standards, NBS Special Publication 361, pp. 227-235.
Preiser, W.F.E. (1988) *Post-Occupancy Evaluation*, Van Nostrand Reinhold Co., New York.
Shibley, R.G. (1982) "Building Evaluation Services", *Progressive Architecture* **12**, 64-66.
Social Services Buildings Research Team (SSBRT) (1976) The Appraisal of Buildings: An Analysis and Review of Existing Architectural and Sociological Approaches, Department of Architecture, Oxford Polytechnic, Oxford.
Yin, R.K. (1989) *Case Study Research: Design and Methods*, Applied Social Research Methods Series, Vol 5, Sage Publications, London.
Zunde, J.M. (1982) *Design Procedures: Level 4*, Longman, New York.

MIX
Papier aus verantwortungsvollen Quellen
Paper from responsible sources
FSC® C105338

If you have any concerns about our products,
you can contact us on
ProductSafety@springernature.com

In case Publisher is established outside the EU,
the EU authorized representative is:
Springer Nature Customer Service Center GmbH
Europaplatz 3, 69115 Heidelberg, Germany

Printed by Libri Plureos GmbH
in Hamburg, Germany